1842 · ALLE · ZEIT · WACH

Herbert Fröhlich (Ed.)

Biological Coherence and Response to External Stimuli

With 97 Figures

Springer-Verlag
Berlin Heidelberg New York
London Paris Tokyo

Prof. Dr. HERBERT FRÖHLICH
Department of Physics
Oliver Lodge Laboratory
Oxford Street
Liverpool L69 3BX
England

ISBN-13:978-3-642-73311-6 e-ISBN-13:978-3-642-73309-3
DOI: 10.1007/978-3-642-73309-3

Library of Congress Cataloging-in-Publication Data. Biological coherence and response to external stimuli/Herbert Fröhlich (ed.). p. cm. Includes bibliographies. 1. Biomolecules – Effect of radiation on. 2. Coherent states. 3. Biophysics. I. Fröhlich, H. (Herbert), 1905– [DNLM: 1. Biophysics. 2. Electromagnetics. QT 34 B604] QP514.2.B575 1988 574.19′15 – dc19 DNLM/DLC 88-4524

Softcover reprint of the hardcover 1st edition 1988

2131/3130-543210

Preface

The existence of coherent excitations in active biological systems has been established in recent years. The present book aims at presenting a survey on the many features that such excitations can exhibit. This does not mean that a "theory of biology" has been established but it implies that such a theory will make use of such excitations. It is hoped that the present book will help in this direction.

April 1988 H. Fröhlich

Contents

Contributors

You will find the addresses at the beginning of the respective contribution

Theoretical Physics and Biology

H. FRÖHLICH [1]

This is the title given to the first Versailles meeting arranged by the Institut de la Vie in 1967 (cf. Marois 1969). Its history goes back to about 1938, when my late friend Max Reiss told me that biological membranes had been found to maintain a small, 100 mV, electric potential difference. On hearing of its thickness of 10^{-6} cm, I found that this corresponds to the enormous electrical field of 10^5 V/cm which ordinary layers will sustain only when special precautions are taken. Also then with an elastic constant corresponding to a sound velocity of 10^5 cm/s, is a frequency of 10^{11} Hz found corresponding to millimetre electric waves. Such frequencies were not available at the time, but I asked the help of Willis Jackson to produce them and hence investigate their properties, in particular biosystems, proposed by Victor Rothschild. This research was abandoned through the outbreak of the war, but I remembered it when approached by M. Marois. Meanwhile I had learned that a single idea does not produce a theory. As a second idea I then considered the possible excitations of coherent modes in this frequency region. Long-range phase correlations had been found to describe the order present in superconductors, superfluid helium, lasers, and others. This then could provide the order present in active biological systems (Fröhlich 1968).

At the same Versailles meeting, I. Prigogine presented his ideas about "dissipative structures" (cf. Prigogine 1980).

The suggestion of the importance of 10^{11} Hz frequencies for biological activities was strongly supported by the publication of research in the Soviet Union, where mm wave spectroscopy had been developed (Devyatkov 1974). They found that a great number of different biological systems have the following properties in common: (a) the effect of irradiation depends strongly on the frequency of the microwaves; (b) in certain microwave-power ranges, the effect of exposure depends weakly on variation of power through several orders of magnitude; (c) the effects depend significantly on the time of irradiation. This confirms the basic results of the theory.

Subsequently a great number of investigation followed. They will not be quoted here in detail but instead reference to relevant review articles will be made; Fröhlich (1977, 1980, 1986a,b). A survey on relevant developments is also presented in the "Green Book" (Fröhlich and Kremer 1983b).

The developments described in this book can be considered as particular cases of Prigogines dissipative structures or of Hakens Synergetics.

1 Department of Physics, Oliver Lodge Laboratory, Oxford Street, P.O. Box 147, Liverpool, L69 3BX, GB

1 Introduction

The present century has witnessed a complete understanding of the structure, interaction, and dynamical properties of atoms and molecules. It was often concluded then that the properties of complex systems will follow once their molecular structure has been found. This idea led to the development of molecular biology.

From the point of view of physics, biological materials are extremely complex and complicated systems. Yet, once they are activated, they function in a most systematic way. In fact on occasions they show sensitivities equal to the highest available in modern technology.

As an example we mention the sensitivity at low light intensities of the human visual system which, according to a careful analysis by Rose (1970), is close to the theoretical limit. The system can thus be considered as an image converter of the highest possible sensitivity. Yet it uses materials of a quite different nature from those used by technologists.

Another example for the extraordinary sensitivity of biological systems is provided by the sensitivity of certain fish to electric signals. They make use of such signals in various ways, as discussed by Bullock (1977). The lowest electric field that has been observed to evoke a response is of the order 10^{-8} V/cm. Again, the fish possesses none of the materials that would be used technically for the required purpose.

Many other examples of the extraordinary physical properties of biological systems exist. It is the task of Theoretical Physics to tentatively introduce appropriate physical concepts. Experimental collaboration is then needed to test these. Close collaboration of theory and experiment is thus required. Existence of the required properties in biological systems may be the result of long evolutionary processes. It must be asked, therefore, whether the relevant properties are possible, rather than whether they are probable.

It must be remarked at once that the relevant properties relate to certain activities of systems that are large from an atomic point of view, and the question of their derivation from micro (atomic) physics arises at once.

It should thus be assumed that from the point of view of physics, biological systems possess a certain order. The structure investigations of molecular biology have demonstrated, however, the absence of spatial order, and the conclusion has often been reached that no physical order exists in biomolecules such as enzymes. The arrangement of amino acids from which they are built was then considered as random (cf. Monod 1972).

The fallacy of this conclusion is evident at once, as the replacement of a particular amino acid in an enzyme by another "wrong" one reduces the efficiency of this particular enzyme. Clearly it is insufficient to consider spatial arrangements only. Physical order does, however, express itself not only in the spatial order. Thus superfluid helium and the electrons in superconductors exhibit an order that might be termed "motional order". It is expressed in terms of macroscopic wave functions and refers to certain phase correlations yielding a "coherence". In fact at very low temperatures when the entropy vanishes, i.e. no disorder exists, the atoms of superfluid liquid helium show the same type of disordered correlation as do other fluids (at higher temperatures). Yet the macroscopic wave function imposes a very subtle correlation in the motion of the helium atoms such that no disorder exists. Stimulated by these features, it will be assumed, as a working hypotheses, that phase correlations of some kind, coherence, will play a decisive role in the description of biological materials and their activities.

2 General

From the point of view of physics, active biological systems may be characterized by three properties:

i. they are relatively stable but far from equilibrium,
ii. they exhibit a non-trivial order,
iii. they have extraordinary dielectric properties.

Furthermore, in dealing with causes for activities, a multicausal approach is required, dealing, e.g. in the case of two interacting systems such as enzymes and substrates, not only with their interaction when they have joined, but also with the cause that brings them together.

To give a trivial but instructive example, consider a cup on a table, partly filled with water. If we tilt it sufficiently, then the water will flow out. The causes are (a) that we have tilted it, and (b) that the free energy is lower after the water has flown out. This also provides an example for a metastable state, for if we tilt the cup only a little, and then return it, water will not flow out, but after few oscillations will return to the original state.

Excitation of metastable states is of basic importance for biological activities. They arise through non-linear restoring forces and may have trivial or non-trivial consequences. To illustrate this, consider a particle with coordinates x bound elastically to $x = 0$ thus having a restoring force proportional to x, and potential energy, proportional to x^2,

$$f = - ax, V = \frac{a}{2} x^2 \geqslant 0, a > 0 . \tag{2.1}$$

The oscillation is harmonic and all displacements are linear. Add now a non-linear term yielding

$$V = \frac{a}{2} x^2 + \frac{b}{4} x^4 , f = - \frac{\partial V}{\partial x} = - ax - bx^3 . \tag{2.2}$$

If $b > 0$, the system will still oscillate, though no longer harmonically. This does not alter the qualitative behaviour as an oscillator as long as $b > 0$. The lowest energy still lies at $x = 0$, and the restoring force is still directed towards $x = 0$.

If, however, $b < 0$, then a qualitative change arises. The restoring force becomes zero not only at $x = 0$, but also at x_0,

$$x_0 = \pm \left(\frac{a}{|b|} \right)^{1/2} . \tag{2.3}$$

The potential now rises with increasing $|x|$ below (x_0), and above it, falls. At x_0, the potential energy has a maximum $3/4\ a^2/b$. At higher energies, thus, the system no longer oscillates.

If on the other hand, $b > 0$ but $a < 0$, then while the restoring force remains zero at $x = 0$ and $x = \pm x_0$, the potential has a maximum at $x = 0$ but two minima with negative values

$$V = -\frac{1}{4}\frac{a^2}{b}, b > 0 \tag{2.4}$$

at $\pm x_0$.

These possibilities, depending on the signs of the parameters a and b, give rise to simple illustrations for concepts relevant in many biological activities. They are:

oscillations when $a > 0, b > 0$
metastable state when $a > 0, b < 0$
bifurcation and chaos when $a < 0, b > 0$.

Thus in the case $a > 0$, $b < 0$, the system is relatively stable near $x = 0$ until the energy exceeds $3/4\ a^2/b$, i.e. at $x = 0$ it is in a metastable state. When $b > 0$, but $a < 0$, then the relative energy maximum at $x = 0$ is unstable, and the system can fall with equal probability into the two minima (2.4) at $\pm x_0$. Hence bifurcation arises connected with the possibility of chaotic behaviour as the properties at $+(x_0)$ and $-(x_0)$ may be very difference. We thus see that already the very simple non-linear potential (2.2) contains important possibilities.

While this very simple model already shows the appearance of metastable states, it also demonstrates the appearance of chaos, arising when two very different states occur with equal probability.

The possibility of deterministic chaos, probably, is the reason for the difficulty arising from an apparent lack of reproducibility in some experiments. This may, however, be apparent only when an insufficient number of experimental data is made available, a feature of bad experimentation. To give a very simple example, consider an assembly of non-interacting harmonic electric oscillators with frequencies ω_c, that are not exactly equal, but all lie close to a ω_0; some are slightly larger, some slightly smaller.

We now investigate this system by measuring the contribution of single oscillations to the real dielectric constant, which as is well known is proportional to $1/(\omega_c^2 - \omega^2)$, where ω is the measurement frequency, also close to ω_0. The result thus is positive or negative, depending on whether ω is smaller or larger than ω_0. If the result of a measurement is denoted by x^1, then the average over many (n) measurements will result in $\bar{x} \approx 0$ arising from a cancellation of positive and negative results. The uneducated investigator will thus conclude "no effect". Actually, however, one should consider not only x, but also $\overline{x^2}$. No effect, i.e. random scattering will differ in a characteristic way from our case by the value of the mean fluctuation $(\overline{x^2} - \bar{x}^2)$ for large n. In fact while each measurement gives a large magnitude which may be positive or negative, i.e. while $\bar{x}$ vanishes, $\overline{x^2}$ will be nearly independent of n, the case of deterministic chaos.

The theory of the properties of large systems, like biological ones, poses, of course, a question arising in the theory of the connection of macro- and microphysics, as will be discussed in the next section. A simple example shows already that the two approaches, reductionist, working with individual particle properties, and holistic, working with properties of large systems, do supplement each other. Thus consider a sound wave in air. Near thermal equilibrium, with thermal disorder, individual particle behaviour is chaotic. Excitation of a sound wave changes this only minimally. Yet, collective properties like the mean density show systematic space-time variation. Thus a large region containing many molecules exhibits properties that are not (or only minimally) noticeable in individual particle motion. Quantitative description requires, however, the knowledge of the compressibility which follows from individual particle motion and interaction.

One should expect, then, that large systems that are not in thermal equilibrium – arising from energy supply – may exhibit properties due to fluctuations which are not strictly reproducible when non-linear processes arise. Thus it has been shown on a particular model (Arnold et al. 1978) that noise may cause phase transitions in excited non-linear systems.

An assembly of subsystems, say growing cells, started initially in equal stage of development, may thus after some time show a distribution of stages of development. The behaviour of a single cell thus is not reproducible. Investigation of many cells is required to demonstrate the distribution (cf. Grundler et al., this Vol.).

In various chapters in this Volume, the great sensitivity of biological systems to very weak electromagnetic radiation at sharp frequencies will be reported. In technology, as is well known, a radio receiver may amplify a very weak signal provided it has sharp frequencies and it is coherent. Here we shall generalize the concept of coherence such that some of the properties of a system at a space time (x_2, t_2) follow from those at another (x_1, t_1). For the ground states of a material this is trivial, of course. For highly excited states as we find them in biology this implies that a certain – though not specifically defined – organization exists.

The concept of coherence originally arose in the context of emission of light from a source. The phases of the light waves emitted from different regions differ and are distributed randomly. As a consequence there exists no interference between them. Theoretically this lack of coherence was the reason for the long delay in establishing the wave nature of light. In lasers, on the contrary, the phases of different regions are correlated exactly; they act coherently.

Three kinds of coherent excitations will arise in the present context; (A) excitation of a single mode of vibration; (B) excitation of a metastable highly polar (ferroelectric) state; (C) excitation of limit cycles or Lotka-Volterra oscillations in complex systems. (A) requires supply of energy s above a certain rate $s > s_0$; (B) requires single supply of energy above a critical value; (C) arises in complex situations such as periodic enzyme reactions. Examples for all three will be given below. (A) will be found to lead to long-range coherent interaction; (B) permits non-chemical storage of energy; solitons are a particular case.

Long-range interaction is required in the control of growth in multicellular biological systems, so that the state of its development can influence every region. Absence of such control leads to cancer.

Long-range correlations are also required in treating the genetic code as language (F. Fröhlich, this Vol.) for there conditions exist that restrict the possible assembly of words.

Returning now to the great sensitivity of certain biological systems to weak electric fields, the opposite property, use of very high electric fields, must be emphasized.

Thus biomembranes maintain fields of the order of 10^5 V/cm, a field in which ordinary materials would break down electrically unless special care were taken. Molecules subjected to such fields will, in general, exhibit non-linear reactions such as change of structure. It will also be noticed that to orient a dipole with a moment μ by a field F against thermal motion at temperature T requires $\mu F > kT$, with $\mu \approx 100$ Debye units = 10^{-16} electrostatic units, typical for a protein, at room temperature this requires $F > 10^5$ V/cm. In comparison, we note that in vacuum, an electromagnetic current with intensity 10 m W/cm^2, which in some instances causes notable changes in biomaterials, corresponds to a field F = 2 V/cm only. Note that the intensity is proportional to F^2.

We also remark in this context that the high field in a membrane causes strong electric polarization. Oscillations will then lead to electric vibrations whose frequencies may be estimated in the order of $10^{10}-10^{11}$ Hz, representing oscillating electric dipoles with frequencies in the millimeter wave region. This frequency follows as the thickness of a membrane is of the order 10^{-6} cm and the elastic properties correspond to a sound velocity of the order 10^5 cm/s. The theory thus suggests the possibility of biological effects in this frequency region. A great number of such effects have actually been found, as discussed in various articles in this Volume.

Finally to return to the three characteristics expressed at the beginning of this section

1. requires that various excitations are stabilized, pointing to metastable states;
2. requires a motional order such as is found in the existence of macro-wavefunctions in superconductors and superfluids, but it also exits in non-equilibrium systems such as lasers, or in maintained particular excitations such as sound waves. Its generalisation leads to coherence;
3. the extraordinary dielectric properties arise from the high electric fields maintained in membranes, in conjunction with the sensitivity to very low field mentioned in the previous sections. Sensitivity to very weak electromagnetic fields with sharp frequency response will be presented in various articles in this Volume.

3 The Connection Between Macro- and Microphysics (cf. H. Fröhlich 1973b)

The non-relativistic many-body problem is governed by the Schrödinger equation in terms of a state function and an appropriate Hamiltonian operator that depends on the interaction between particles, or between particles and external fields. The state function depends on the coordination of all the particles involved, say N particles. State functions can be classified according to the energy involved, and there exist an enormous number of them, even in a limited range of energies, increasing more or less exponentially with N.

When we are dealing with a crystal near its ground state, a systematic classification of the states is possible in terms of small linear displacements of the particles that form the crystal. The lowest excitations then can be expressed in terms of non-interacting normal mode vibrations.

Larger displacements, however, do involve non-linear properties. They lead to interactions between normal modes and their treatment causes difficulties. Powerful methods have been developed, however, to deal with an N-body problem, and the opinion is frequently expressed that the derivation of all solutions of the N-body problem – in a certain energy range – would be desirable, for from these solutions would then follow all macroscopic properties.

Not only is this aim impossible in principle, but it is also highly undesirable, for macrophysics deals with concepts quite different from those used in microphysics, and the task of selecting the interesting features from the immense number of micro-solutions would be more formidable than finding these solutions in the first place.

To illustrate the enormous number of micro states, consider a linear chain of 200 amino acids forming a protein, and treat these amino acids as single particles. Because there exist 20 different amino acids, 20^{200} different arrangements (states) exist, each containing 200 amino acids.

Or to present a very different example, consider an ideal gas of N particles in thermal equilibrium at room temperature. According to a well-known theorem by Boltzmann, the number of micro states W involved is given by the entropy $S = sN$ as $W = \exp(S/k) = \exp N(s/k) \approx 10^N$, where use is made of the entropy s per particle. Suppose a super computer that calculates them takes τ seconds per state, then a time $t \approx 10^N \tau/N$ will be required. If the computer has been running since the beginning of the Universe, $t \approx 10^{17}$ s, it can have dealt with about $N \sim 30$ only, even for very small τ.

Clearly it is not possible to derive macroscopic properties from the solution of the N-body problem, but this does not mean that macrophysics can not be derived from microphysics. The method to be used is to express macroscopic quantities in terms of particles; then, instead of solving the N-body Schrödinger equation, use it to derive relations between these macroscopic quantities. Thus, as an example, the Navier-Stokes equations, which govern the flow of liquids, can be derived. The first, holistic, step then is the definition of relevant macroquantities, namely charge density $\rho(x, t)$ and current density $j(x, t)$ both space-time-dependent. Both can be expressed in terms of averages of individual particle contribution; introducing these into the Schrödinger equation, and carrying out appropriate averaging then yields the Navier-Stokes equations. These also contain space-time-independent parameters like compressibility. Their derivation requires the use of reductionist concepts like two-particle interaction.

The success of the procedure rests in the definition of relevant multiparticle quantities, ρ and j, which after averaging give rise to the macroscopic densities and their space-time connection. To use a similar method in biology requires definition of appropriate quantities. This is not a systematic step. It is the task of theoretical physics to propose such appropriate concepts, and to derive consequences that can be tested experimentally. Clearly, as mentioned before, close collaboration between theory and experiment is thus required.

In general terms, active biological systems exhibit an organization which, however, is not based on spatial order. They receive random energy through metabolism (or light) and use a part of this energy to maintain and develop a complex organization. In simple physical systems, in general, a supply of random energy leads to an increase in temperature. Exceptions arise in the case of machines which require complex arrangements. More simple, in principle, is the case of the laser where the supply of energy yields the well known laser states exhibiting long-range coherence – it has been pointed out above that more general types of coherence are available, and will be discussed below in more detail. They can be maintained through supply of energy in spatially disordered systems, and are concerned with relatively few degrees of freedom only.

4 Coherent Excitations

In the present and following section, models for the three types of coherent excitations will be presented. It must be realized that the derivation of these excitations makes use of particular material properties. Other models will exist. They have in common that non-linear interactions are relevant such that supply of random energy does not (only) result in heating of the particular system, but in the excitation of certain coherent (ordered) states that are stabilized.

4.1 Metastable Highly Polar States (cf. Fröhlich 1980, 1986a,b)

A material polarized electrically with polarization P possesses a self-energy proportional to P^2. This energy depends on the shape of the material. As a consequence, deformation of the material will cause a change of the electric self-energy which may be positive or negative. A change in elastic energy, due to the deformation, will also occur, but this is always positive if the material was in equilibrium before it was deformed. As an example we note that the ratio of the electric energies of a slab and of a sphere, both having equal volume and polarisation, is $3\,\epsilon/(\epsilon + 2)$, where ϵ is the dielectric constant (cf. Fröhlich 1958).

As an explicit example, consider a sphere whose electro static energy is $\frac{1}{2}\gamma P^2 > 0$. If deformed into an ellipsoid with eccentricity η, this energy is proportional to η, and higher powers, as the electrostatic energy does not have an extremum for the sphere, $\eta = 0$. The elastic energy arising from the displacement will, however, have no linear term in η, but will be proportional to η^2, as elastically the sphere is the equilibrium state. The total energy U, therefore, for small η will have the form

$$U = \frac{1}{2}\gamma P^2 (1 + c\eta) + \frac{1}{4} S\eta^2 , \tag{4.1}$$

where c is a positive or negative constant and $S > 0$ is an elastic constant. Minimizing with regard to η yields

$$\eta = - c\gamma P^2 /S , \tag{4.2}$$

and hence

$$U = \frac{1}{2}\gamma P^2 - \left(c \frac{1}{2}\gamma P^2\right)^2 /S . \tag{4.3}$$

Thus U behaves like the example (2.2) with $a > 0$, $b < 0$, i.e. as discussed in the context it has a metastable state at $P^2 = 0$, but decays for sufficiently large excitations.

At such excitations, however, one would expect higher, i.e. P^6 terms, which would then provide a stabilization at high polar excitation.

To present a more formal model, consider a homogeneous polarization field P with energy density $\frac{1}{2}\omega^2P^2$, ω being proportional to the frequency of the free field oscillation. This field interacts with an elastic field with potential energy density $\frac{1}{2}{}^2\rho$ for longitudinal elastic displacements yielding a density ρ. Their interaction should then be proportional to ρP^2 and further terms proportional to ρP^6 should be included to provide stabilisation at high P^2.

The total potential energy then has the form

$$W = \frac{1}{2}\omega^2P^2 + \left(\frac{1}{2}S\rho^2 + cP\rho\right)(1 - d^2P^2). \tag{4.4}$$

Here the coupling constant c may be positive or negative

Minimising with regard to ρ yields

$$\rho = -\frac{c}{S^2}P^2, \tag{4.5}$$

and hence from (4.4),

$$W = \frac{1}{2}\omega^2P^2 - \frac{1}{2}\frac{c^2}{S^2}P^4 + \frac{1}{2}\frac{c^2d^2}{S^2}P^6, \tag{4.6}$$

as required above. This potential energy thus increases proportional to P^2 for small P^2, and P^6 for large P^2. In between it has a maximum followed by a minimum.

For detailed discussion introduce

$$y = 3\,d^2P\,,\ \lambda = 3\,s^2d^2\omega^2/c\,, \tag{4.7}$$

then

$$\frac{6\,d^2}{\omega^2}W = y - \frac{1}{\lambda}y^2 + \frac{1}{3\lambda}y^3. \tag{4.8}$$

Minimizing with regard to P^2, i.e. y, requires $y = y_0$,

$$0 = 1 - \frac{2}{\lambda}y_0 + \frac{1}{\lambda}y_0^2\,, \quad \text{i.e. } y_{0\pm} = 1 \pm \sqrt{1-\lambda}\,, \tag{4.9}$$

and hence, using

$$y_0^3 = 2\,y_0^2 - \lambda y_0\,, \tag{4.10}$$

or inserting (4.9)

$$\frac{6\,d^2}{\omega^2}W_0 = 1 - \frac{2}{3\lambda} + \frac{2}{3}\frac{(1-\lambda)^{3/2}}{\lambda}. \tag{4.11}$$

We now note that when $\lambda > 1$, we have no real solutions. Otherwise, however, the potential energy has a minimum at P = 0, then rises to a maximum when $y_0 = 1 - \sqrt{1-\lambda}$, followed by a minimum at $y_0 = 1 + \sqrt{1-\lambda}$ and then rises proportional to P^6 when P^2 is sufficiently large. The minimum and maximum coincide when $\lambda = 1$ at $y_0 = 1$. When $\lambda = 3/4$ then y_0 is $\frac{1}{2}$ and 3/2, and $W_0 = 0$ at $y_0 = 3/2$, and $6\,d^2\,W_0/\omega^2 =$

2/9 at $y_0 = \frac{1}{2}$. In this case thus the minimum coincides with the minimum at P = 0. For $\lambda < 3/4$ the potential energy of this highly polar state is negative, and this state then is the ground state – the material is ferroelectric. Thus when $3/4 < \lambda < 1$, the material possesses a highly polar metastable state. This condition on λ, in particular $\lambda < 1$, indicates the conditions the material must satisfy. They are elastic softness (small S), high polarisability (small ω) in conjunction with large coupling between the fields (large c^2). These conditions tend to be satisfied by biological materials.

It has thus been shown that if certain material conditions are satisfied, interaction between highly polar and elastic displacements leads to the establishment of metastable highly polar states, i.e. ferroelectric states.

This model may, of course, be generalized to space-time-dependent conditions, by introducing space-time-dependent polar and elastic fields. Contact can also be made with the excitation of solitary waves, solitons, which have been widely investigated for both biological and non-biological materials (Bilz et al. 1981; Sobell 1985; Davydov 1978). This also includes the Davydov soliton, based on the polar stretched C = 0 group stabilized by elastic displacements.

4.2 Coherent Excitation of a Single Polar Mode (cf. Fröhlich 1977)

Biological materials contain many ions. Their linear displacements above the ground state can be developed into normal modes with frequencies, say ω_j, in a range

$$0 < \omega_1 \leqslant \omega_j \leqslant \omega_2 \,, \quad j = 1,2,\ldots,z \,, \tag{4.12}$$

if there exist z modes. The lowest one, ω_1, is different from zero.

The long-range Coulomb interaction establishes that the normal modes extend over large inhomogenuous regions, a cell, or large units.

Assume now our system to be embedded in a heat bath of temperature T with which it interacts strongly, and that energy is supplied to it randomly at a rate s_j to ω_j. It will then be shown that when this supply exceeds a critical rate, a single mode will be excited very strongly, similar to the phenomenon of Einstein Bose condensations.

To derive this on a simple model, non-linear interaction terms must be considered.

Let n_j be the number of quanta by which mode ω_j is excited, and let s_j be the rate of metabolic energy supplied to it. Linear interaction with the heat bath will then impose a Planck distribution for n_j. This interaction is described in terms of emission and absorption of single quanta $h\omega_j$ between the modes and the heat bath. When non-linear interactions are taken into account, however, then energy exchange in terms of two quanta will occur. Most important amongst these are simultaneous emission and absorption of quanta $h\omega_j$ and $h\omega_l$. The rate equation for n_j then becomes

$$\begin{aligned} n_j &= S_j - \Phi_j \left(n_j\, e^{\beta\omega_j} - (1+n_j)\right) - \\ &- \sum_l X_{jl} \left[n_j\, (1+n_e)\, e^{\beta\omega_j} - n_l\, (1+n_j)\, e^{\beta\omega_l}\right. , \end{aligned} \tag{4.13}$$

where $\beta = \hbar/hT$ and Φ and X are transition possibilities. The ratio of absorption and emission terms, proportional to n and 1 + n, respectively, follows from the principle of detailed balance.

In the stationary case, $\dot{n}_j = 0$.

Summing over all states ω_j, the X_{jc} terms vanish, and the total rate of energy supply

$$\Sigma S_j = \Sigma \Phi_j \left[n_j e^{\beta\omega_j} - (1 + n_j)\right] \tag{4.14}$$

depends on the Φ-terms only.

While solution of the rate equation (4.13) can be easily provided, it will be noted that for large X, these X-terms alone largely determine the stationary solutions, i.e. they require

$$n_j (1 + n_l) e^{\beta\omega_j} = n_l (1 + n_j) e^{\beta\omega_l} . \tag{4.15}$$

Solutions of this equation involves an unknown term μ,

$$n_j = \frac{1}{e^{\beta(\omega_j - \mu)} - 1} , \tag{4.16}$$

i.e. a Bose distribution.

Now the total number of quanta is through (4.14) determined by Φ. Hence μ must be determined from the known total number of quanta, $\Sigma n_j = N$. Also note that n_j must be positive or zero so that $\mu \leqslant \omega_1$ is required. When the total number of quanta exceeds a critical value, then μ must closely approach ω_1, yielding a large n_1, as in Bose condensation. The frequency ω_1 is then very strongly (coherently) excited. Detailed considerations (Pokorny 1982) show that in certain circumstances other frequencies might be strongly excited.

To understand the possibility of this type of excitation, the strong (non-linear) interaction with the heat bath must be emphasized. This attempts to impose its temperature on the distributions. But while in a Bose gas the number of particles is fixed, so that Einstein condensation arises only when the temperature is sufficiently lowered, in our case the temperature is fixed but the number of particles (quanta) is increased by supply of energy.

It should be emphasized now that other interaction terms could be introduced which would prevent condensation (Fröhlich 1982), though other possibilities for coherent excitation may also arise, as, e.g. in a computer calculation (Fermi et al. 1965) on the fate of energy supply to anharmonically coupled oscillators.

Exact solution of Eq. (4.13) shows that the energy supply s_j may occur at any or all frequencies of the polar modes provided the total, Σs_j, exceeds the critical s_0. In general, the subsequent excitation of ω_1 then requires some time. The system thus possesses storing ability until sufficient energy is pumped into ω_1. Let t_1 be the time required for this process. We expect it to be the faster the larger Σs_j but it will never occur ($t_1 = \infty$) as long as $\Sigma s_j < s_0$. Thus if a and n are positive integers we may assume

$$t_1 = a/(\Sigma s_j - s_0)^n , \quad \Sigma s_j \geqslant s_0 . \tag{4.17}$$

We now consider a complex biological process consisting of many steps, only one of which depends on the excitation, and if t_2 is the time required for all these processes, then the total time t for the process is $t = t_1 + t_2$ and hence the total rate is

$$\frac{1}{t} = \frac{(\Sigma s_j - s_0)^n}{a + t_2 (\Sigma s_j - s_0)^n} , \quad \Sigma s_j > s_0 . \tag{4.18}$$

It thus depends in a step-like manner on the rate of energy supply.

Finally it should be remarked that the time t_1 will be expected to be particularly short if the main energy supply occurs at the frequency ω_1 which will be excited; resonance behaviour with regard to the applied frequency must thus be expected.

This energy supply need not only arise from biological metabolism but may also be due to externally applied radiation. Resonance effects due to the influence of applied radiation may thus be expected, and in fact, are often observed, as discussed in various papers in this Volume.

5 Periodic Enzyme Reactions (cf. Fröhlich 1980, 1986a,b)

High catalytic power in conjunction with strict specificity constitutes a most striking feature common to all enzymes. High catalytic power requires a reduction of activation energies, and the metastable state with its high internal electric field may activate this. Furthermore, in the following section it will be shown that frequency selective long-range interactions arise from the excitation of coherent vibrations. This might then yield a selective attraction of substrates.

The importance of the metastable state for the reduction of activation energies points to the importance of electrostatic interaction within the enzyme-substrate complex, as has also been confirmed by a more detailed study (Wada and Nakamura 1981). Clearly water and its ions must play an important role in a quantitative analysis.

Consider now a model in which enzyme molecules are arranged spatially, as suggested in Clegg (1984), and when excited they interact with substrate molecules. In the course of this interaction, substrate molecules are destroyed, and some of the liberated energy is used for the activation of other enzymes. These have a finite lifetime so that some of them fall back into the non-activated state. In the course of this process, substrate molecules are coherently excited and hence attract other substrates into the region. The activation-deactivation of enzymes gives rise to a corresponding activation-deactivation of the correlated electric dipoles, and hence to an electric signal. This model, originally developed for brain waves, may be relevant for other periodic enzyme processes as well.

Let Z be the number of non-activated enzyme molecules, considered to be large compared with the number N of activated enzymes, so that approximately Z is time-independent. The rate of change of N is then proportional to NZS, where S is the number of substrates, a triple collision with reaction rate α. Also if β is the rate constant for spontaneous decay

$$\frac{dN}{dt} = \alpha NZS - \beta N \,, \tag{5.1}$$

$$\frac{dS}{dt} = - \alpha NZS + \gamma S \,, \tag{5.2}$$

where γ is the rate constant for the attraction of substrates.

These equations permit an implicit solution

$$N^{\gamma} S^{\beta} = c e^{\alpha Z (N+S)} \,, \tag{5.3}$$

where c is a constant. Also a time-independent solution N_0, S_0 exists

$$N_0 = \frac{\gamma}{\alpha Z}, \quad S_0 = \frac{\beta}{\alpha Z}, \tag{5.4}$$

so that

$$\nu = N - N_0, \quad \sigma = S - S_0 \tag{5.5}$$

satisfies the well-known Lotka-Volterra equations

$$\frac{d\nu}{dt} = \gamma\sigma + \alpha Z + \nu, \quad \frac{d\sigma}{dt} = -\gamma\nu - \alpha Z\sigma\nu, \tag{5.6}$$

which have time-periodic solutions. The system thus does not develop into the time-independent state, $\nu = 0, \sigma = 0$.

In fact, if σ and ν are so small that $\sigma\nu$ can be neglected, the solutions are

$$\nu = c\sqrt{\gamma}\cos\omega t, \quad \sigma = c\sqrt{\beta}\sin\omega t \tag{5.7}$$

$$\omega^2 = \beta\gamma. \tag{5.8}$$

We thus find a periodic enzyme reaction whose frequency is determined by the rate constants for recombination and attraction.

As a consequence, electric vibrations with the same frequency are generated at each enzyme site. The spatial arrangements of these sites determine to what extent these signals are observable at a distance.

The polarised active enzymes might exhibit a tendency towards forming a ferroelectric state. The total polarisation, P, of the region is then proportional to ν if we assume that the contribution of the time-independent N_0 is screened through mobile ions. The interaction between excited enzymes will then cause a tendency towards the ferroelectric state, hence to a contribution to $\dot{\nu}$. If we consider the relevant activation energy to be proportional to $(P_{max}^2 - P^2)$, then $\dot{\nu}$ has the form

$$\frac{d\nu}{dt} = \gamma\sigma + Z\sigma\nu + (c^2 e^{-\Gamma^2\nu^2} - d^2)\,\nu, \tag{5.9}$$

where Γ^2, c^2, d^2 are positive constants ($-d^2\nu$ is an ordinary damping term). Γ^2 will be proportional to $1/T$, T = temperature. Developing the exponent and neglecting $Z\sigma\nu$,

$$\frac{d\nu}{dt} = \gamma\sigma + (c^2 - d^2 - c^2\Gamma^2\nu^2)\nu \tag{5.10}$$

or differentiating and using (5.6)

$$\frac{d^2\nu}{dt^2} = -\beta\gamma\nu + ((c^2 - d^2 - 3)\,\Gamma^2\nu^2)\,\frac{d\nu}{dt}. \tag{5.11}$$

This is the equation of a limit cycle, provided $c^2 > d^2$. This means that oscillations, whatever way they start, approach at a later time a state with the same definite amplitude. The system thus stores energy and is stable against certain perturbations. Solutions of the more exact equation (5.10) have been shown to be limit cycles (cf. Kaiser, this Vol.). An appropriate relatively small perturbation may, however, cause their collapse and the liberation of the stored energy.

The present model of periodic enzyme reactions has led to the excitation of two different coherent electric vibrations, one arising from excitation of molecular vibrations, say ω_1, leading to the attraction of substrates; the second, say ω_2, determined by certain rate constants, being the frequency of a limit cycle. It must be expected that ω_1 is very much larger than ω_2, $\omega_1 \gg \omega_2$. Superposition of the two electric vibrations then results in a high frequency ω_1, modulated by a low frequency ω_2.

It is of interest in this context that application of a high frequency radiation to various biological systems has led to much greater effects after low frequency modulation of the high frequency. Systematic investigations on the ω_1 and ω_2 frequency dependence have, however, not yet been performed.

6 Resonance Interaction (cf. Fröhlich 1972)

Consider the interaction of two electric harmonic oscillations with frequencies ω_1 and ω_2 respectively, at a distance R. In appropriate units for their amplitudes q_1 and q_2, the equations of motion are

$$\ddot{q}_1 + \omega_1^2 q_1 = -\gamma q_2 \;;\quad \ddot{q}_2 + \omega_2^2 q_2 = -\gamma q_1 \tag{6.1}$$

and the energy H is

$$H = \frac{1}{2}(\dot{q}_1^2 + \omega_1^2 q_1^2) + \frac{1}{2}(\dot{q}_2^2 + \omega_2^2 q_2^2) + \gamma q_1 q_2 , \tag{6.2}$$

where the dot represents the time derivative. For given amplitudes, the electric interaction is measured in terms of γ, which is proportional to $1/R^3$, and depends on various angles. The properties of the system depend on the frequencies $\Omega\pm$ of the normal modes. Oscillations proportional to $\cos\Omega t$ or to $\sin\Omega t$ exist, and for both there are two types of oscillations, considering the angles as fixed, and treating R as a given parameter.

From (6.1), inserting $q_i \to q_i \cos\Omega t$, we find

$$(\omega_1^2 - \Omega^2)\, q_1 = -\gamma q_2 , \quad (\omega_2^2 - \Omega^2)\, q_2 = -\gamma q_1 \tag{6.3}$$

and hence

$$(\Omega^2 - \omega_1^2)(\Omega^2 - \omega_2^2) = \gamma^2 . \tag{6.4}$$

Solving for Ω^2 gives the normal modes

$$\Omega_\pm^2 = \omega_+^2 \pm (\omega_-^4 + \gamma^2)^{1/2} , \tag{6.5}$$

where

$$\omega_\pm^2 = \frac{1}{2}(\omega_1^2 \pm \omega_2^2) . \tag{6.6}$$

Note from (6.4) $\Omega^2 = 0$ when $\omega_1^2\omega_2^2 = \gamma^2$; the system is then unstable.

The case of resonance is defined by

$$\omega_1 = \omega_2 = \omega_+ , \quad \omega_- = 0 . \tag{6.7}$$

Hence from

$$\Omega_\pm^2 = \omega_+^2 \pm \gamma \tag{6.8}$$

and from (6.3)

$$q_1 = \pm q_2 \,, \tag{6.9}$$

which provide the well-known normal modes, $q_1 \pm q_2$. Also clearly, if $\omega_+^4 \gg \gamma^2$, then

$$\Omega_\pm = \omega_+ \pm \frac{1}{2}\frac{\gamma}{\omega_+} + \ldots \,. \tag{6.10}$$

In the absence of resonance, however, when $\omega_-^4 \gg \gamma^2$, then

$$\Omega_\pm^2 = \omega_+^2 \pm \omega_-^2 \pm \frac{1}{\gamma}\gamma^2/\omega_-^2 + \ldots \,. \tag{6.11}$$

Note $\omega_+^2 \pm \omega_-^2$ is ω_1^2 or ω_2^2. No linear term in γ exists in this case.

The resultant interaction of the two systems is determined by the γ-dependent part of the two frequencies Ω_+ and Ω_-, and their excitations. Consider first the case that $\Omega_\pm \gg kT$. In thermal equilibrium, both energy and free energy are then determined by the zero point energy $\frac{1}{2}\hbar\Omega$ of the oscillations. The interaction energy I then using (6.5) is given by

$$\begin{aligned}\frac{2\,I}{\hbar} &= (\Omega_+ + \Omega_-) - (\Omega_+ + \Omega_-)_{\gamma=0} \\ &= (\omega_+^2 + (\omega_-^4 + \gamma^2)^2)^{1/2} + (\omega_+^2 - (\omega_-^4 + \gamma^2)^{1/2})^{1/2} \\ &\quad - (\omega_+^2 + \omega_-^2)^{1/2} - (\omega_+^2 - \omega_-^2)^{1/2} \,.\end{aligned} \tag{6.12}$$

In the particular case of resonance, when $\omega_-^2 = 0$ we find for I_R,

$$\frac{2\,I_R}{\hbar\omega_+} = (1 + \gamma/\omega_+^2)^{1/2} + (1 - \gamma^2\omega_+)^{1/2} - 2 \,, \tag{6.13}$$

which is even in γ. If in particular $\gamma^2 \ll \omega_+^4$ then developing

$$\frac{2\,I_R}{\hbar\omega_+} \cong -\frac{1}{4}\gamma^2_{\omega_+^4} \propto -\frac{1}{R^6} \,, \tag{6.14}$$

which is the London interaction. Note that the short-range linear in $\gamma \propto 1/R^3$ interaction cancels.

A similar result is found in thermal equilibrium in the classical limit $KT \gg$ to Ω when the interaction I is given by

$$\frac{I}{KT} = \log(\Omega_+\Omega_- - \omega_1\omega_2) = \frac{1}{2}\log \Omega_+^2\Omega_-^2/\omega_1^2\omega_2^2 \,. \tag{6.15}$$

Using (6.5) and (6.6) thus

$$\frac{I}{KT} = \frac{1}{2}\log(\omega_+^4 - (\omega_-^4 + \gamma_1^2))/\omega_1^2\omega_2^2 = \frac{1}{2}\log(1 - \gamma^2/\omega_1^2\omega_2^2) \,, \tag{6.16}$$

i.e. negative (attractive) proportional to $\gamma^2 \propto 1/R^6$ as obtained in the London interaction.

Thus even in the resonance case when the frequencies contain a term linear in γ [cf. (6.8)] the residual thermal equilibrium interaction I is proportional to γ^2, i.e. no long-range interaction proportional to γ is retained.

Coherent excitations, however, arise outside thermal equilibrium by excitation of low-lying particular modes. In the resonance case, strong excitation of the frequency Ω involves the negative $-\gamma$; we use $\gamma > 0$ otherwise interchange + and − in (6.10). Let Ω_- be excited with n_- quanta; then the interaction, using (6.10), is

$$I = \left(n_- + \frac{1}{2}\right)\hbar\,[\Omega_+ - \Omega_-(\gamma = 0)] = -\left(n_- + \frac{1}{2}\right)\hbar\gamma/2\,\omega_+ . \tag{6.17}$$

It is of long range proportional to $1/R^3$.

Clearly this arises from resonance, for in the absence of resonance the result has been found to be proportional to $\gamma^2 \propto 1/R^6$.

Models for strong, coherent, excitations have been discussed in Sect. 5. They thus lead to long-range frequency selective interaction, i.e. to resonance interactions as required in the case of periodic enzyme reactions, cf. Sect. 5.

In the interaction between two large molecules, one of them may be positioned in a biomembrane and hence under the influence of a large electric field. It will be shown now that this has a strong influence on the interaction. Let F be proportional to the external field and hence replace the equations of motion (6.1) by

$$\ddot{q}_1 + \omega_1^2 q_1 = -\gamma q_2 + F\,, \quad \ddot{q}_2 + \omega_2^2 q_2 = -\gamma q_1\,. \tag{6.18}$$

Particular solutions when q_1 and q_2 are time independent q_1^0, q_2^0 provide

$$q_2^0 = -\frac{\gamma}{\omega_2^2} q_1^0\,, \quad \left(\omega_1^2 - \frac{\gamma^2}{\omega_2^2}\right) q_1^0 = F\,. \tag{6.19}$$

Hence

$$q_1^0 = \frac{F\,\omega_2^2}{\omega_1^2\omega_2^2 - \gamma^2}\,, \quad q_2^0 = \frac{-\gamma F}{\omega_1^2\omega_2^2 - \gamma^2}\,. \tag{6.20}$$

As mentioned after Eq. (6.6), the system is unstable when $\omega_1^2\omega_2^2 = \gamma^2$.

Note that if we set

$$q_1 = q_1^0 + Q_1\,, \quad q_2 = q_2^0 = Q_2 \tag{6.21}$$

with time-dependent Q_1 and Q_2, these Q_1 and Q_2 satisfy the F independent Eq. (6.1). From (6.2) the q_1^0, q_2^0 give rise to the energy

$$H = \frac{1}{2}\,\omega_1^2 q_1^{0\,2} + \frac{1}{2}\,\omega_2^2 q_2^{0\,2} + \gamma q_1^0 q_2^0 - F q_1 \tag{6.22}$$

or inserting from (6.20)

$$H = -\frac{1}{2}\,\frac{\omega_2^2 F^2}{\omega_1^2\omega_2^2 - \gamma^2}\,. \tag{6.23}$$

We note in particular the singularity when the system is unstable at $\omega_1^2\omega_2^2 = \gamma^2$. This defines a critical distance, say R_0, of the two molecules, as γ depends on their distance. Depending on whether their distance is slightly smaller or larger than R_0, the magnitude of the energy is large but jumps from positive to negative. This can give rise, of course, to chaotic behaviour.

Models of the interaction of two molecules, one of them in the membrane, the other outside it, must, of course, be considered in detail. It may be of interest that the electronic charge e, the proton mass M, and the width of the membrane of 10^{-6} cm permit definition of a frequency ω,

$$\omega^2 = e^2/d^3 M \cong 10^{23}\ (\mathrm{Hz})^2\ , \tag{6.24}$$

leading to a frequency in the 10^{11} Hz region, which is of the order of the frequency of membrane vibrations.

7 Quantization of Magnetic Flux

The integral of a magnetic field H over a cross-section perpendicular to it is defined as the magnetic flux Φ. Introducing a vector potential A in the usual way, $H = \mathrm{curl}\, A$, we find with the help of Stoke's law

$$\Phi = \int \mathrm{curl}\, A \; d\sigma = \oint A_s ds\ , \tag{7.1}$$

where the line integral ds represents the borders of the surface σ.

In superconductors it has been found that the magnetic flux is quantized. F. Bloch (1968) has shown, however, that flux quantization is not restricted to superconductors, but in fact is a completely general property of all materials. The reason that it had not been observed is due to the small energy differences in neighbouring flux quantum numbers, so that in the usual energy distribution single flux quanta do not arise.

For the proof we use the Schrödinger equation for interacting particles, i.e. we disregard the spin

$$\sum_j \left\{ \frac{\hbar^2}{2\, m_j} \left(\frac{2}{i}\, \partial_j - \frac{e_j}{\hbar c} A\,(x_j) \right)^2 + \sum_k V\,(\nu_j - x_k) \right\} \Psi = E\, \Psi\ . \tag{7.2}$$

Here m_j and e_j are mass and charge of the particle with coordinate x_j; V is the potential energy. We consider the case in which the magnetic field is restricted to a certain area σ such that the line integral in (7.1) can be taken in a region in which the magnetic field vanishes. In this region the vector potential A is a gradient, say

$$\frac{e_i}{\hbar c} A\,(x_j) = \partial_j\, X\,(x_j)\ , \tag{7.3}$$

so that with the gauge transformation

$$\Psi = e^{i \sum_j X_j (x_j)}\, \Psi_0\ . \tag{7.4}$$

A can be eliminated from (7.2) with the help of (7.3). Using the closed path for integration (7.1) in this region we find by introducing (7.3)

$$\Phi = \frac{\hbar c}{e_j}\, \Delta X_j\ , \tag{7.5}$$

where ΔX_j is the jump of X_j when the s-path is closed. Uniqueness of Ψ then requires

$$\Delta X_j = 2\, \pi n_j\ , \tag{7.6}$$

where n_j is any positive or negative integer, or zero. Thus

$$\Phi = \frac{\hbar c}{e_j}\, n_j \tag{7.7}$$

for all particles, i.e. independent of j, as $A(x)$ depends on the space coordinate x only.

For electrons which are always present, $e_j = e$, and the smallest non-vanishing n_j becomes $n_j = n = 1$, i.e.

$$\Phi = \Phi_0 = \frac{\hbar c}{e} . \tag{7.8}$$

This implies, of course, that for other particles with a different charge, e_j, n_j/e_j must be the same as for electrons, which is always possible as long as integer charges only exist. The flux Φ is thus an integer multiple of Φ_0.

In the theory of superconductivity where flux quantization was first noticed, a flux quantum

$$\Phi = \frac{1}{2}\Phi_0 = \frac{hc}{2e} = 2 \times 10^{-7} \text{ electrostatic cgs units} \tag{7.9}$$

is found. The missing factor 2 in our derivation probably arises from neglecting the spin.

Apart from this it has been shown, however, that magnetic flux quantization is not restricted to superconductors but is a completely general phenomena.

Clearly in a homogeneous field, H, the flux quantum defines the surface of σ. If in particular σ is a circle with radius R then we have

$$R^2 \pi = \Phi = 2 \times 10^{-7} , \tag{7.10}$$

i.e.

$$R^2 = \frac{2 \times 10^{-7} \text{ cm}^2}{\pi H} , \tag{7.11}$$

if H is measured in Gauss. In the earth magnetic field, $H \approx 0.5$ Gauss and hence $R \approx 3 \times 10^{-4}$ cm, i.e. of the order of many biological cells. In a field of z times the earth's magnetic field, we would have

$$R \cong 3 \times 10^{-4} \sqrt{z} . \tag{7.12}$$

To appreciate the energies involved we consider a cylinder, radius R, height ℓ, a magnetic field H parallel to the axis, and assume that its strength gives rise to n flux quanta, so that

$$HR^2 \pi = n\Phi_0 . \tag{7.13}$$

The energy within the cylinder is then

$$U = \frac{H^2}{8\pi} R^2 \pi\ell = \frac{(HR^2 \pi)^2 \ell}{8 R^2 \pi} = \frac{\Phi_0^2 n^2 \ell}{8 \pi^2 R^2} , \tag{7.14}$$

or using $R = 10^{-4}$ cm,

$$U \cong \frac{4 \times 10^{-14} n^2}{8 \pi^2 10^{-4}} \frac{\ell}{R} \cong \frac{1}{2} 10^{-11} n^2 \frac{\ell}{R} \text{ erg} , \tag{7.15}$$

or as at room temperature $kT \approx 5 \times 10^{-14}$ erg,

$$\frac{U}{KT} \cong n^2 \frac{\ell}{R} 10^3 , \tag{7.16}$$

quite considerable.

On the other hand, macroscopic sizes, say R = 1 cm, according to (7.11) require only about 10^{-7} Gauss to establish a single flux quantum. Observation of flux quanta must thus deal with small systems as they are found in biology.

8 Multicomponent Systems and the Cancer Problem (cf. Fröhlich 1978, 1980)

Bacterial cells divide as long as nutrient and space are available. Normal tissues and organs are subjected to a control, however, which prevents them from growing beyond their appropriate size, or of mutating. This control is absent in cancer. From a naive point of view of physics, thus, the cancer problem requires finding this control. Clearly it requires a long-range interaction between different cells, i.e. it must involve all, or most of the cells of a particular tissue or organ. It will be suggested, then, that the required control can rest in the excitation of a particular mode of vibration of the relevant tissue or organ.

Consider first an assembly of equal units capable of polar oscillations, with equal frequency, e.g. a polar crystal. In crystal dynamics it is shown that this gives rise to a band of polar modes whose lowest frequency is different from zero. These modes can be divided into longitudinal and transverse modes; furthermore three modes with zero wave number, i.e. giant dipole vibrations, exist, whose frequencies depend on the macroscopic shape of the material. This arises from the long range of the interaction between polar vibrations.

If one of these equal units with equal frequencies is replaced by another one with sufficiently different frequency, then a localised oscillation with frequency outside the band of polar modes arises.

A non-crystalline arrangement of equal polar units does not basically change these conclusions, i.e. zero wave number polar modes exist; their frequencies depend on the macroscopic shape.

Consider now that our units are cells forming a tissue or organ, and that through the supply of metabolic energy a single mode is strongly, coherently, excited as discussed in Sect. 4.2, requiring an energy supply at a rate larger than a critical s_0. At this rate the system exhibits Einstein-Bose excitation of a single mode.

Alternatively, we might assume excitation of a limit cycle as discussed in connection with Eq. (5.11).

Both these excitations arise from non-linear interactions. Their dynamic consequences have not yet been fully explored, but even without such details important consequences can be drawn.

Thus in either case a coherent electric vibration carried out by the individual cells extends through the whole tissue (organ), and the vibrations of each individual cell responds to it, i.e. each cell is held in an appropriate phase and vibrational frequency.

Consider now that changes in some of the cell's physical characteristics have taken place; if strong enough, this cell will no longer be in appropriate resonance with the excited collective mode. In turn, this cell will no longer contribute to the collective excitation. Energy must be made available to perform such a change in an individual cell; its interaction with the excited mode thus acts as an inhibition on such changes. Forces may be expected to exist which tend to counteract the change in the particular cell. If, however, a considerable number of cells have undergone such changes, then the intensity of the excited normal mode will be strongly reduced and it will no longer be strong enough to inhibit changes in individual cells.

Thus a situation prevails which in some respect corresponds to order in physical systems described in terms of an order parameter (Haken 1977). In our case the strength of the coherent vibration represents the order parameter. Changes in certain physical properties of individual cells, which remove them from taking part in the collective vibration, cause disorder. An order-disorder transition may thus be anticipated. Disorder means that the coherent excitation is no longer maintained.

Two types of forces now arise from the excited mode: (i) those exerted by the whole system on a single cell outside it; (ii) forces acting within the whole system, which together with standard short-range forces determine the mean distance between cells, and possibly also the macroscopic shape. They also determine the frequency of the particular tissue (organ). Particular frequencies would then be characteristic for particular cell differentiation, as first proposed by F. Fröhlich (1973).

Now in cell division, and in cell mutation, certain changes in the structure of cells take place. The control exerted by the excited mode can then be expected to constrain these, i.e. it will inhibit cell division in regions in which the excitation is sufficiently strong. Arising from the complex structure of tissues and organs this may involve certain regions, but not others. Thus if we consider a sheet with in-phase oscillations perpendicular to the surface, but with a boundary condition of vanishing amplitude at the edges, then the inhibition of cell division would not hold for cells at the edges.

Now in cancer, the control exerted by the excited mode is no longer active. In terms of the model described by an order parameter, a phase transition from order to disorder has taken place. A critical number of cells must thus have undergone a transition which changes their frequencies.

Detailed quantitative treatment of the relevant non-linear problem has not been given yet, but will no doubt be possible in a number of ways, resembling, for example, those given by Haken (1977) or Arnold et al. (1978). The main feature they must describe is the existence of an order in terms of a non-linearly excited coherent electric vibration which controls cell division and mutation, and which stabilizes the relevant frequency, in conjunction with a possible order-disorder transition in which these controls are absent. The frequency region may either be based on membrane vibrations, order 10^{11} Hz, or be much lower as in the case of limit cycles (5.11).

An important conclusion is then that in the disordered (cancer) state order might be restored by external irradiation with the correct frequency. This requires, of course, extensive experimental investigations to establish the critical frequency.

9 Experiments

Three types of coherent excitations have been introduced from the point of view of theoretical Physics. They have particular characteristics and experiments can be devised to prove their existence. It must be mentioned at once, however, that this does not yet constitute a theory of biology – such a theory will rather make use of these excitations. Thus to give a trivial example for such a statement, to show that metallic electric conduction is carried by electrons does not yet constitute a theory of conductivity. A theory of biology combining a great complexity with a systematic activity may well require a new concept. Meanwhile, experimental evidence may yield some further information. Two types of investigations should be considered, one in which the concept of coherent excitation is used to interpret known biological features, a second in which experiments are carried out to establish the relevant excitations. Most of the chapters in this Volume deal with these problems, and as a result of these, and other investigations, the existence of coherent excitations can be considered as established.

Many of the experiments deal with monocellular materials. Interaction between cells, on the other hand, may well be an important feature, making use of collective excitations. This holds, for instance, in the model of control of cancer, presented in the previous section, and even the discussion of the linguistic of the genetic code, by F. Fröhlich, this Volume, makes use of long-range coherent excitations.

Further use of the possibility of collective coherent excitations is presented in the chapter by Kell and Hameroff in this Volume as well as by Huth et al. (1984). Evidence for collective excitations also arises in B. Goodwins continuous phase transitions Goodwin and Trainor (1980) and in Sobell (1985), as well as in various Chapters in the book Modern Bioelectrochemistry, ed. F. Gutmann and H. Keyser (1986). Evidence for long-range interaction based on the excitation of coherent vibrations is given by Rowlands in this Volume. Hasted's chapter in this Volume presents direct evidence for the existence of metastable excited states, as do the investigations by Mascarenhas (1975). Furthermore the so-called Davydov (1979) soliton also represents a particular case of metastable state. The existence of slow electric vibrations in active cells, as required by periodic enzyme reactions discussed in Sect. 5 is shown in the chapter by Pohl and Pollock in this Volume.

Fairly up-to-date references on these types of investigations are available in the review articles by Fröhlich (1977, 1980, 1983b, 1986a,b). Explicit reference should be made to the importance of the properties of cell water as shown by Clegg (1984). Also the possible importance of O^{2-} for coherent electric vibrations had been stressed by Bilz et al. (1981).

By far the most extensive investigations refer to the interaction of biological systems with electromagnetic radiation, many of them mentioned in the above quoted reviews. Some of them show very striking effects, but difficulties with their reproducibility have arisen some times. It is only in recent years that this has been shown to be due to the occurrence of deterministic chaos. The mathematical treatment of the occurrence of chaos in non-linear excitations is presented in Kaiser's chapter in this Volume. Its importance in biology has been realized at an early date by Tomita (1984). Roughly speaking, it implies that in a non-linear system, a very small change in the initial conditions may result in a large change of final states. A trivial example has been presented in Sect. 2.

Another important feature that can lead to difficulties in reproductibility arises in an assembly of cells that interact but are not close enough to form bound states. Synchronization of their vibrations, and hence of properties of scattering may then require special treatments that have not yet been standardized. As an example consider the Raman effect. Here when vibrations are non-thermally excited, then the ratio of anti-Stokes to Stokes lines should be larger than in thermal equilibrium. A very great number of investigations on active bacterial cells have been carried out by Webb (1980). Under the particular conditions it can be shown (cf. Fröhlich 1986a) that a high state of synchronization is required to find the lines, but unfortunately the details of preperation have not been given. It is of interest, therefore, that Bannikov et al. (1980) find a great sensitivity to the cultivation conditions for successful results.

Drissler and Santo (1983) were successful only after 20 failures. They were informed by synchronization experts that with their method they cannot expect anything better. A highly reproducible method was then proposed which successfully, reproducibly, confirmed a non-thermal excitation, as discussed on p. 254 of Fröhlich (1986a). Further evidence for excitations are obtained in photosynthesis, cf. Drissler is article in this book.

The excitation of coherent electric vibrations invites, of course, investigations of the interaction with electromagnetic waves. It must be pointed out in this context that in an infinite ionic crystal most of the frequencies are optically inactive as the wave numbers of the vibrations and the radiation must coincide. Activity can arise, however, from the effects of surface and defects. Emission of radiation arising from coherent excitations in the system must then be expected to be very weak, largely due to internal surfaces. Nevertheless it should reflect the sharp frequency distribution and the coherence of the excitation. It seems likely that the observed "ultraweak" photon emission from active biosystems by Li et al. (1983) is due to such effects.

A considerable number of experiments have been carried out on the effect of low intensity, sharp frequency and irradiation on biological activity as discussed by Kremer and by Grundler et al. in this Volume. They do, in fact, confirm the predictions of the theory as discussed in some detail by Fröhlich (1980). Thus if the rate of energy supply arising from random biological supply is just below the critical s_0 at which coherent oscillations are excited, then a small external supply will lift the total above s_0, especially when it is supplied at the frequency that will be excited. Alternatively, the random energy supply may be increased through a thermal current arising from the application of a small temperature difference as shown by Fröhlich (1985).

The work by Grundler et al. presented in this Volume provides further evidence for the relevance of deterministic chaos in biologically active materials. This, of course, is to be expected as from the point of view of physics such materials are energetically highly excited and thus contain a distribution over a large number of physical states which individually may react differently to irradiation.

Finally in this section it must be mentioned that interests different from scientific ones are relevant in research on microwave radiation – cf. the November 1986, number 72, issue of the Bioelectromagnetic Society Newsletter.

References

Arnold L, Horsthemke W, Lefever R (1978) White and coloured noise and transition phenomena in nonlinear systems. Z Physik B29:367–373

Bannikov VS, Bezruchko SM, Grishankova EV, Kuzmin SB, Mituagin Yu A, Orlov R Yu, Roskkov SB, Sokolina VA (1980) Investigation of *Bacillus megaterium* cells. Dokl Akad Nauk SSSR 253:No 2:479–480

Bilz H, Büttner H, Fröhlich H (1981) Electret model for the collective behaviour of biological systems. Z Naturforsch 36b:208–212

Bloch F (1968) Simple interpretation of the Josephson effect. Phys Rev Lett 21:1241–1243

Bullock TH (1977) Electromagnetic sensing in fish. Neurosci Res Prog Bull vol 15, 1:17–22

Clegg JS (1984) Intracellular water and the cytomatrix. J Cell Biol 99:167s–171s

Davydov AS (1978) Solitons bioenergetics and mechanism of muscle contraction: Int J Quantum Chem 16:5–17

Devyatkov ND (1974) Sov Phys-Usp (engl. transl.) 16:568

Drissler F, Santo L (1983) Raman effect. In: Fröhlich H, Kremer F (eds) Coherent excitations in biological systems. Springer, Berlin Heidelberg New York, pp 6–9

Fermi E, Pasta JR, Ulam SM (1965) Collected works of Fermi, vol 2. University of Chicago Press, pp 978–980

Fröhlich F (1973) Life as a collective phenomenon. In: Haken H, Wagner M (eds) Cooperative phenomena. Springer, Berlin Heidelberg New York, pp VII–XII

Fröhlich H (1958) Theory of dielectrics, 2nd edn. Oxford Press, Oxford

Fröhlich H (1968) Long range coherence and energy storage in biological systems. Int J Quantum Chem 2:641–649

Fröhlich H (1972) Selective long range dispersion forces between large systems. Phys Lett 39A: 153–154

Fröhlich H (1973a) Collective behaviour of non-linearly coupled oscillator fields. Collect Phenom 1:101–109

Fröhlich H (1973b) The connection between macro- and microphysics. Riv Nuovo Cim 3:490–534

Fröhlich H (1975) The extraordinary dielectric properties of biological materials and the action of enzymes. Proc Natl Acad Sci USA 72:4211–4215

Fröhlich H (1977) Long-range coherence in biological systems. Riv Nuovo Cim 7:399–418

Fröhlich H (1978) Coherent electric vibrations in biological systems and the cancer problem: IEEE Trans Microwave Theory Tech 26:613–617

Fröhlich H (1980) The biological effects of microwaves and related questions. In: Advances in electronics and electron physics, vol 53. Academic Press, New York, pp 85–152

Fröhlich H (1982) Conditions for coherent excitations in biological systems. Phys Lett 93A: 105–106

Fröhlich H (1983a) Evidence for coherent excitation in biological systems. Int J Quantum Chem 23:1589–1595

Fröhlich H, Kremer F (eds) (1983b) Coherent excitations in biological systems. Springer, Berlin Heidelberg New York

Fröhlich H (1985) Further evidence for coherent excitations in biological systems. Phys Lett 110A:480–481

Fröhlich H (1986a) Coherent excitations in active biological systems. In: Gutmann F, Keyser H (eds) Modern bioelectrochemistry. Plenum, New York, pp 241–261

Fröhlich H (1986b) Coherence and action of enzymes. In: Rickey Welch G (ed) The fluctuating enzyme. John Wiley, New York, pp 421–449

Goodwin BC, Trainor LEH (1980) Cleavage lines of holoblastic eggs. J Theor Biol 86:757–761

Haken H (1977) Synergetics. Springer, Berlin Heidelberg New York

Huth GC, Bond JD, Tove PV (1984) Nonlinear Tunneling barriers at high frequencies and their possible logic processing function in biological membranes. In: Ross Adey N, Lawrence AF (eds) Nonlinear electrodynamics in biological systems. Plenum, New York, pp 227–241

Li KH, Popp FA, Nagl W, Klima H (1983) Indications of optical coherence in biological systems and its possible significance. In: Fröhlich H, Kremer F (eds) Coherent excitations in biological systems. Springer, Berlin Heidelberg New York, pp 117–122

Marois M (ed) (1969) Theoretical physics and biology. North Holland, Amsterdam, p 13
Mascarenhas S (1975) Electrets in biophysics. J Electrastat 1:141–146
Monod J (1972) Chance and necessity (transl.). Collins, London, p 95
Pokorny J (1982) Computer analysis of coherent electric vibrations in biological systems. Czech J Phys B321:928–935, J Theor Biol 981:21–27
Prigogine I (1980) From being to becoming. Freeman, San Francisco
Rose (1970) Image Technol 12:8
Sobell HM (1985) Kink-antikink bound states in DNA structure. In: Jumak and McPherson (eds) Biological macromolecules and assemblies, vol 2. Wiley, New York, pp 172–232
Tomita K (1984) The significance of the concept chaos. Prog Theor Phys Suppl 79:1–25
Wada A, Nakamura H (1981) Nature of the charges distribution in proteins. Nature 293:757–758
Webb S (1980) Laser Raman spectroscopy of living cells. Phys Rep 60:201–224

Theory of Non-Linear Excitations

F. Kaiser [1]

1 Introduction

The description of order and function in biological systems has been a challenge to scientists for many years. From the point of view of theoretical physics, biological function must be treated in terms of dynamic properties. A consideration of the motion of individual particles (atoms, molecules, ...) is meaningless because of the enormous number of possible states. In addition, the problem of the interaction between electromagnetic fields and biological systems has gained increasing interest both through theoretical considerations and through rather exciting experimental results (Fröhlich 1980, 1986a; Kaiser 1981, 1983a,b, 1984; Grundler, this Vol.).

As early as 1967, it was emphasized (Fröhlich 1969) that biological systems exhibit a relative stability for some modes of behaviour. In vivo, these modes remain very far from thermal equilibrium. One manifestation of such behaviour is a coherent excitation, which means that a single mode is strongly excited. The stabilization of such a mode requires non-linear interactions between the relevant biological subunits. Since the functional complexity of biological materials requires the application of macroscopic theories, the application of the concepts on non-linear dynamics seems appropriate. Fröhlich's concept of coherent excitations lies within this general frame work (Fröhlich 1980, 1986b). It incorporates the necessity that these specific modes be thermally decoupled from the remaining part of the system, which will act as a randomizing heat bath.

The existence and stabilization of far from equilibrium states by non-linear interactions within at least some subunits of a physical, chemical or biological system is intimately linked with cooperative processes. Besides the well-known strong equilibrium cooperativity, thermodynamic metastable states and nonequilibrium transitions in cooperatively stabilized systems can occur, provided a certain energy input is present. In equilibrium, an entire subunit or domain of a macromolecular system reacts as a unit, which means that it transforms as a whole.

Examples are chemical transformations and physical changes of macrovariables (equilibrium phase transitions). In non-equilibrium situations, non-linearities create dissipative elements which lead to new states including trigger action, threshold and hysteresis. The additional interaction of these non-equilibrium states with external excitations increases in a dramatic way the number and types of non-equilibrium states.

1 Institut für Angewandte Physik – Theorie –, Technische Hochschule Darmstadt, Hochschulstraße 2, 6100 Darmstadt 1, FRG

Two ways of non-linear excitation are possible. Strong internal or external fields require the application of non-linear laws (material laws) for the internal dynamics, which in turn can result in coherent excitations. Examples of this type of excitations are laser systems and instabilities and waves in plasmas. Weak or extremely weak external fields can create non-linear effects by coupling to cooperative processes in non-equilibrium systems, for direct processes (e.g., interaction field-molecule) the fields are too weak. This situation is assumed to play an essential role in biological systems. Particularly, non-thermal effects in irradiated biosystems, which are still the matter of highly controverse discussions, receive a physical basis within this concept. To be more specific, the interaction field – biosystem – is reduced to the interaction field – cooperative process. A possible realization is the coupling of electromagnetic fields to coherent excitations.

Two types of coherent excitations have been suggested within the realm of non-equilibrium cooperativity (Fröhlich 1968a,b, 1986b). (1) In the *high polarization model* the supply of energy beyond a certain threshold generates metastable highly polar states. (2) In the *coherent vibration model*, oscillations are produced and stabilized by a certain rate of energy flux through the system. Both types of excitations are based on definite physical models and they have received some experimental support (Fröhlich and Kremer 1983). The possible existence of static and dynamic coherent excitations can have far-reaching consequences for the temporal and spatial order and its function in biological systems. It is well known that a large number of oscillating phenomena in biological systems exists, with periods ranging from milliseconds (biochemical oscillations, neuronal activity) to hours (circadian rhythms) and even to months and years (population cycles) (Kaiser 1980; Rapp 1986).

In recent years it has become clear that non-linear phenomena and their interactions with fields and forces is abundant in nature. Examples range from mechanics (forced anharmonic oscillators), hydrodynamics (pattern formation and turbulence), electrics and electronics (non-linear circuits, Josephson junctions), non-linear optics (laser, optical bistability), acoustics (cavitation bubbles), chemistry (oscillating reactions), biochemistry (glycolytic oscillations) to biology (stimulated heart beat, EEG, neuronal activity). Only some disciplines with a few typical examples have been mentioned.

From a theoretical point of view, non-linear phenomena require the investigation of non-linear dynamical systems with only a few relevant degrees of freedom. These studies include the internal dynamical behaviour and its perturbation by external influences. For example, a non-linear oscillator that is subjected to an external periodic forcing can exhibit stable and unstable periodic and almost periodic responses, coexisting subharmonic states and onset of a completely irregular motion. This complex dynamical behaviour seems to be an aspect of biological regulation, but, rather surprisingly, it is found in very simple non-linear systems. The essential non-linear phenomena are (1) waves that behave like particles (e.g. solitons), (2) transitions within regular motions and from regular motions to irregular ones, (3) the formation of spatial or temporal-spatial structures from an isotropic medium.

As a result of a large and rapidly increasing number of theoretical and experimental studies with non-linear systems one can state: a new physics is developing, which can be qualified as non-linear, whose temporal evolution (evolution in time) is similar in all non-linear systems and permits a unified description. It comprises fascinating phenomena: turbulent, irregular or chaotic motion originating from simple regular oscillations ranging from single mechanical oscillators to complex biological rhythms.

Without becoming lost in the mathematical details, the relevant dynamical non-linear phenomena are described and the way in which they appear in analytical and numerical investigations and in experiments is investigated. To keep the present contribution self-contained, only the temporal evolution of excited non-linear systems will be presented. The occurrence of irregular shapes and pattern (fractals) in both geometrical and dynamical descriptions and their relevance or connection to temporal and spatial chaos is beyond the scope of this article. Finally, the possible relevance of non-linear excitations in biological systems will briefly be discussed. Some remarks on spatio-temporal structures are given in Appendix B.

2 Non-Linear Dynamics

The theoretical investigation of non-linear excitations is performed with the help of non-linear evolution equations. These equations describe the dynamic behaviour of the state of the underlying model. Dynamical systems change their state with time. Systems of this type are continuous systems

$$\dot{\underline{x}} = \underline{F}(\underline{x}, \mu, t) \tag{1}$$

or discrete systems

$$\underline{x}_{n+1} = \underline{f}(\underline{x}_n, \mu) . \tag{2}$$

The state of the system is $\underline{x} = (x_1, \ldots, x_m)$ and $\underline{x}_n = (x_{1,n}, \ldots, x_{m,n})$ (i.e. $\underline{x}$, $\underline{x}_n \in R^m$, m-dimensional system). The non-linear functions $\underline{F}$ and $\underline{f}$ describe the development of the system with time in a continuous or discontinuous way ($\dot{x}$ means derivative of $\underline{x}$ with respect to time). The state at time n + 1 is given by the state at time n by the function $\underline{f}$. μ is a parameter that determines the actual state of the system. System (1) consists of m ordinary non-linear differential equations of first order, system (2) represents m non-linear one-dimensional maps. Differential-delay equations may occur as well in non-linear dynamical problems.

2.1 Some Historical Background

Non-linear dynamics is a very old problem. It originates in planetary motions. Henri Poincaré was the first to investigate the complex behaviour of simple mathematical systems by applying geometrical methods and studying topological structures in the phase space (Poincaré 1892). He discovered that strongly deterministic equations for the motion of planets and other mechanical systems could display an irregular or chaotic motion. Some years later, a mathematical basis for this behaviour was given (Birkhoff 1932). Only in 1963 in a numerical "experiment" on a model of boundary layer convection was it discovered that a system of three first-order non-linear differential equations can exhibit a chaotic behaviour (Lorenz 1963). Contrary to Poincaré's example (deterministic chaos in conservative or Hamiltonian deterministic systems), Lorenz discovered deterministic chaos in dissipative systems.

The Lorenz model may be viewed as the prototype example for many non-linear dynamical systems, e.g. for the biologically motivated studies on populations (May 1976). The development of the digital computer offered an additional tool to study aspects of non-linear dynamics which were previously considered to be too complex.

The essential results of Lorenz are: (1) oscillations with a pseudo-random time-behaviour (now called chaotic), (2) trajectories that oscillate chaotically for a long time before they run into a static or periodic stable stationary state (preturbulence), (3) some trajectories alternate between chaotic and stable periodic oscillations (intermittency), (4) for certain parameter values trajectories appear chaotic although they stay in the neighborhood of an unstable periodic oscillation (noisy periodicity).

It took another 10 years before the importance of these results was recognized. Since then the number of both theoretical and experimental studies on the complex behaviour of simple systems has rapidly increased (for a review see, e.g. Hirsch 1976; Gurel and Rössler 1979; Collet and Eckmann 1980; Ott 1981; Schuster 1984; Bergé et al. 1984; Cvitanović 1984; Hao 1984; Fischer and Smith 1985; Holden 1986; Thomson and Stewart 1986).

Some problems remain unsolved. For example, the identification of the chaotic attractor in the mathematical theory (Birkhoff) and in the simulation "experiment" (Lorenz) has not yet succeeded, i.e. the chaoticity of the numerically computed results has not been proven in the strong mathematical sense. Furthermore, it should be emphasized that chaos is only a part of the fascinating behaviour that non-linear systems can exhibit. Since already for regular motion a huge number of states and bifurcations exists, the following of some trends and the concern only with chaotic motion should be avoided. Especially for externally excited systems, regular motion and its bifurcation need to be considered.

2.2 Standard Methods and Regular Motion

Many essential features of non-linear systems are already contained in systems described by Eq. (1). Furthermore, continuous systems are in many cases closer to the physical situation. In order to characterize a stationary and periodic motion, only finite dimensional continuous systems are considered. Without loss of generality, the system (1) is reduced to a non-linear, driven, damped oscillator system (Appendix A):

$$\ddot{x} + g(x,\dot{x})\dot{x} + f(x)x = F(t)\,, \tag{3}$$

which for $F(t) = F_0 + F_1 \cos \lambda t$ can be written as

$$\dot{x} = y \tag{4a}$$

$$\dot{y} = -\, g(x,y)y - f(x)x + F_0 + F_1 \cos z \tag{4b}$$

$$\dot{z} = \lambda\,. \tag{4c}$$

System (3) is a two-dimensional non-autonomous system, system (4) the corresponding three-dimensional autonomous system. These systems are sufficient for a discussion of non-linear excitations, although they contain the weakness of all modelling approaches: a real world phenomenon originally modelled by partial differential equations is described by finite-dimensional ordinary differential equations. The question remains, whether the results from a mathematical theory or from a computer-generated numerical output coincide with experimental results. Fortunately, for many problems this reduction seems appropriate. The field or force F(t) in Eq. (3) is assumed to be an external driver which is composed of a static part (dc field or bias) and a periodic component with fixed frequency λ. Both internal or stochastic drivers might occur as well, but they are neglected, since a restriction to systematic stimuli seems necessary.

With the help of Eq. (3), the behaviour of externally excited non-linear systems is discussed, based on a rather conceptual basis. Equation (3) is non-linear, it cannot be integrated in most cases by analytical methods, i.e. no analytic solutions are found. In order to obtain information on the motion of system (3), one first of all looks for stationary solutions of the unexcited system. Since non-linear systems exhibit stable and non-stable solutions for the same set of internal and external parameters, a stability analysis is requested. Only those solutions that persist under small perturbations are considered stable. Standard methods are available for these investigations (Andronov et al. 1970; Marsden and McCracken 1976).

In Fig. 1 the essential features to be considered for an investigation of regular motions in non-linear and excited systems are given in a rather qualitative and schematic way:

Oscillation diagram: the amplitude x is given as it develops with time t (for concrete situations, this follows from numerical integrations).

Phase plane diagram: the velocity $\dot{x}$ is given as a function of the amplitude x, the arrow represents the motion with increasing time t. In the example given, the motion is periodic, it repeats for t = T (T = period of the oscillation).

Power-spectrum: square of the Fourier coefficient $c(\omega)$ as a function of the frequency ω. The signal or amplitude x(t) is Fourier-transformed

$$c(\omega) = \lim_{T\to\infty} \int_0^T x(t)\, e^{i\omega t} dt \,. \tag{5}$$

For periodic motion the power spectrum consists of discrete lines only.

Bifurcation diagram: steady state amplitude x as a function of a bifurcation parameter μ (μ is either an internal parameter or one of the external ones (λ, F_1). For a certain value of μ, the steady state $\tilde{x}$ becomes unstable, two stable amplitudes $x \gtrless \tilde{x}$ coexist.

Bifurcations are the most important phenomena occurring in non-linear systems. They are responsible for a large series of different states that the system occupies for different excitations. Often bifurcating systems exhibit hysteresis. The system fails to return to its original state after a temporary change of its control.

Fixpoint: simplest stable steady state solution. For increasing time, all trajectories terminate in this point (provided no further stable steady state exist). Stable fixpoints are static attractors.

Limit cycle: closed trajectory in phase space. All trajectories terminate in this cycle, no other closed cycle lies in its neighborhood. Without external drive [F(t) = 0], the limit cycle corresponds to a periodic stable solution of the non-linear system, whose

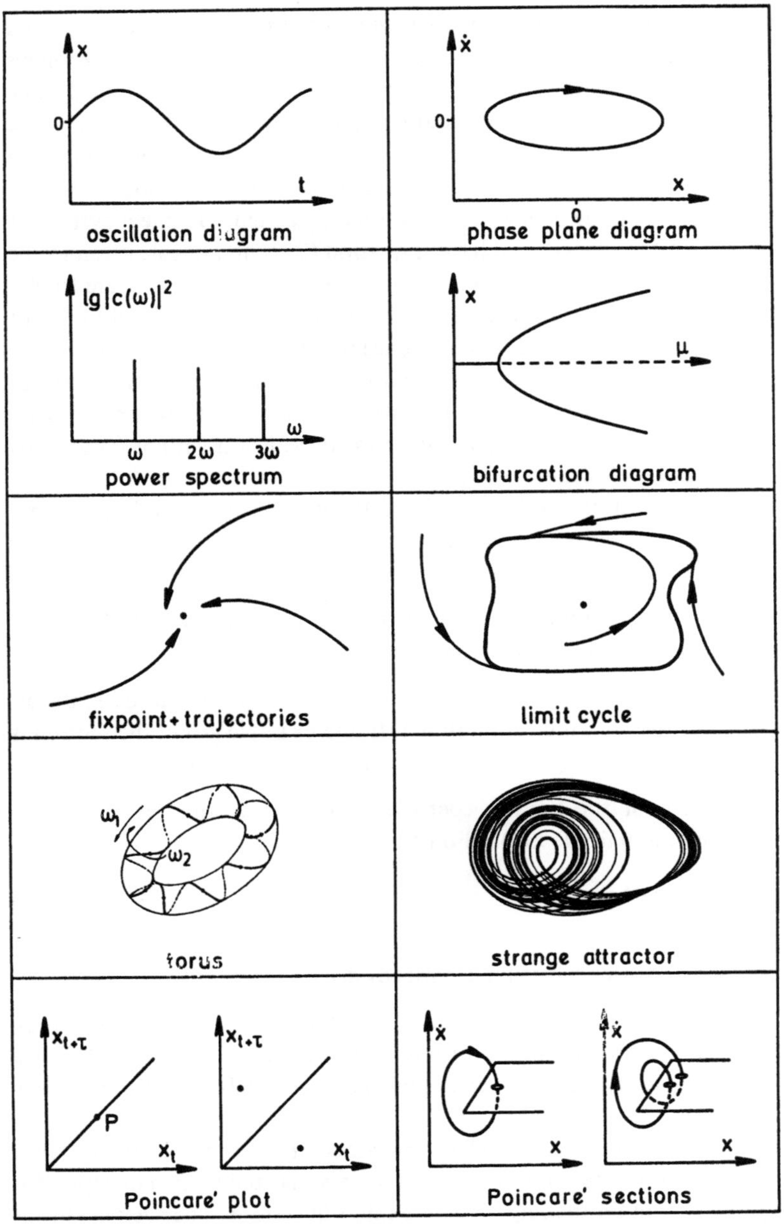

Fig. 1. Schematic view of the essential constituents coming from an analysis of the regular motion of non-linear excited systems. The details are discussed in Sect. 2.2

amplitude and frequency is determined by internal parameters (self-sustained oscillation). Stable limit cycles act as periodic attractors.

Torus: the system's trajectories move on a two-dimensional torroidal surface. Two frequencies are present, oscillations around the torus (ω_2) and along the torus (ω_1). For rational $p/q = \omega_2/\omega_1$, a periodic motion with period q results, the cycle closes after q cycles. For ω_2/ω_1 irrational, a quasi periodic motion is given. The trajectory never closes and covers the whole torus. $\rho = p/q$ is the rotation number.

Strange attractor: the trajectory never closes, but its motion is restricted to a finite region (basis of attraction). The trajectory does not cross, the present diagram is a projection of a three-dimensional $x-\dot{x}-t$ diagram in the $x-\dot{x}-$ plane. The motion is called irregular or chaotic. Its chaoticity stems from a permanent non-linear folding and stretching of trajectories. Separating trajectories are again brought together, close trajectories are separated. These properties explain the content of the word "strange attractor". Quite general, strange attractors (limit cycles) represent a mapping of non-periodic (periodic) self-oscillations in the phase-space of dynamical systems.

From topological considerations it can be proven that only certain types of stable attractors can exist for a fixed dimension m. For systems of m first order ordinary differential equations one finds:

m = 1	m = 2	m = 3	stable fixpoints	/ static attractor
	2	3	+ stable limit cycles	/ periodic attractor
	2	3	+ torroidal phase space	/ motion on a torus
		3	+ strange attractor	/ chaotic motion .

Chaotic motion is only possible in three- or higher-dimensional non-linear autonomous systems or equivalently, in two- or higher dimensional non-autonomous systems, as given by Eqs. (3) and (4). However, the situation changes completely for discrete systems [Eqs. (2)]. Chaos in the sense of strange attractors can be present in one-dimensional iterative non-invertible maps and in two-dimensional iterative maps. All the statements are necessary requirements.

Further information can be obtained by the following procedures:

Poincaré section: the trajectories of the 3-d phase space $(x-\dot{x}-t)$ or $(x-y-z)$ are cut with a 2-d plane ($t = 2\pi i$ or $z = 0$ for example). The projection of the strobe points where the trajectories intersect the plane from top to bottom are redrawn in a $x-\dot{x}$ plane. For periodic states one obtains exactly one point, for double-period oscillations two points, etc. With this method, one can separate periodic, quasi-periodic and aperiodic motions.

Poincaré plot: the x-values of the Poincaré sections are plotted in a $x_{t+\tau}-x_t$ diagram, where $\tau = \frac{2\pi}{\lambda}$ for externally driven systems. Thus the points of intersection are ordered in time, the 2-d diagram corresponds to a stroboscopic plot, the actual amplitude at times $t + n\tau$ ($n = 0,1,2,...$) is drawn. Again, periodic, quasi-periodic and aperiodic states can be distinguished and the stability can be shown.

It should be emphasized that the standard methods of Fig. 1 are not restricted to three-dimensional systems. The transformation to Poincaré sections and to Poincaré plots reduces the continuous dynamics of differential equations as given by (1), (3) and (4) to interated maps [discrete system (2)]. Quite generally, one intersects a R^m phase space with a R^{m-1}-dimensional surface.

For externally driven non-linear systems, some further characteristics yield informations on the system's behaviour. In Fig. 2 the steady state amplitude as a function of either the external frequency λ or the external driver amplitude F_1 are drawn in a rather qualitative and schematic way. For linear systems, the resulting frequency is given by λ. The system oscillates with the driver frequency and with an amplitude that is directly proportional to the external field F_1. A broad resonance peak in the frequency response diagram (x_s-λ) results. The behaviour changes drastically in the non-linear case. Assuming a stable limit cycle with amplitude X_{LC} and frequency ω_{LC} for $F(t) = 0$, one obtains for an increasing driver amplitude F_1 the transitions (bifurcations):

internal frequency ω_{LC} $\rightarrow$ quasiperiodic states $\rightarrow$ frequency λ ,
internal free amplitude x_{LC} $\rightarrow$ quasiperiodic states $\rightarrow$ increasing amplitude ,

i.e. a transition from the free to the entrained oscillation. For a fixed F_1 and a variable driver frequency, one finds:

internal frequency ω_{LC} $\rightarrow$ synchronization to λ $\rightarrow$ internal frequency ω_{LC} ,
internal amplitude x_{LC} $\rightarrow$ sharp resonance peak $\rightarrow$ internal amplitude x_{LC} .

Only for a certain range of the frequency does the internal oscillation become entrained. Below and above this region, the free oscillation dominates.

Depending on the external parameters λ and F_1, three essential types of behaviour result: free oscillation (unperturbed limit cycle); oscillation into a state which is a non-linear mixture of free and a forcing oscillation; entrained oscillation (synchronized, phase-locked). Bifurcations to sub- and superharmonic oscillations can occur, as

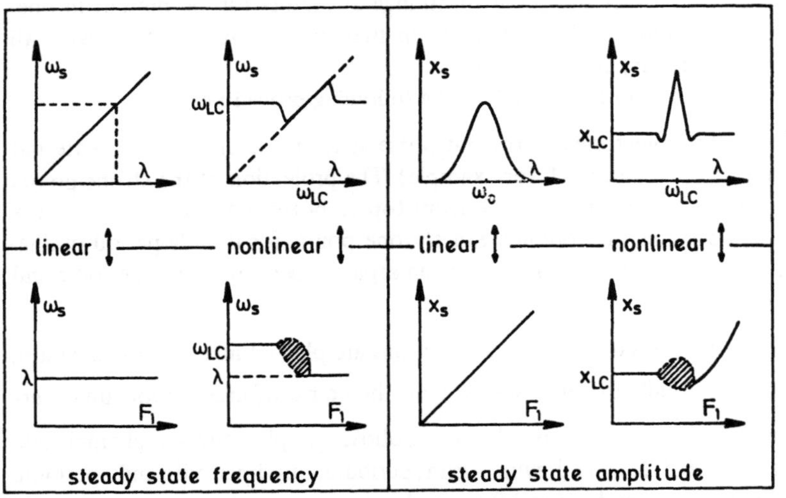

Fig. 2. Schematic view of the steady state frequency and steady state amplitude of a linear and a non-linear oscillator (shown are the results for oscillators A8 and A10). The oscillators are driven by $F(t) = F_1 \cos \lambda t$ (subscript s = steady state, subscript LC = limit cycle; see Sect. 2.2)

well as bifurcations to some other periodic and to chaotic states. These states have been omitted in Fig. 2.

Figures 1 and 2 contain the essential features of excited non-linear systems, which follow from standard methods for the investigations of non-linear continuous systems. Some concrete results for a specific system (coherent oscillation model, Appendix B) are shown in Figs. 4–8. Further information on driven non-linear systems can be found in many textbooks (e.g. Hayashi 1964; Nayfeh and Mook 1979; Guckenheimer and Holmes 1983).

2.3 Characterization of Chaotic Motion

More refined methods are required to discuss the complete behaviour of driven non-linear systems. A series of problems and questions arises. Some of them can be stated as follows: (1) is an apparently aperiodic state found in numerical calculations or in experiments really chaotic, or is it quasi-periodic or periodic with an extremely long period? (2) how can one separate noise and uniform randomness from deterministic irregularities and structured chaos? (3) what is the origin of deterministic chaoticity and how can one measure the strength of chaos? (4) how can one construct attractors from measured signals (i.e. time series of some variables)?

Meanwhile, some procedures have been developed which allow for some answers to these questions. Since these methods in many instances are a direct continuation of the standard methods, a close connection to the concept of Sect. 2.2 and to Fig. 1 is indispensable. Again, only systems are considered whose time-dependent motion is deterministic, i.e. there exists a set of differential equations which determine the future behaviour of the system from given initial conditions [see Eqs. (1)]. In the following, deterministic chaos denotes a nonperiodic or irregular motion which is sometimes called disordered or turbulent. It is generated in non-linear systems by deterministic laws for the time evolution of the system.

To keep the discussion within a reasonable range, again a schematic drawing is given for some of the essential methods (Fig. 3). Some of the definitions are already given in Sect. 2.2.

Autocorrelation function: this function measures the correlation between subsequent signals or amplitudes $x(t)$. For regular motion it oscillates or remains constant, for chaotic motion it decays exponentially, i.e. no long-term correlations for two motions exist. In the diagram, only the positive envelope is drawn. The autocorrelation function is the Fourier transform of the power spectrum.

Power spectrum: for a chaotic motion, one obtains an exponential decay of the different frequency contributions for increasing frequency λ. Some broad band peaks are superimposed at low frequencies, which is a typical characteristic of deterministic chaos.

Divergence of trajectories: a deterministic motion is identical only for exactly the same starting conditions. For slight deviations, no exact repetitions occur, nearby starting trajectories separate extremely.

Bifurcation diagram: a whole bifurcation series is shown for the steady state amplitude. The external field strength F_1 is the bifurcation parameter. Period doubling se-

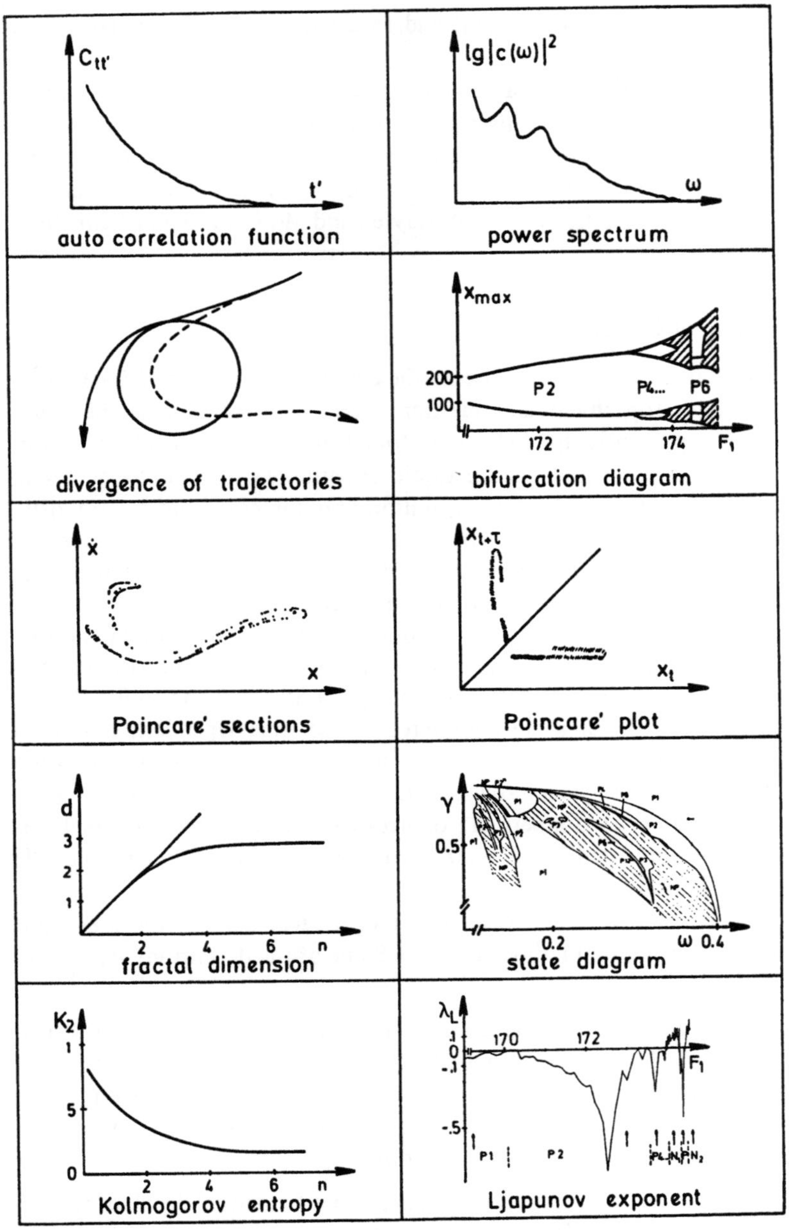

Fig. 3. Extension of Fig. 1 for chaotic motion. The bifurcation diagram, the Ljapunov exponent and the Poincaré plot are calculations for model B1, the Poincaré section and the state diagram are results for a non-linear driven pendulum [Eqs. (A12)]. The details are discussed in Sect. 2.3

quences are followed by chaotic states (2-band chaos in this example, see the model of Appendix B). Bifurcation diagrams are restricted to low dimensions, e.g. one steady state as a function of a single parameter is considered. For a multivariable and a multi-parameter system, these diagrams are not applicable.

Poincaré section and Poincaré plot: both diagrams show large series of points, which are space-filling in many cases. Contrary to a random motion, the occupied space is restricted to a certain amount of the whole available space. Again the reduction of dimension and the transition from continuous to discrete dynamics should be mentioned.

State diagram: the bifurcation manifold in parameter space is given. In the present example (driven torsion pedulum) an internal parameter (damping γ) and an external one (driver frequency λ) determine the state. P_n refers to periodic states with period $T_n = n \frac{2\pi}{\lambda}$, the shaded regions are chaotic motions. Different bifurcation routes are exhibited.

Some criteria have been presented for a chaotic motion. In all the examples it is demonstrated that the occurrence of chaos corresponds to a qualitative change of the behaviour. More refined techniques exist which allow for an investigation of chaotic attractors for both numerical and experimental systems. In many experiments one observes signals which seem random, and one wants to know if the underlying motion is chaotic in the deterministic sense. Furthermore, the measured time signal is one-dimensional and discrete in most cases, i.e. the trajectory for the time development is projected to one coordinate. However, it is well known that this single coordinate contains information from the other variables. Certain procedures have been developed to construct the attractor from a data set and to extract the inherent information from the measured signal (Schuster 1984). Some examples are given in Fig. 3.

Fractal dimension: different dimensions for strange attractors (deterministic chaos) can be defined, e.g. Hausdorff dimension, information dimension, correlation dimension. All these dimensions have non-integer values for chaotic states (fractal dimensions). By embedding procedures, these dimensions can be calculated. The embedding dimension n is the lowest integer dimension that contains the whole attractor. In the diagram, the fractal dimension d is shown for increasing n. Already for small n, d ceases to change. For random systems d is proportional to n. It should be emphasized that to a certain extent noise can be extracted from the irregularities which originate from deterministic motion. In addition, fractal dimensions provide some indication of how many relevant degrees of freedom are involved in the dynamics of the system.

Kolmogorov entropy: K is a fundamental measure for a chaotic motion. It gives the average rate at which information about the state of a dynamical system is lost with time. For regular motion K becomes zero, for random systems it is infinite. Deterministic chaos exhibits a small, positive K-value. It gives some insight into the predictability of the system.

Ljapunov exponent: the exponent λ_L changes with the internal or external parameters. It represents a generalization of the linear stability analysis, the latter is only applicable to fix points and limit cycles. Systems with at least one $\lambda_L > 0$ are chaotic.

λ_L measures the average divergence of nearby trajectories, chaotic states exhibit an exponential divergence for at least one λ_L. For an m-dimensional system, m exponents exist. For m = 3 one obtains (only the sign of λ_L is given):

stable fixpoint $(-,-,-)$; stable limit cycle $(0,-,-)$
stable torus $(0,0,-)$; strange attractor $(+,0,-)$.

In the diagram, the transitions $\lambda_L < 0$ to $\lambda_L > 0$ indicate bifurcations from periodic states (P1, P2, P6 etc.) to non-periodic or chaotic states (N_1, N_2) for the model in Appendix B). $\lambda_L > 0$ (chaotic states) implies that the system is characterized by a restricted long-time predictability. Furthermore, chaotic states exhibit asymptotic stability, though they possess an internal instability. It originates from a finite basin of attraction (see Sect. 2.2: strange attractor).

Fractal dimensions, Kolmogorov entropy and Ljapunov exponents are closely related to each other. They are the relevant measures for an identification of deterministic chaotic states, particularly for an analysis of experimental signals.

3 Theoretical and Experimental Implications

3.1 Non-Linear Excitations and Bifurcations

Externally driven non-linear systems exhibit an enormous variability of behaviour. If one changes at least one internal or external parameter, the system undergoes continuous or discontinuous changes from one attractor to another at some critical values. The determining parameter is called the control or bifurcation parameter. The transitions are bifurcations of steady-state solutions of the dynamical system.

There are three types of bifurcations. (1) *Hopf Bifurcation:* typical examples are transitions from a static attractor (fixpoint) to a periodic attractor (limit cycle) and from the latter one to the motion on a torus. (2) *Saddle-node or tangent bifurcations:* transitions from a limit cycle to a new one, discontinuous transitions on a hysteresis and transitions from quasi-periodic to phase-locked periodic states are the dominating bifurcations. (3) *Period doubling bifurcations:* a limit cycle of period T bifurcates into an oscillation with period 2T, which in many cases is followed by a whole cascade of further period doubling bifurcations to states with periods 4T, 8T ... 2^nT. Period tripling and multipling bifurcations can also occur.

Hopf and saddle-node bifurcations can create new frequencies, e.g. subharmonics of the driver frequency. Furthermore, all three types of bifurcations can terminate in a chaotic motion due to the following scheme:

1. A finite sequence of Hopf bifurcations (as a rule three are sufficient) leads to a chaotic attractor (Ruelle-Takens-Newhouse route).
2. Chaoticity can result from saddle-node bifurcations (Manneville-Pomeau or intermittency route).
3. An infinite cascade of period doubling bifurcations terminates in a chaotic state (Grossmann-Feigenbaum route).

Besides these three main routes to chaos, which have been found in many non-linear dissipative systems and non-linear maps and in more than 100 different experimental systems, some further and specific transition to strange attractors exist (Schuster 1984). From a theoretical point of view, the period doubling route has received a dominating position. This route exhibits a universal behaviour and obeys several scaling laws. These universal features are also found in externally driven non-linear oscillators. They demonstrate that dissipative and non-linear systems are the necessary and essential constituents for the dynamic systems. A detailed knowledge of the specific internal dynamics is of minor importance. Under this aspect, the modelling approaches for non-linear phenomena, e.g. in biological systems, become less problematic. Many speculative suggestions can be taken into consideration provided a physical basis for the models is retained.

In Figs. 4–7 the numerical results for an externally driven non-linear oscillator (coherent oscillation model, Appendix B) are given. The system exhibits two period doubling routes to chaos for a certain fixed frequency λ and increasing driver field strength F_1 with a fixed static field F_0 (see Ljapunov exponent of Fig. 3). The route consists of: $P2^n \rightarrow$ chaotic $\rightarrow P3 \cdot 2^n \rightarrow$ chaotic. For different static fields F_0, some other routes or only parts of them occur, including distontinuous transitions (hysteresis) and periodic windows within an irregular regime (Kaiser 1985a,b). In Fig. 8 the steady state amplitude is given for three different ways of an external excitation in model B1: (1) $F(t) = F_0$, the amplitude decreases to zero, the oscillation bifurcates into a static attractor ($x_s = 0$), which for further increasing drive is shifted to values $x_s > 0$, non-oscillating. (2) $F(t) = F_1 \cos\lambda t$. A periodic field leads to a strong increase until for certain F_1 the oscillation becomes unstable (collapse and onset of travelling waves). (3) $F(t) = F_0 (1 + \cos\lambda t)$. The combination of a dc and an ac field stabilizes the oscillations. A large number of bifurcations results, similar to those shown in Figs. 4–7 for $F(t) = F_0 + F_1 \cos\lambda t$.

It should be emphasized that the superposition of a static field (dc) to a periodic field (ac) breaks an existing inversion symmetry. Additional bifurcations are possible. For example, a period doubling bifurcation can only occur if the inversion symmetry has been broken, since in inversion symmetric systems only odd harmonics and subharmonics are allowed.

3.2 Specific Effects in Excited Systems

Non-linear dynamics in general and non-linear oscillations in particular are suitable representatives for cooperative processes and coherent excitations, respectively. Non-linear oscillations and vibrations participate as a non-monotonous evolution in the inanimate and animate world. The interaction of these states with external fields creates a rich variety of non-linear responses. From the point of view of weakly irradiated biological systems, some results of the foregoing sections are partially reviewed. Especially processes or mechanism that cannot be described or explained by direct high intensity processes or by heating reveal a new interpretation. Non-thermal effects in irradiated biosystems belong to these systems.

Very briefly, the most exciting results for driven systems and their relevance for biological systems are summarized (cf. Figs. 1–3):

1. Sharp resonance peaks in the frequency response refer to extremely frequency-sensitive processes found in experiments (e.g. Grundler and Keilmann 1983; Fröhlich 1986a,b).

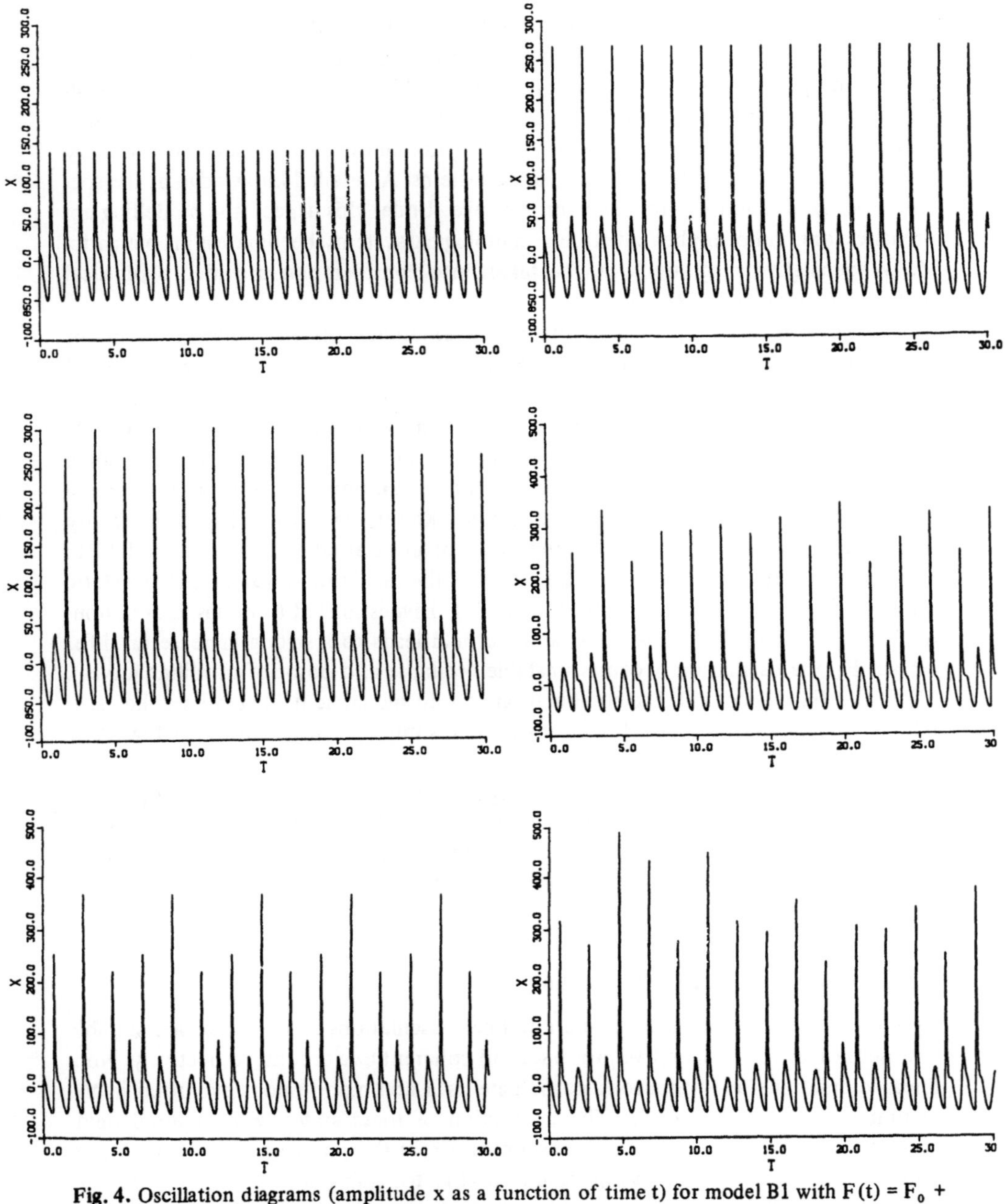

Fig. 4. Oscillation diagrams (amplitude x as a function of time t) for model B1 with $F(t) = F_0 + F_1 \cos\lambda t$. With increasing driver strength F_1, a series of bifurcations occurs. From left to right and from top to bottom six different states are drawn:

periodic states P1, P2, P4 non-periodic state N_1,
periodic state P6 non-periodic state N_2 $(Pn = nT = n \frac{2\pi}{\lambda})$.

The critical Ljapunov exponent for the whole bifurcation sequence is shown in Fig. 3

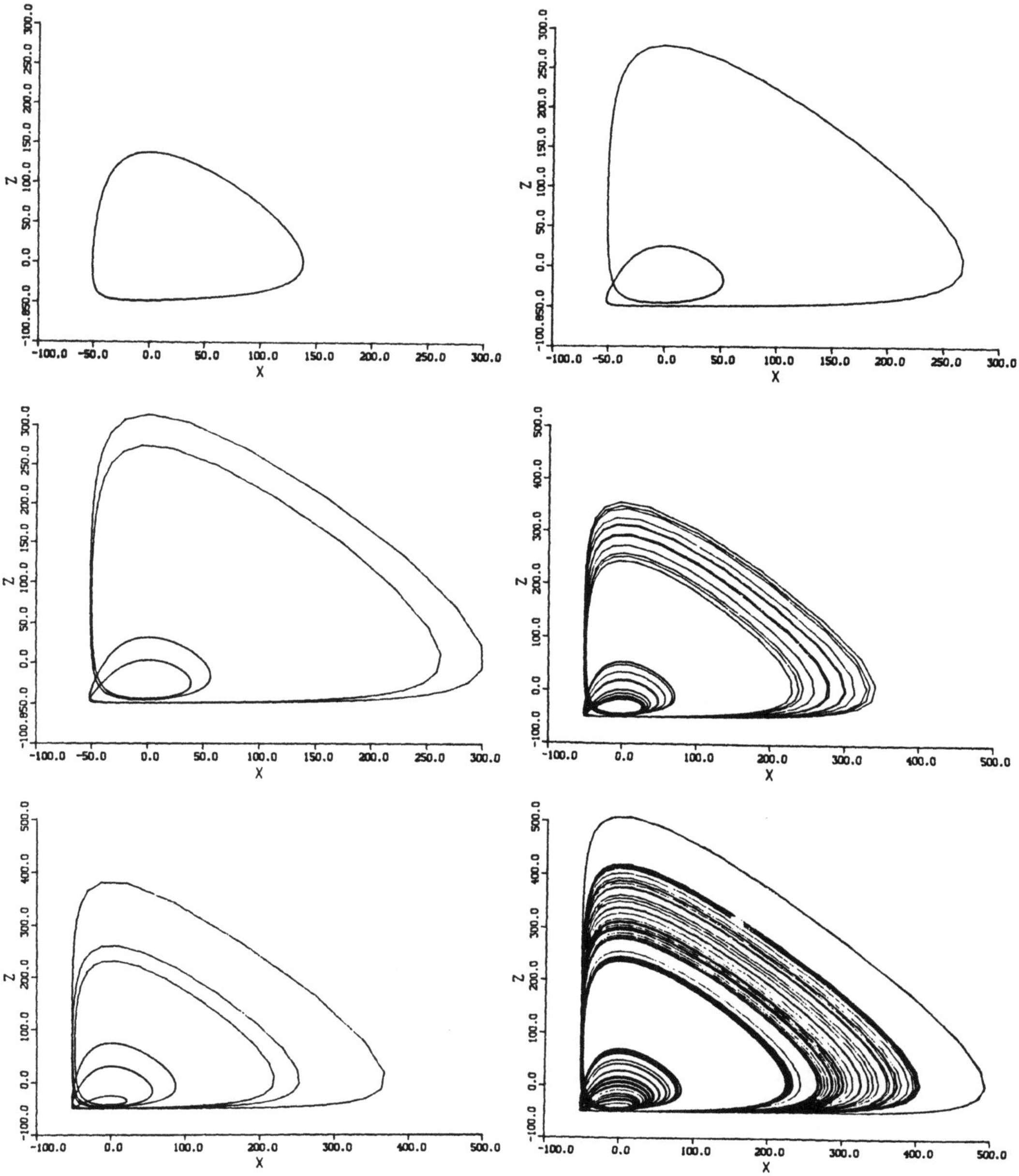

Fig. 5. Phase space diagrams for the oscillations of Fig. 4

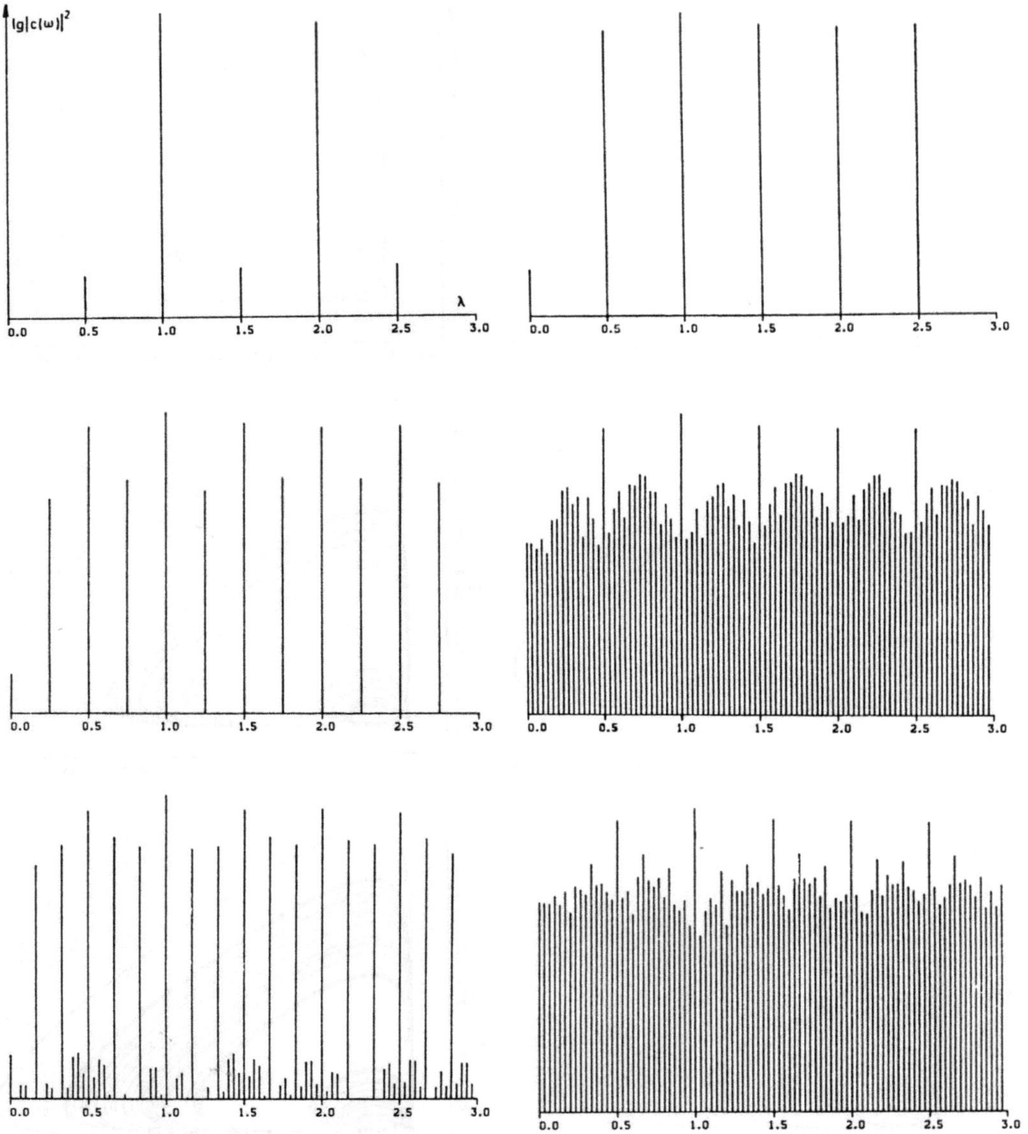

Fig. 6. Power spectra for the oscillations of Fig. 4. The frequency is drawn in units of λ

Fig. 7. a Poincaré sections for the six different states of oscillations of Figs. 4–6. *Left:* P1-P2-P4-N_1 ; *right:* P6 and N_2 (z is equivalent to x). **b** 3-d phase space x–$\dot{x}$–t for the periodic states P1 and P2 and the chaotic state N_2 of Fig. 4 (*from left to right*). **c** Oscillation diagrams for different types of excitations in model B1. From *left to right* and from *top to bottom*:

$\lambda \ll \omega_{LC}$	$F_0 = 0$		→	partial quenching,
$\lambda \approx \omega_{LC}$	$F_0 = 0$	F_1 small	→	non-linear beat,
$\lambda < \omega_{LC}$	$F_0 \neq 0$	F_1 large	→	transition chaotic-periodic,
$\lambda < \omega_{LC}$	$F_0 \neq 0$	F_1 large	→	intermittency

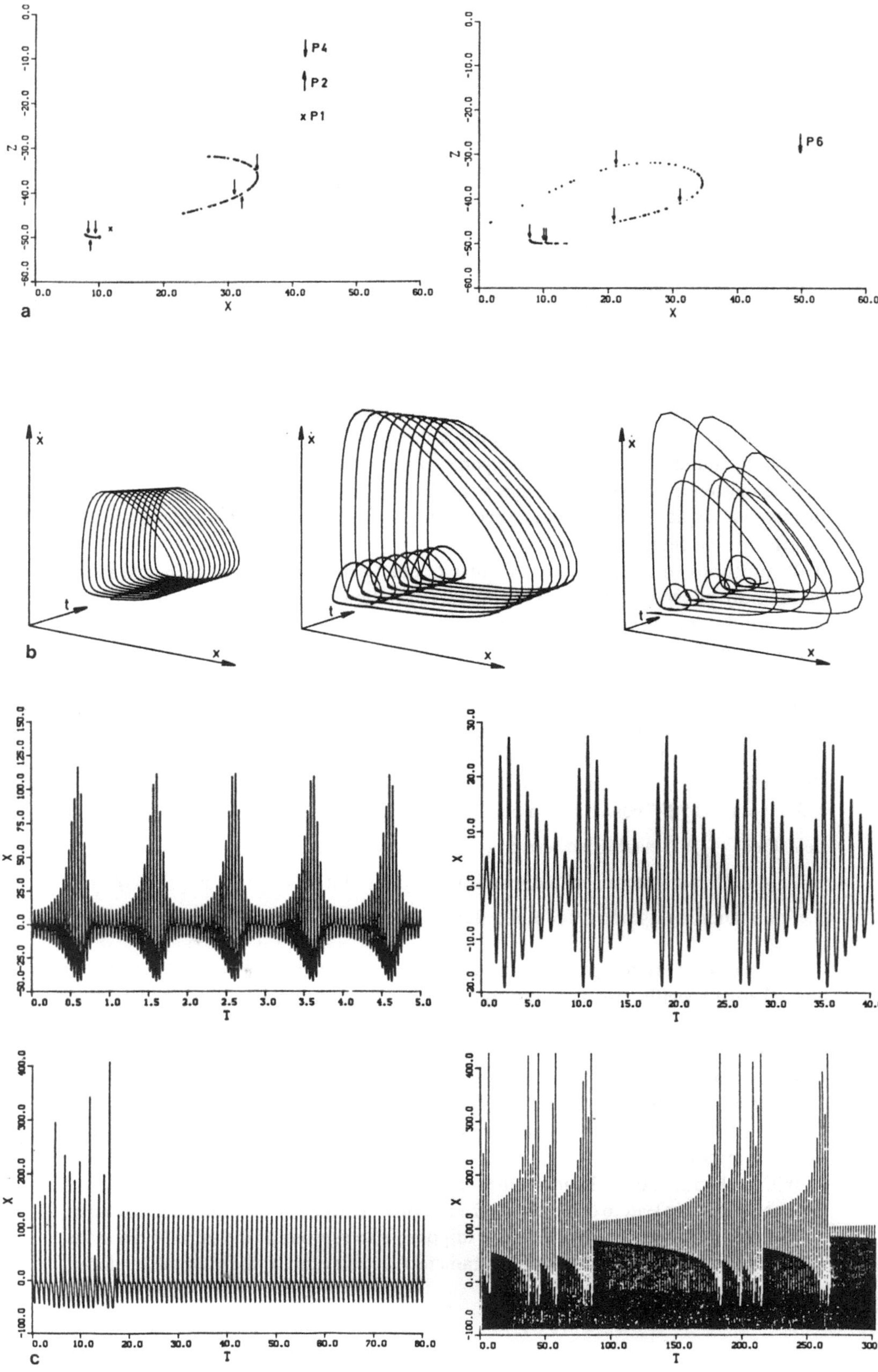

a
P4
P2
x P1
P6
Z
X
b
ẋ
t
x
c
X
T

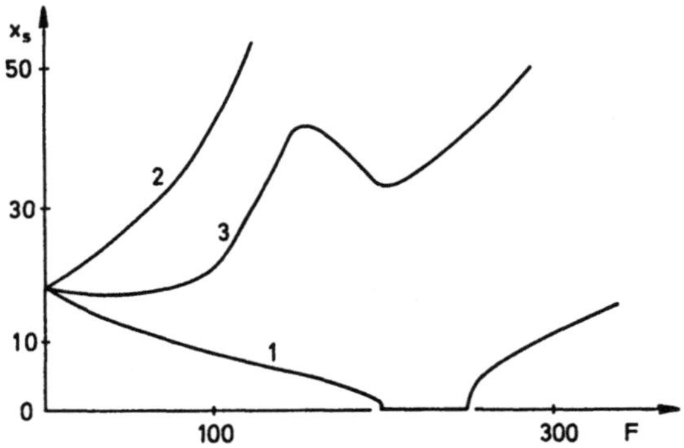

Fig. 8. Steady state amplitude x_s of model B1 as a function of the external driver field F for three different excitations:
1. $F(t) = F_0$
2. $F(t) = F_1 \cos\lambda t$
3. $F(t) = F_0 (1 + \cos\lambda t)$

2. Self-excited oscillations (limit cycles) display for certain λ- and F_1-regions a phase-locked or synchronized behaviour. Outside these regions, the free oscillations prevail. Threshold and saturation found in experiments receive a simple explanation.
3. The existence of extremely large transients before the bifurcation to a new state occurs may explain why in some experiments long time irradiations are required for an effect to be detected.
4. Bifurcations to a rich variety of other frequencies of oscillations (sub- and super-harmonics) allow for effects at different irradiation frequencies. The existence of both positive and negative effects in a very near frequency range (e.g. in growth rate experiments, Grundler and Keilmann 1983) becomes plausible.
5. Below, but very close to a bifurcation, the driven system exhibits an extreme sensitivity to that frequency which determines the motion above the bifurcation point. This leads to the amplification of small signals in bifurcating dynamical systems. Examples are well known from both experiments and theoretical considerations (Wiesenfeld and McNamara 1986). If, for example, a system oscillates with frequency λ due to a field $F_1 \cos\lambda t$ near a period doubling bifurcation, a very weak additional field $\epsilon \cos(\frac{1}{2}\lambda t)$ establishes a strong oscillation with half the original frequency, i.e. with $\frac{1}{2}\lambda$. Thus a system near a dynamical instability might strongly amplify coherent (i.e. non-random) signals. Effects at different frequencies can be either evoked or suppressed.

Bifurcations between periodic states and to quasi-periodic oscillations have offered a number of specific effects in driven systems. The additional possibility of running the system into chaotic states opens a new insight into more subtle experiments and also into some disagreeable experimental results (e.g. lack of reproduceability). Including chaotic states, the above list of effects can be continued:
6. Deterministic chaotic states display an extreme initial state sensitivity and loss of final state predictability. Very small changes of initial conditions can cause a complete different motion.
7. Within chaotic states a large number of periodic states exists. Some extremely small parameter changes lead to bifurcations from regular to irregular motions and vice versa.

In experiments with biosystems it is impossible to reproduce the same initial state. Furthermore, it is not known which parameters are relevant for the internal dynamics.

The failure to reproduce some specific effects appears in a new light as well as the existence of different effects under the same conditions.

8. The possibility to construct attractors from time signals that look irregular makes it possible to gain information on the internal dynamics. The number of relevant variables can be estimated provided the irregularity has a deterministic basis. Furthermore, the influence of always present noise can be studied separately.

It should be emphasized that chaotic states can also be created by internal means. The irregularities that originate from a deterministic basis must be viewed as an essential functional component of biosystems (e.g. in EEG signals, heart rhythms, gene reproduction etc.). They are in these cases not a pathological sign (Kaiser 1987).

It seems that chaos plays an important role in biological systems, at least in certain stages of development. Its existence necessitates a reinterpretation of both biological function and measured effects. There is increasing evidence that parts of active biological systems are in a chaotic state or can be driven into a chaotic one by external stimuli. Examples are periodic and chaotic activity in a molluscan neuron (Holden et al. 1982), chaos in sinusoidally stimulated neurons (Hayashi et al. 1982), chaotic activity in enzyme proteins (Mandell et al. 1982) and alternating periodic and chaotic sequences observed in periodically forced giant axons of squid (Aihara et al. 1985). These examples belong to the class of externally driven periodic systems where the bifurcations from periodic to deterministic chaotic states occur on different routes via quasiperiodicity, period-doubling sequences, intermittency etc. In these systems, deterministic chaos seems conclusive. However, there are experiments which in almost all cases exhibit only irregular motions. In order to classify these chaotic states, a detailed analysis must be performed. In Rapp et al. (1985) and Rapp (1986), a correlation analysis on the dynamics of neural activity in the simian motor cortex has been done. It shows that the spontaneous activity may be chaotic in the deterministic sense, exhibited by a strange attractor of very low dimensionality. Evidence of chaotic dynamics of brain activity during certain sleep cycles is reported in Babloyanz et al. (1985). Again a correlation analysis yields a low-dimensional chaotic attractor.

Some modelling approaches describing rhythmic activity of the brain and some experimental facts are reviewed in Kaiser (1985b). However, some care must be taken into account when a dimensional analysis of experimental data is performed (Layne et al. 1986). For noisy signals, large data sets are required. Furthermore, many different methods of calculating the dimension of attractors have been introduced; in some cases the results disagree.

Finally, some new aspects of bifurcations in driven non-linear systems are presented in an extremely speculative fashion. In driven systems the actual steady state results from a balance of the internally dissipated energy with the external energy input. For a number of models the energy input and the dissipated energy E_D has been calculated (Kaiser 1987 and unpubl.). Figure 9 shows some results for E_D as a function of either the field strength F_1 or frequency λ. In all models investigated so far near a period-doubling bifurcation, the system bifurcates into a state with decreasing E_D. This means that at the bifurcation point the system chooses a state which can be maintained by a smaller amount of energy input per period. Discontinuous transitions are in many cases transitions to large amplitude oscillations. The new value for the dissipated energy in these new states is much smaller compared to the small amplitude

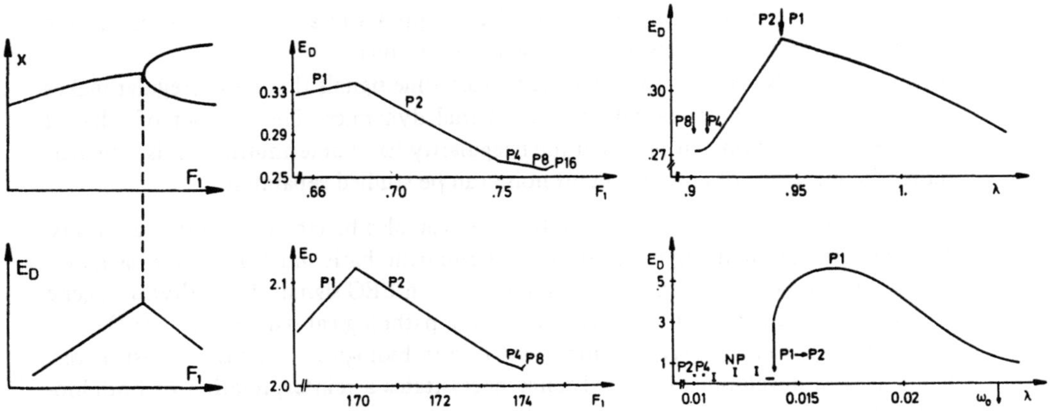

Fig. 9. Dissipated energy E_D as a function of driver frequency λ or driver field strength F_1. *Left:* steady state amplitude x and E_D as a function of F_1 for a period doubling bifurcation (schematic view). *Center:* E_D as a function of F_1 for models A12 (*top*) and B1 (*bottom*). *Right:* E_D as a function of λ for model A12 (*top*) and for a periodically pumped one mode laser model (*bottom*)

oscillations. Period doubling sequences for different models show a similar behaviour. Some rules seem to exist that represent certain physical criteria for bifurcations to occur. We conjecture that presumably a new universal behaviour lies behind these very first results.

Appendix A: Excited Non-Linear Oscillators

The m-dimensional continuous system of m first-order ordinary differential equations is transformed into one ordinary differential equation of the order m. Supposing that this transformation can explicitly be performed on obtains for m = 2

$$\ddot{x} + g(x,\dot{x})\dot{x} + f(x)x = F(t)\,. \tag{A1}$$

An external forcing term F(t) has been added.

x(t) is taken as the amplitude of an oscillation of a physical, chemical or biological system. The first term describes the acceleration of the system, the second one its damping, and the last one the stiffness of the system. Assuming that the non-linear functions $g(x,\dot{x})$ and f(x) can be expanded

$$g(x,\dot{x}) = a + b_1 x + b_2 x^2 \ldots + c_1 \dot{x} + c_2 \dot{x}^2 + \ldots$$

$$f(x) = \alpha + \beta x^2 + \gamma x^4 + \ldots$$

and restricting to some dominating terms, A1 is reduced to

$$\ddot{x} + (a + bx^2 + \ldots)\dot{x} + (\alpha + \beta x^2 + \ldots)x = F(t)\,. \tag{A2}$$

Equation (A2) contains many oscillating systems, the most important ones are:

1. *Unforced, undamped, linear:* harmonic oscillator

$$\ddot{x} + \alpha x = 0\,. \tag{A3}$$

The analytic solution is a constant sine wave for all times. Its phase and amplitude are determined by initial conditions, its period by the stiffness parameter α.

2. *Undamped, unforced, non-linear:* anharmonic oscillator

$$\ddot{x} + (\alpha + \beta x^2)x = 0 . \tag{A4}$$

Analytic solutions exist. The period depends on the amplitude of the steady wavy motion. Beat modes, higher harmonics and multi-stability can occur.

3. *Damped, unforced, linear:* damped harmonic oscillator

$$\ddot{x} + a\dot{x} + \alpha x = 0 . \tag{A5}$$

The analytic solution is an exponentially damped sine-wave or a non-oscillatory exponential decay of the amplitude. The period depends on the damping parameter a, for long times a resting state results.

4. *Damped, unforced, non-linear:* damped, anharmonic oscillator

Duffing oscillator $\quad \ddot{x} + a\dot{x} + (\alpha + \beta x^2)x = 0 \quad$ (A6)

Van der Pol oscillator $\quad \ddot{x} + (a + bx^2)\dot{x} + \alpha x = 0 . \quad$ (A7)

No analytic solutions are known. Transients to an asymptotically stable equilibrium state (resting state, point attractor) in the Duffing case and to an asymptotically stable periodic state (limit cycle, periodic attractor) in the Van der Pol case occur.

A necessary requirement for the existence of a limit cycle is a non-linear damping term, a non-linear stiffness is neither sufficient nor necessary.

For externally driven oscillators, at least a linear damping is required. However, damping is always inherent in real dynamic systems. Some typical examples are briefly discussed for $F(t) = F_1 \cos\lambda t$.

5. *Damped, forced, linear:* driven harmonic oscillator

$$\ddot{x} + a\dot{x} + \alpha x = F(t) . \tag{A8}$$

This is the classical resonance problem (see Fig. 2). The analytic solution is a superposition of the solution of the undriven oscillator (damped sine-wave as a transient) and a particular solution of the whole system, a steady state oscillator with the external frequency.

6. *Damped, forced, non-linear:* driven anharmonic oscillator

driven Duffing oscillator $\quad \ddot{x} + a\dot{x} + (\alpha + \beta x^2)x = F(t) \quad$ (A9)

driven Van der Pol oscillator $\quad \ddot{x} + (a + bx^2)\dot{x} + \alpha x = F(t) . \quad$ (A10)

No closed form analytic solutions exist. One obtains transients and stable steady-state oscillations, whether the non-linearity is in the stiffness or in the damping or in both terms. Besides harmonics, sub- and subharmonics occur, as well as quasi-periodic and chaotic states. Many different bifurcation routes exist. Both oscillators are excellent representatives for non-linear excited systems (Guckenheimer and Holms 1983; Thomsen and Stewart 1986). Some other well-known examples, exhibiting a similar behaviour, are (Beckert et al. 1985):

driven pendulum $\quad \ddot{x} + a\dot{x} + A \sin x = F(t) \quad$ (A11)

driven torsion pendulum $\quad \ddot{x} + a\dot{x} + Ax - B\sin x = F(t) . \quad$ (A12)

Appendix B: A Model For Coherent Oscillations

The combination of the two coherent excitations (highly polar metastable state, coherent vibrations, see Sect. 1) has been the basis for a low-frequency periodic chemical reaction giving rise to slow electrical oscillations which are coherent. Interactions between excited states create additional processes. The model equations read (Fröhlich 1974, pers. commun.; Fröhlich 1977):

$$\dot{x} = \gamma y + \alpha xy + (c^2 e^{-\Gamma^2 x^2} - d^2)x + F(t)$$
$$\dot{y} = -\beta x - \alpha xy \; . \qquad \text{(B1)}$$

γy, βx and αxy result from a non-linear chemical reaction. The third term of the first equation describes collective gain processes and dielectric losses (Kaiser 1977a,b). System (B1) is a two-dimensional non-autonomous system and can be viewed as an anharmonic, damped and driven oscillator. It has been investigated in great detail for $F(t) = F_0 + F_1 \cos\lambda t$ (Kaiser 1978a,b, 1983a,b). Some results are:

For $F(t) = 0$ a stable limit cycle exists, provided $c^2 > d^2$. For $c^2 > d^2 + \beta$ three fixpoints exist in addition (2 are unstable, 1 is stable). For $F(t) \neq 0$ a competition between the internal oscillation (self-sustained oscillation) and the external periodic field starts. Without a static component ($F_0 = 0$), only one period-doubling bifurcation occurs. Beyond a certain threshold $F_{1,c}$, the oscillation is unstable, leading to the onset of propagating pulses (Kaiser 1978b).

The results of a temporal stability analysis are not necessarily identical with spatial stability. However, a deep connection exists between the irregular motion of dynamical systems (i.e. temporal behaviour in non-linear ordinary differential equations) and wave propagation (i.e. spatiotemporal behaviour in non-linear partial differential equations). Typical non-linear waves are solitary waves, including transition waves, travelling periodic waves, wave trains and singular pulses. A specific and very important solitary wave is the *soliton*. A soliton is a solitary wave which preserves its shape and speed in a collision with another solitary wave. This extraordinary behaviour exhibits a loss-free transport of energy, possibly accompanied by a mass or charge transfer. The excitation of a metastable state can be performed by this mechanism.

Most soliton models belong to a certain class of non-linear partial differential equations in one space and one time dimension. Three-dimensional solitons (topological solitons) occur in models that describe dislocations and disclinations in crystals. As an important result, solitary waves and solitons transmit a variety of messages encoded in the wave form (see Del Giudice et al., this Vol., for a possible relevance of solitons in biological systems).

Besides the general concept of coherent excitations, it was the possible collapse of an internal oscillation that led Fröhlich to the above model. The idea is as follows: an internal oscillation stores a certain amount of energy. A weak external field and correspondingly a very small energy input can destabilize this oscillation after some bifurcations.

Internally stored energy is set free, which in turn can create effects within the system although the externally applied energy is far too weak to perform this effect. An extreme sensitivity to external fields as found in some experiments can be explained (Fröhlich 1977). However, since system (B1) exhibits for $F_0 \neq 0$ a rich variety of bifurcations, the limit cycle collapse is only one aspect of its behaviour. Different routes to chaos, continuous and discontinuous transitions and large hysteresis effects are the essential phenomena. Therefore, B1 not only models coherent excitations, it is an excellent tool for intensive studies of bifurcations and related questions (Kaiser 1985a).

References

Aihara K, Matsumoto G, Ichikawa M (1985) An alternating periodic-chaotic sequence observed in neuronal oscillators. Phys Lett 111A:251–255
Andronov AA, Vitt AA, Khaikin SE (1970) Theory of oscillators. Pergamon, Oxford
Babloyanz A, Salazar JM, Nicolis C (1985) Evidence of chaotic dynamics of brain activity during the sleep cycle. Phys Lett 111A:152–156
Beckert S, Schock U, Schulz CD, Weidlich T, Kaiser F (1985) Experiments of the bifurcation behaviour of a forced nonlinear pendulum. Phys Lett 107A:347–350
Bergé P, Pomeau Y, Vidal C (1984) L'ordre dans le chaos. Hermann, Paris
Birkhoff GD (1932) Sur quelques courbes fermées remarquables. Bull Soc Math Fr 60:1–26
Collet P, Eckmann JP (1980) Iterated maps on the interval as dynamical systems. Birkhäuser, Boston
Cvitanović P (ed) (1984) Universality in chaos. Adam Hilger, Bristol
Fischer P, Smith WR (eds) (1985) Chaos, fractals and dynamics. Marcel Dekker, New York
Fröhlich H (1968a) Long range coherence and energy storage in biological systems. Int J Quantum Chem 2:641–649
Fröhlich H (1968b) Bosé condensation of strongly excited longitudinal electric modes. Phys Lett A26:402–403
Fröhlich H (1969) Quantum mechanical concepts in biology. In: Marois M (ed) Theoretical physics and biological. North-Holland, Amsterdam, pp 13–22
Fröhlich H (1977) Possibilities of long- and short-range electric interactions of biological systems. Neurosci Res Program Bull 15:67–70
Fröhlich H (1980) The biological effects of microwaves and related questions. Adv Electronics Electron Phys 53:85–152
Fröhlich H (1986) Coherent excitations in active biological systems. In: Gutman F, Keyzer H (eds) Modern bioelectrochemistry. Plenum, New York, pp 241–261
Fröhlich H, Kremer F (1983) Coherent excitations in biological systems. Springer, Berlin Heidelberg New York
Grundler W, Keilmann F (1983) Sharp resonances in yeast prove nonthermal sensitivity to microwaves. Phys Rev Lett 51:1214–1216
Guckenheimer J, Holmes P (1983) Nonlinear oscillations, dynamical systems, and bifurcations of vector fields. Springer, Berlin Heidelberg New York
Gurel O, Rössler OE (eds) (1979) Bifurcation theory and applications in scientific disciplines. New York Academy of Sciences, New York
Hao B-L (1984) Chaos. World Scientific, Singapore
Hayashi C (1964) Nonlinear oscillations in physical systems. McCraw-Hill, New York
Hayashi H, Yshizuka S, Ohta M, Hirakawa K (1982) Chaotic behaviour in the onchidium giant neurone under sinusoidal stimulation. Phys Lett 88A:435–438
Hirsch MW (1976) Differential topology. Springer, Berlin Heidelberg New York
Holden AV (ed) (1986) Chaos. Manchester University Press
Holden AV, Winlow W, Haydon PG (1982) The induction of periodic and chaotic activity in a molluscan neurone. Biol Cybern 43:169–173
Kaiser F (1977a) Limit cycle model for brain waves. Phys Lett 62A:63–64
Kaiser F (1977b) Limit cycle model for brain waves. Biol Cybern 27:155–163
Kaiser F (1978a) Coherent oscillations in biological system I. Z Naturforsch 33a: 294–304
Kaiser F (1978b) Coherent oscillations in biological systems II. Z Naturforsch 33a:418–431

Kaiser F (1980) Nonlinear oscillations in physical and biological systems. In: Uslenghi PLE (ed) Nonlinear electromagnetics. Academic Press, New York, pp 343–389, Mir, Moskau, pp 250–281 (in Russian)

Kaiser F (1981) Coherent modes in biological systems. In: Illinger KH (ed) Biological effects of nonionizing radiation. ACS Symp Ser 151:219–241

Kaiser F (1983a) Theory of resonant effects of rf and mw energy. In: Grandolfo M, Michaelson SM, Rindi A (eds) Biological effects and dosimetry of nonionizing radiation. Plenum, New York, pp 251–282

Kaiser F (1983b) Specific effects in externally driven self-sustained oscillating biophysical model systems. In: Fröhlich H, Kremer F (eds) Coherent excitations in biological systems. Springer, Berlin Heidelberg New York, pp 128–133

Kaiser F (1984) Entrainment, quasiperiodicity, chaos, collapse: bifurcation routes of externally driven self-sustained oscillating systems. In: Adey WR, Lawrence AF (eds) Nonlinear electrodynamics in biological systems. Plenum, New York, pp 393–412

Kaiser F (1985a) Coherence, synchronization, chaos: cooperative processes in excited biological systems. In: Chiabrera A, Nicolini C, Schwan HP (eds) Interactions between electromagnetic fields and cells. Plenum, New York, pp 131–155

Kaiser F (1985b) Cooperative behaviour in brain activity: response to external drives. In: Mishra RK (ed) World Scientific, Singapore, pp 467–491

Kaiser F (1987) The role of chaos in biological systems. In: Barett W, Pohl H (eds) Energy transfer dynamics. Springer, Berlin Heidelberg New York, pp 224–236

Layne SP, Mayer-Kress G, Holzfuss J (1986) Problems associated with dimensional analysis of EEG data. In: Mayer-Kress G (ed) Dimensions and entropies in chaotic systems. Springer, Berlin Heidelberg New York, pp 246–256

Lorenz EN (1983) Deterministic nonperiodic flow. J Atoms Sci 20:130–141

Mandell AJ, Russo PV, Knapp S (1982) In: Haken H (ed) Evolution of order and chaos in physics, chemistry and biology. Springer, Berlin Heidelberg New York, pp 270–282

Marsden JE, McCracken M (1976) The Hopf bifurcation and its application. Springer, Berlin Heidelberg New York

May RM (1976) Simple mathematical models with very complicated dynamics. Nature 261:459–467

Nayfeh AH, Mook DT (1979) Nonlinear oscillations. Wiley, New York

Ott E (1981) Strange attractors and chaotic motions of dynamical systems. Rev Mod Phys 53: 655–671

Poincaré H (1892) Les méthodes nouvelles de la mécanique céleste. Gauthier-Villars, Paris

Rapp PE (1986) Oscillations and chaos in cellular metabolism and physiological systems. In: Holden AV (ed) Chaos. Manchester University Press

Rapp PE, Albano AM, Zimmerman ID, Deguzman GC, Greenbaum NN (1985) Dynamics of spontaneous neural activity in the simian motor cortex, the dimension of chaotic neurons. Phys Lett 110A:335–338

Schuster HG (1984) Deterministic chaos. Physik Verlag, Weinheim

Thomson JMT, Stewart HB (1986) Nonlinear dynamics and chaos. J Wiley, New York

Wiesenfeld K, McNamara B (1986) Small-signal amplification in bifurcating dynamical systems. Phys Rev A33:629–642

Structures, Correlations and Electromagnetic Interactions in Living Matter: Theory and Applications

E. Del Giudice[1], S. Doglia[1], M. Milani[1] and G. Vitiello[2]

1 Introduction

Modern physics has elucidated many problems about the structure of complex systems by connecting the apparent macroscopic features to the collective properties of microscopic components. The bridge is provided by the Quantum Field Theory (QFT), which has been recognized as equivalent to a statistical mechanics of assemblies with infinite degrees of freedom. Moreover the quantum theory has been able to account for the emergence of ordered systems from non-ordered sets of microscopic components. Crystals, ferromagnets and superconductors have been successfully described by this approach.

Living systems exhibit more complex kinds of order than these systems. Components seem to be correlated in a dynamic (space-time) rather than in a static (purely spatial) sense. On the other hand, in a living system there are so many types of microscopic components (in a cell there are several hundreds of molecular species) that one absolutely cannot pretend to produce a dynamical description of the system by following the story of each component. A description of the system in terms of a Lagrangian formalism seems therefore unattainable; this is one of the main reasons for introducing a conceptual (and formal) framework other than a purely mechanical one, namely a description in terms of field theory. In such a framework the Lagrangian formalism can be partially recovered. We could assume the general symmetry features of a possible Lagrangian, neglecting its explicit analytic form. Finally, the living system can only exist far from thermodynamic equilibrium, so that a correct description of the relationship living system – environment is as fundamental as the internal description of the system itself to understand life. Actually, living systems can survive in many different environments, so that the relationship system – environment cannot be given by a sharply defined coupling such as could be written in Lagrangian terms. We could then split physical information into a Lagrangian description of the internal components of the system and their interactions, and a non-Lagrangian description of the influence of the environment on the system.

As a consequence of this complexity, theoretical physics has had a harder life with living systems than with the much simpler systems investigated in solid-state physics. Computational techniques are hampered by the lack of an explicit Lagrangian.

---

1 Dipartimento di Fisica, Università degli Studi di Milano and INFN, Via Celoria, 16, I-20133 Milano, Italy

2 Dipartimento di Fisica, Università degli Studi di Salerno and INFN, I-84100 Salerno, Italy

General theorems and group theoretic approaches could then be explored. It is scarcely necessary to justify this choice further, since a satisfactory dynamic description of living systems has not yet been provided by the more traditional approaches in biology. These approaches have so far provided detailed information about the number, structure and biological function of the different components of the living system, but have not accounted for the cooperation of these components into a working biological system.

The standard procedure of theoretical physics in bridging microscopic structure and macroscopic features is to identify the physical variables which, in a given system, play the role of "order parameters", namely of the macroscopic fields which describe the correlation among the microscopic components. In a ferromagnet, for instance, this variable is the density of magnetization.

The proposal of Herbert Fröhlich in 1968 (Fröhlich 1968) to assume the density of electric polarization as the "order parameter" relevant for biological systems was then a fundamental step. Living systems can be affected by many agents in many different ways, but these influences add up to modifications of this basic parameter, the density of electric polarization.

In the following we shall elaborate the basic proposal of Fröhlich (1980) in the framework of QFT and shall derive some general consequences.

2 The QFT Theoretical Framework

Here let us briefly summarize the main aspects of a QFT for a system exhibiting an order parameter (Umezawa et al. 1982). The basic objects are a set of field operators (the Heisenberg fields) globally denoted by ψ, in terms of which the Lagrangian of the system is given: $\mathscr{L} = \mathscr{L}(\psi)$. The motion equations which are derived from $\mathscr{L}$ are denoted by

$$\Lambda(\partial)\,\psi = J(\psi)\,. \tag{1}$$

$\Lambda(\partial)$ is a differential kinetic operator and $J(\psi)$ is a non-linear functional of ψ describing the interaction. The invariance of the theory under a symmetry group G is assumed, so that

$$\begin{aligned} \mathscr{L}(\psi)\,(\mathscr{L}) &= \mathscr{L}(\psi') \\ \Lambda(\partial)\,\psi' &= J(\psi') \\ \psi \rightarrow \psi' &= g\psi\;;\quad g \in G\,. \end{aligned} \tag{2}$$

The dynamics is completely specified when a realization of the field operators ψ is given. In other words, the assignment of the Lagrangian is not sufficient for a complete description of our system and the space $\mathcal{H}$ of the states must also be assigned. In QFT, however, the assignment of $\mathcal{H}$ requires a choice among many different possibilities (Friedrichs 1953) each one corresponding to a different value of some relevant functional of the ψ fields, say $F(\psi)$. The order parameter is defined as the expectation value of $F(\psi)$ in the ground state $|0\rangle$ of the space $\mathcal{H}$:

$$\langle 0\,|F(\psi)|\,0\rangle = v\,. \tag{3}$$

In a different representation, say $\mathcal{H}'$, the ground state expectation value of F (ψ) would be v'. It has been shown that it is impossible to relate $\mathcal{H}$ and $\mathcal{H}'$ by a unitary transformation (Friedrichs 1953; Itzykson 1980). Different representations are then said to be unitarily inequivalent representations of the canonical commutation relations and there are infinitely many of them in QFT. When in Eq. (3) $v \neq 0$, it is said that there is spontaneous breakdown of symmetry (when $v = 0$ the symmetry is not broken). When $v \neq 0$, although the Lagrangian is invariant under the symmetry group G [see Eq. (2)], the ground state of the physical system is not invariant under the same symmetry group G. We observe that invariance of $\mathcal{L}$ under G corresponds to the possibility of implementing transformations $\psi \to \psi' = g\psi$, $g \in G$, without supplying energy. When there is spontaneous breakdown of symmetry and G is a continuous group, the Goldstone theorem and the Anderson-Higgs-Kibble (AHK) mechanism predict that long-range forces (i.e. gauge fields, as for instance the electromagnetic field) are described by massive fields (Itzykson and Zuber 1980). When long-range forces are absent, then gapless modes explicitly appear in the physical spectrum (they are called the Goldstone modes). These gapless modes are a dynamical manifestation of the order characterizing the ground state, a measure of which is given by the value v of the order parameter. In the AHK mechanism these gapless modes do not appear and in the regions where the gauge fields propagate (the so-called normal regions) the order is destroyed. A set of asymptotic fields, say φ^{in} (or equivalently φ^{out}), must be introduced which are more directly related to observable modes of the system. The physical state representation, say $\mathcal{H}^{in}$, is a realization for the set φ^{in} and therefore a realization of the Heisenberg fields ψ $\mathcal{H}^{in}$ is obtained through a mapping

$$\psi \stackrel{w}{=} \Psi\,[\varphi^{in}]\,, \tag{4}$$

where w (weak equality) denotes the fact that the equality holds when expectation values in $\mathcal{H}^{in}$ are computed for both members of Eq. (4). Since the set of φ^{in} fields includes the gapless Goldstone modes, say χ^{in}, the set of equations for φ^{in} include the equations for χ^{in}

$$K\,(\partial)\,\chi^{in} = 0\,. \tag{5}$$

K (∂) is a differential operator without a mass term since χ^{in} are gapless. Thus the transformation

$$\chi^{in} \to \chi^{in} + c \quad (c = \text{const}) \tag{6}$$

is a symmetry transformation for Eq. (5) [K (∂) $\cdot$ c = 0). Therefore we see that the set of equations for φ^{in} are invariant under a symmetry group G^{in} which in general is the contraction of G. A group A is called a contraction of a group B when the algebra of A can be derived as a limit of the algebra of B when the commutators of the broken generators are made vanishing. This fact simply reflects the above statement that the ground state is not invariant under G. Moreover, the transformation (6) describes the coherent condensation of Goldstone modes in the ground state, which is a dynamical description of the order of the system. The transformation (6) indeed induces a change in the expectation value of the number operator $(\chi^{in+}\,\chi^{in})$:

$$0 = \langle 0\,|\chi^{in+}\,\chi^{in}|\,0\rangle \to \langle 0\,|(\chi^{in+} + c^{*})\,(\chi^{in} + c)|\,0\rangle = |c|^2\,.$$

and also we see that the state $|0>$ appears as a (Glauber) coherent state for $\chi^{in'} = \chi^{in} + c$ since

$$<0\,|\chi^{in'} + \chi^{in'}|0> = |c|^2 \,. \tag{7}$$

3 A Scheme for the Living System

Let us now schematize a living system in the above QFT framework. We assume that the living system is the final step of a dynamical evolution which originates from basic interactions within a set of electric dipoles (Del Giudice et al. 1982c). An electric dipole is just the *physical* schematization of a biomolecule. Interactions among dipoles are: (a) of short range, such as Van der Waals and H-bonding; (b) of long range, electromagnetic (Coulomb and radiative) forces.

According to Sect. 2, the short-range part of the Lagrangian in the presence of unsymmetric vacuum undergoes a Goldstone mechanism (Del Giudice et al. 1985b), whereas the electromagnetic part undergoes an Anderson-Higgs-Kibble mechanism (Del Giudice et al. 1986). Let us discuss at first the short-range part of the Lagrangian. We assume that the short-range part of the Lagrangian is invariant under the SU (2) group of dipole rotations. The non-invariance of the ground state under that group amounts to the assumption that the system exhibits an electret (a long-lasting non-vanishing electric polarization along a preferred direction) in the ground state.

The definition of the ground state of the system requires some care (Del Giudice et al., submitted). It is not easy to define the minimum energy state out of thermodynamic equilibrium. In this case the system could span over many different states with (approximately) the same energy. It has recently been recognized that most biomolecules exhibit many low-lying states whose energy spacing is very small and it is difficult to trace a clear dynamic law of evolution among them (Frauenfelder 1985). The attempts to describe the biological system as a spin glass are simply the consequence of that feature.

On the other hand, if we prescribe a unique ground state (vacuum) to the living system, we would assume the obligation to prescribe a definite value to its electric polarization value, but the living system is an open system depending upon the environmental influence as much as on the influence of internal forces. The actual value of the electric polarization would then depend not only on the internal dynamics controlled by the Lagrangian, but also on the external influences which are not taken into account by the Lagrangian. We must then assume as initial input not just a unique value of electric polarization but a wide range of possible values. Each of these values will then define a possible ground state for the living system. In the framework of Sect. 2, the invariance of Eq. (5) under a constant shift implies the possibility of adding a constant to the order parameter without any expense of energy.

All the unitarily inequivalent representations of the group E (2) – which is the contraction of the group SU (2) – can actually describe the system. Within each representation the ground state is a true vacuum in the sense of the usual mathematical properties of Hilbert space, so that the standard formalism of symmetry breaking, which requires the uniqueness of the vacuum, can be safely used (Del Giudice et al., submitted).

Moreover the invariance under the transformation (6) implies that the infinitely many vacua corresponding to the different values of the constant have the same energy. Inequivalence under unitary transformations among these vacua means that the transition among them cannot be induced by any unitary operator. The physical system has then a twofold dynamics; the system evolves within each Hilbert space according to the Lagrangian and moves also among different spaces driven by external interactions. The physical system is then described by a physical space which is the sum of all the inequivalent Hilbert spaces, each one corresponding to a different electric polarization value of the vacuum. The transition among these different spaces is performed by changing the numerical value of the order parameter, the vacuum electric polarization P in the formulas. Straightforward application of the symmetry-breaking mechanism brings us to the following conclusions:

a) The physical states of the system are coherent states of the quanta of the generators of the broken physical transformation.

b) If we assume a non-vanishing electret along a direction, the Goldstone theorem prescribes the emergence of two Goldstone fields b and b^+. There would remain only a U (1) rotational invariance around the electret direction. But, as discussed in Del Giudice et al. (1986), we have also a local U (1) invariance, i.e. a local phase invariance, and this would imply the necessary appearance of an electromagnetic field which would destroy the dipole ordering. In order to avoid this inescapable consequence, we must break the global U (1) invariance and correspondingly we would have a third Goldstone field.

c) The massless Goldstone modes induce in the system a correlation which makes all the dipoles oscillate coherently. This correlation appears macroscopically as a sharply defined polarization oscillation (Del Giudice et al. 1985b).

Let us now introduce the electromagnetic field (Del Giudice et al. 1986). The correlation among the dipoles actually spoils the system invariance under local phase transformation, which would break its coherence. Electromagnetic field is consequently prevented from entering in the ordered region. It would actually open its way within the array of ordered dipoles along particular singular lines (vortices) where correlation is removed (normal region). In terms of fields along these singular pathways, the Goldstone boson corresponding to the breakdown of the global U (1) invariance does not contribute to the physical spectrum.

A third component for the photon field appears, producing a photon mass. The acquired mass depends upon the symmetry breaking parameter. The relationship is

$$M^2 = \text{const } P \,, \tag{8}$$

where P is the vacuum expectation value of the electric polarization P.

Electrodynamics in a living system is no longer governed by the Maxwell equations

$$\Box A_\mu = J_\mu \,, \quad \Box \equiv \frac{1}{c^2}\frac{\partial^2}{\partial f^2} - \frac{\partial^2}{\partial x^2} - \frac{\partial^2}{\partial g^2} - \frac{\partial^2}{\partial t^2} \tag{9}$$

but by the equations

$$\Box A_\mu + M^2 A_\mu = J_\mu \,. \tag{10}$$

This can be seen as a particular case of the equations governing the propagation of electromagnetic fields in non-linear media.

Equations (10) describe a field A_μ which is not allowed to spread throughout the space, but that behaves as a massive particle, describing a network of trajectories within the normal regions. The above trajectories are not the geometrical lines of the classical physics approach, but are thin filaments whose radius, R, is given on a quantum mechanical basis by

$$R = \frac{\hbar}{c}\frac{1}{M}\,. \tag{11}$$

In the next section we will discuss the consequences of the above results for the self-organization of living systems. We conclude this section by a final consideration. All the above conclusions are valid at infinite volume. At finite volume, a general theorem of Von Neumann prescribes that all the representations of the commutation relations are unitarily equivalent. This means that all the different Hilbert spaces of the above scheme become unitarily equivalent. Moreover, since the Goldstone bosons are trapped into a finite volume, they would acquire a mass inversely proportional to the finite size. In our formalism the requirement of a finite polarization is naturally achieved by introducing a finite size for the system and this in turn leads to nonzero effective mass for the Goldstone boson. It is easy to estimate that for a size of 25 microns we would obtain a mass which at room temperature would be approximately 0.3 kT.

Actually we obtain

$$M = \frac{\hbar}{c}\frac{1}{R} \Rightarrow Mc^2 = 1.26 \times 10^{-21}\ \mathrm{J} = 0.3\ \mathrm{kT}\,. \tag{12}$$

The infinitely many vacua which are degenerate at infinite volume would then transform themselves into a rich family of low-lying states whose spacing has the same order of magnitude of thermal fluctuations. Transitions among them, and consequently transitions among different degrees of microscopic order, can be triggered by thermal fluctuations, as pointed out by the Prigogine school (Prigogine and Nicolis 1977). As a matter of fact, in our approach, order is not created by thermal fluctuations but by dynamics. Thermal fluctuations push the system through its different possible orderings. This is clearly realized if we consider that at infinite volume no energy at all is required for such transitions. We also observe that when thermal fluctuations come into play because of the finite size of the system, the separation of the internal dynamics from the environmental influence is no longer possible. These contributions mix together at finite volume, and it is then impossible to establish simple causal relationships between the observed biological evolution and dynamics.

4 Finite Size and Temperature

In this section we show that the problem of finite size of the system is tightly connected with its non-vanishing temperature. The third principle of thermodynamics, formulated by Nernst in 1906, states that a physical system (i.e. a finite size system) cannot reach the temperature T = 0 K. This statement is the macroscopic consequence of the quantum uncertainty principle at microscopic level.

Let us consider a component of such a system with a non-zero effective mass, m. Suppose, for simplicity, the system to be one-dimensional.

Since this component is in a bound state, its average momentum $\overline{p}$ is zero. Let us apply the Heisenberg uncertainty relationship

$$\frac{\hbar^2}{4} \leqslant (\Delta p)^2 (\Delta q)^2 = \overline{(p - \overline{p})^2} (\Delta q)^2 = \frac{\overline{p^2}}{2\,m} 2\,m\,(\Delta q)^2 \,. \qquad (13)$$

The factor $\overline{p^2}/2$ m is the average kinetic energy of each component. Temperature is proportional to the average kinetic energy of the population of components; the average of the population cannot be lower than the average of each component. Consequently this last average is the minimum temperature T_{min} that the system can assume without varying the value of the effective mass of the component, namely without varying its internal structure. This temperature is

$$T_{min} = \frac{\hbar^2}{4\,m\,k(\Delta q)^2} \,. \qquad (14)$$

If the system volume is finite, $(\Delta q)^2$ is a finite number so that T_{min} cannot be zero.

Let us now discuss the temperature of the living system. At infinite volume, the ordered system we have described in Sect. 3 quite obviously has T = 0 K. A non-zero temperature could be obtained only by introducing thermal fluctuations. Actually the working temperature of a living system ($T \cong 300$ K) is quite high.

Moreover a temperature decrease does not improve the ordering of the system as would be expected by the fact that fluctuations also decrease. On the contrary, in living systems the dynamical order disappears when temperature falls below the room temperature. This striking feature suggests an alternative meaning of the temperature of the living system.

Let us assume that Goldstone bosons behave as an ideal gas

$$\frac{p^2}{2\,M_B} = \frac{3}{2} kT \qquad M_B : \text{Goldstone boson mass}\,, \qquad (15)$$

introduce the de Broglie relation

$$P = h/\lambda$$

and the system size

$$R = \frac{\hbar}{c} \frac{1}{M_B}\,,$$

and moreover impose the stationarity condition

$$R = n\,\frac{\lambda}{2}$$

with n = integer. We obtain

$$R = \frac{\pi hc}{6k} n^2 \frac{1}{T} \,. \qquad (16)$$

At T = 300 K and n = 1, R = 25 micron.

This length would correspond to the sum of the cell diameter and the width of the correlated medium around it.

The above rough estimate suggests that the observed biological parameters are not inconsistent with the assumption that the living system at room temperature is almost free of thermal fluctuations and its temperature depends mainly on zero point quantum oscillations.

We introduce a simple physical picture which accounts for room temperature as the minimum temperature for a metabolically active system.

Let us make a rough estimate, by assuming water as the unique biocomponent. Actually, water accounts for the 70%–80% of the whole living system. The correlation among the water dipoles amounts to a correlation among the electron pairs that in each molecule are responsible for the dipole. We can consider the dipole oscillation as produced by the oscillation of its electron pair. Let us consider a chain of correlated dipoles. The correlation along this chain can be kept stationary if the oscillation of the electron pairs of the i-th dipole of the chain spans the interval between the (i−1)-th and the (i+1)-th one. A shorter span would not sustain the correlation, while a longer span would produce fluctuating correlations among many different dipoles as it occurs in ordinary water. Consequently in our case we can choose Δq to be twice the average distance among water molecules which under ordinary conditions is 3.1 Å. Moreover the mass of the electron pair can be assumed to be twice the electron mass, since the correlation is possible only when the electron pair is not so tightly connected to its molecule. Actually this explains why the correlation occurs in an excited metastable state of the system as pointed out by Fröhlich (1968). In Eq. (14) then $m = 2, m_e = 1.8 \times 10^{-27}$ g and $\Delta q = 6.2$ Å. We obtain $T = 288$ K = 15 °C.

Thermal effects can be added to the above scheme. It is to be expected that a thermalization of the system would be in competition with the order arising from the symmetry breakdown. It is actually well known that a transition to the disordered configuration (the symmetric phase) can be obtained for some critical temperature T_c (symmetry restoration). As the appearance of the mass M of the field A_μ is a dynamical effect since the AHK mechanism is produced by the symmetry breakdown [see also the M dependence on P, Eq. (8)], one would expect that as $T \to T_c$ not only the order parameter but also the mass M of the electromagnetic field vanishes. However, it can be shown (R. Mańka and G. Vitiello, in preparation) that M(T) does not vanish as $T \to T_c$ because of non-zero contributions coming from thermal fluctuations of the A_μ field. In our opinion, this situation is a very interesting one, both from the theoretical and from the phenomenological point of view. From a theoretical point of view it describes indeed the occurrence of ordered domains in the ground state. A similar situation occurs in ferromagnets, where the ordered state arises from the correlation of ordered domains, each one contributing to the global macroscopic magnetization of the system. As $T \to T_c$ the order among the domains is lost although each domain will exhibit its own magnetization. As a result hysteresis appears. Hysteresis arises since the disordered ground state at T_c is not unitarily equivalent to the disordered ground state at $T = 0$ at infinite volume. This means that spontaneous breakdown of symmetry at $T = 0$ (namely the ordered ground state at $T = 0$) cannot be washed away even at $T = T_c$. In other words $M(T_c) \neq 0$ denotes a *memory* of the symmetry breakdown at $T = 0$. This observation brings us to the phenomenological implication: $M(T_c) \neq 0$. Since the appearance of a mass for the A_μ field expresses the filamentation of the A_μ propagation as discussed in Sect. 3, we can conclude that $M(T_c) \neq 0$ denotes the persistence (at least for some time) of the filamentary structure even in the absence of the underlying ordered dipole array (the order parameter vanishes as $T \to T_c$). The persistence of the filamentary structure (memory of the order at $T < T_c$) can have far-reaching consequences on the biological activity. We mention here some of them: great flexibility of the biological system in adjusting itself to changeable environments; biochemical activity going on in "dead times" of the metabolic history of the system. Contrary to the usual behaviour of non-living matter, temperature effects cooperate, instead of being destructive, to preserve "memory" of the structures hitherto produced. Even for $T < T_c$ thermal effects contribute to M(T) so that if order were accidentally destroyed at $T < T_c$ by any agents, filaments (and the biochemical activity localized along them) would nevertheless survive for some time, maintaining metabolic activity. An interesting application of the above argument may concern brain activity.

5 Filamentary Structures in Biological Matter

As mentioned in Sect. 3, the dynamics of the electromagnetic field A_μ within the biological system is ruled by the equation

$$\Box A_\mu + M^2 A_\mu = J_\mu \, . \tag{10}$$

According to this equation, the photon behaves as a massive particle while travelling across the medium: the field is self-confined within filaments with radius

$$R = \frac{\hbar}{c} \frac{1}{M} \, .$$

Assuming a complete aqueous medium and calling α the fraction of oriented dipoles we obtain the estimate:

$$\frac{M}{\sqrt{\alpha}} = 13.6 \text{ eV}$$

and then

$$R = \frac{146}{\sqrt{\alpha}} \text{ Å} \, .$$

It is interesting to compare this estimate with the measured radius of cytoplasm microtubules, which is about 125 Å (Wolosewick and Porter 1979; Clegg 1981, 1983). This comparison suggests $\alpha \sim 1$; i.e. the water dipoles are all correlated. This is a further indication, together with the considerations on temperature in Sect. 4, that the living system dynamics is dominated by coherent modes.

We discuss now the significance of the electromagnetic field filaments for biological organization (Del Giudice et al. 1982a,b, 1983).

First of all we point out that the confinement of the E.M. field into filaments realizes a shielding of the system components against destabilizing Coulomb interactions. In this way a channeling of the E.M. interaction is realized. The most important consequence of the electromagnetic field confinement is, however, the production of interfaces where strong field gradients are present. The general expression of the force per unit volume produced by an electromagnetic field on a continuous distribution of electric polarization P is (Pavlov 1978)

$$\vec{F} = \varphi \vec{E} + (\vec{P} \cdot \vec{\nabla}) \vec{E} + \frac{1}{c} \frac{\partial \vec{P}}{\partial t} \times \vec{H} . \tag{17}$$

The second term $(\vec{P} \cdot \vec{\nabla})\vec{E}$ is usually disregarded when slowly varying fields are considered, but in highly inhomogeneous fields this term becomes dominant. This fact has been exploited, for instance in laser physics, in order to produce selective movements of different atomic species (Metcalf 1987). Recent progress in laser-induced trapping and cooling of materials has been reported (Chu et al. 1986). In the cases of gas and dilute solutions, this term can be approximated as (Askar'yan 1974):

$$F_{grad} = \text{const}\ \nabla E^2\ \Sigma_K \frac{\omega_K^2 - \omega_0^2}{(\omega_K^2 - \omega_0^2)^2 + \Gamma_K^2} . \tag{18}$$

It is important to realize that the above force is the only one in the general expression of Eq. (17) which allows a recognition of different individual molecules by the field. The gradient term, in contrast with the other terms, does not depend on the charge. It recognizes the peculiar frequencies ω_K of oscillation of the surrounding molecules. Consequently a field of a given frequency ω_0, while exerting an attractive ($\omega_0 > \omega_K$) or repulsive ($\omega_0 < \omega_K$) force, can produce on its side boundaries condensation or rarefaction of different molecular species. The induced inhomogeneities depend upon the steepness of the field jump on the interface. We will discuss in the following some phenomenological implications. In order to produce a significant inhomogeneity, the strength of the Goldstone correlation among dipoles must be contained in a suitable window. Both limits of strong and weak correlation are actually critical for the occurrence of this mechanism. A weak correlation would produce a photon mass too small for the consequent filamentation of the electromagnetic field. This field is then not strong enough to produce gradient forces able to counteract the disruptive effect of diffusive mechanisms on ordering. On the opposite side an excessively strong correlation would prevent the EM field from penetrating into the system. Consequently filaments would not occur, and there would be no interface where the above selective forces could operate. We then obtain the startling result that an increase of microscopic physical order would produce a decrease of macroscopic order (for instance, chemical reactions order). Life then appears to be possible only on the interface between order and disorder or between differently ordered domains. The above prediction can be compared with the finding that tissue affected by carcinogenic transformations or viral infection exhibits a microscopic order higher than that found in healthy tissues (Reid 1987). Transformed cells have a cytoskeleton with few thick filaments instead of the rich network which is found in healthy cells. Transformed cells and also viral materials exhibit an enhanced electric permittivity, which implies a very low response to the electromagnetic signals coming from the environment.

Let us now discuss how a biochemically ordered structure can arise in the framework of the above mechanism. It could occur that the frequency ω of the propagating field matches the oscillation frequency ω_K of a particular molecular species present in solution. The molecules would then become strongly attracted on the external surface of the filament. They would also become oriented by the field and squeezed one against the other by the pressure radiation term in the expansion of Eq. (17). A linear aggregate would be produced; polymerization could then take place or not. In the latter case, energy, which usually would be spent for polymerization in a non-biological situation, could be invested for different purposes. For instance, in the actin filament of the cytoplasm, the ATP unit present on each monomer of G actin, if not employed for polymerization, could be used to produce soliton excitations, which, as we will discuss in Sect. 8, would play a significant role in the system dynamics.

Formation of the molecule aggregate gives rise to a coating of the field filament. This coating modifies the dielectric parameters of the medium and could then determine a change in the propagation regime. This approach resembles the renormalization group approach. We keep the form of Eq. (10), but we vary the numerical value of the parameter $M^2 = f(P)$, according to the sequential steps of the physical process. At the beginning the system is a dilute solution of biomolecules, so that the value of M^2 is mainly controlled by water polarization. The dynamic mechanism produces, through Eq. (18), a condensation of a molecular species on the side boundaries of the filaments. This produces in turn a variation of the polarization P and consequently of the mass M of the photon; the filament radius and all the connected properties are then renormalized, creating the possibility of additional molecular processes. A time-dependent pattern is then likely to emerge. In particular, since the molecules present in solution may change according to the sequence of reactions originated by the gradient force, also the pattern of frequencies ω_k changes with time, giving rise to a time-dependent pattern of molecular processes and of possible chemical reactions.

Let us consider a particular sequence within the above scheme. Chemical reactions occurring along a filament release an energy output on the interface of a correlated region. This heat output (positive or negative) then does not propagate in a diffusive way but in wave form on the correlations, giving rise to a polarization wave. Its frequency depends on the strength and on the range of the correlation and is then time-dependent. This wave in turn produces new attractions and repulsions. Chemical reactions are then sequentially interlocked; the energy output of any particular one, instead of increasing the temperature of the system, provides the energy input at another reaction. A coherent feedback scheme might emerge.

We discuss now the role of the "radiation pressure" term $\frac{1}{c}\frac{\partial \vec{P}}{\partial t} \times \vec{H}$ in Eq. (17). Since the EM field is a massive field and then not purely transverse, this term has both a longitudinal and a transverse component. The latter adds up to the gradient force, contributing to the overall stability of the filament; the former pushes molecules and ions along the filament in the wave direction. It is interesting to note that cytoplasm filaments have been recognized either as the place where most biochemical reactions occur or as the main transport rail system of the cell (Walosewick and Porter 1979; Clegg 1981, 1983).

Finally we note that, since the propagating field breaks dipole correlations within the filament, we could expect an "electrostriction effect", namely a lower water density in the filament respective to the bulk water. It could be the object of further investigation to compare this effect with the hydrophobic properties reported for the inner space of biomolecules.

Filamentation in an extracellular medium has been discussed elsewhere (Del Giudice et al. 1982d, 1984a, and in press).

6 Low Intensity Coherent Electromagnetic Emission from Living Matter

The finite size of the living system produces also one additional feature, namely the emission of low intensity coherent electromagnetic radiation. It has been shown that in the AHK mechanism a current density can be defined in the ground state (Matsumoto et al. 1975):

$$T_\mu(x) = \langle 0 \,|\hat{T}_\mu(x)|\, 0 \rangle = M^2 \left[A_\mu(x) - \frac{1}{q}\frac{\partial}{\partial x_\mu} f(x) \right]. \tag{19}$$

In this definition A_μ (x) is a c-number denoting an electromagnetic classical vector field and $\partial/\partial x\, f(x)$ is the derivative of the local condensation of the quanta of the Goldstone field operator ϑ emerging from the breakdown of the U (1) rotational invariance around the polarization direction. The right hand side of Eq. (19) has the form of a gauge transformation $A'_\mu = A_\mu - \frac{\partial}{\partial x_\mu}\lambda(x)$ that leaves the electric and magnetic field $\vec{E}$ and $\vec{H}$ invariant, provided that $\frac{\partial}{\partial x_\mu}\frac{\partial}{\partial x_\gamma}\lambda(x) = \frac{\partial}{\partial x_\gamma}\frac{\partial}{\partial x_\mu}\lambda(x)$. Then T_μ can be considered equivalent to zero whenever $\left(\frac{\partial}{\partial x_\mu}\frac{\partial}{\partial x_\nu} f = \frac{\partial}{\partial x_\nu}\frac{\partial}{\partial x_\mu} f\right)$. But when f(x) is a singular function $\left[\frac{\partial}{\partial x_\mu}\frac{\partial}{\partial x_\nu} f \neq \frac{\partial}{\partial x_\nu}\frac{\partial}{\partial x_\mu} f\right]$ we obtain $T_\mu(x) = \langle 0 \,|\hat{T}_\mu|\, 0 \rangle \neq 0$. The ground state behaves then as a coherent state and moreover a non-vanishing classical electromagnetic field appears. This field arises as a consequence of a spatial variation of the coherent condensation of the Goldstone bosons and in turn it is coherent. The coherence of the matter field triggers the coherence of the electromagnetic field. Singularities in the local condensation f(x) arise for instance on the boundary surfaces of the system or whenever cracks appear because of damages or diseases. Such singularities have been proven possible only when Goldstone bosons are massless (Umezawa et al. 1982). Therefore a significant Goldstone mass, namely a very small system size, would prevent the appearance of such singularities and the subsequent appearance of the field $\vec{A}$. We have already shown that in the actual biological system the Goldstone boson requires a very small mass. This very small mass is the real threshold for the appearance of the low intensity coherent EM emission. Finally the frequency spectrum of this classical field is expected to be continuous. In conclusion, the presence of singularities in the local condensation of Goldstone bosons produces beyond a very low threshold emission of coherent electromagnetic radiation on a continuous spectrum whose intensity is expected to increase with the number and extension of the space domains where such a condensation is a singular function. In particular, we expect an increase of the intensity when the system is damaged or is crumbling.

The above prediction fits completely into the experimental findings reported by Nagl, Popp and colleagues (Popp 1986).

7 Magnetic Flux Quantization and Josephson-Like Effects

An additional consequence of the AHK mechanism is magnetic flux quantization (Del Giudice et al. 1986). In the region where f(x) is regular and consequently $\vec{J} = 0$, the vanishing of the right hand side of Eq. (10) gives (in the gauge $A_0 = 0$; div $\vec{A} = 0$)

$$\hbar \nabla \vartheta = q\vec{A} \, . \tag{20}$$

Let us integrate this equation along a closed line. We obtain

$$\phi(\vec{H}) = \frac{h}{q} n \, , \tag{21}$$

i.e. the magnetic flux across a surface defined by the closed line is quantized.

The above results appear usually in superconductors (Solymar 1972). On the other hand, a zero resistivity appears also in the biological system under consideration, since screening currents must arise in order to confine the electromagnetic fields outside the correlated region. Non-Maxwellian electrodynamics implied by Eq. (10) is necessarily connected with electric currents flowing without losses on the side boundaries of the filaments where fields are trapped.

We can then expect the emergence in biological systems of phenomena typically associated with superconductors such as peculiar effects of weak magnetic fields, magnetic flux quantization and Josephson effects.

The possibility of the above effects has been suggested by a number of experiments (Japhary-Asl and Smith 1983).

The above considerations are extremely important when the system under investigation is a single living cell, that can be characterized by a simple constant phase.

If we now consider a system composed of two vicinal cells, a Josephson-like behaviour can be expected, since they mimick a Josephson junction, i.e. two superconductors separated by a thin non-conducting material (the barrier). Actually Josephson effects in biological systems have already been predicted on different considerations and experimentally investigated (Japhary-Asl and Smith 1983).

Experiments on living cells at about the time of cell division have shown: emission of electromagnetic radiation from the system, voltage steps in currents-voltage characteristics, and modification of these in the presence of electromagnetic radiation of proper frequencies.

The connection is apparent between the different phenomenological properties of the biological system, which are linked in our approach through simple fundamental relationships.

Moreover Josephson-like behaviour reported in Japhary-Asl and Smith (1983) suggests some insights concerning the microscopic agent of the correlation, since comparison between the usual voltage-frequency conversion law (Barone and Paternò 1982)

$$h\nu = qV \, , \tag{22}$$

and the empirical relationship indicates that two electrons are involved in the correlation.

We conclude this section with two comments.

Josephson junction can be the basic unit for compact computer systems characterized by low energy dissipation. The dynamically sustained Josephson-like elements inside a living system can be a time-dependent network able to perform both activities: information storage and information processing.

Moreover it is interesting to point out that cooperative behaviours arise in arrays of Josephson junctions (that could represent a set of cells in tissue or in a population).

It has been shown (Bonifacio et al. 1984) that by changing the density of elements (number of elements per unit volume) it is possible to induce a phase transition in the behaviour of Josephson elements, from a single-element behaviour to a cooperative one. The coupling agents between these junctions are the resonant photons of the electromagnetic fields that can be absorbed or emitted by the junctions.

Effects such as phase synchronization and voltage locking are typical of the cooperative regime (Clark 1973).

Finally, a relevant role is played by the dissipation term, that is the parameter that contains information about the absorption of the EM field inside the volume containing the array of Josephson junctions and the leakage of the EM field outside the volume occupied by the junctions.

Chaotic behaviour, instead of a correlated one, can also occur depending on the value of the dissipation term.

It can be concluded that the absorption, emission and escape of photons in the volume can give rise to ordered processes (that in our approach could lead to tissue formation and control) or disordered and chaotic processes (that in our approach could lead to uncontrolled proliferation of cells). Suitable variations of electromagnetic fields could induce transitions in one sense or in the other.

The assimilation of a multicellular biological system to an array of Josephson junctions implies that external electromagnetic fields, although very weak, could affect in a significant way the normal assets of the system. Experiments on Josephson devices usually require laboratories carefully screened against unwanted electromagnetic perturbations. It is conceivable that the detection of coherent structures in biological processes requires a comparable care.

8 Solitons on Molecular Chains and the Water Electret

In this section we consider a possible mechanism for the appearance of an electret (i.e. a long-lasting non-vanishing electric polarization) around the molecular components of the biological system. The formation of the electret is the first step of the self-organizing process of the biological system. Experimental evidence (Celaschi and Mascarenhas 1977) suggests that the electret disappears when the biomolecule is dehydrated. Some orientating electromagnetic field could be at the origin of this process. We consider as a candidate the field associated to a soliton slowly travelling on the chain.

A mechanism for soliton formation on biomolecular chains has been proposed by Davydov (Davydov 1979). We list here a number of relevant points:

1. When a localized deformation is produced on a chain of loosely and anharmonically bound monomers, a stable solitary excitation (soliton) appears and propagates on the chain.

2. A necessary condition is also the inability of the chain to dissipate energy outwards. In other terms, the host system must be essentially one-dimensional without two- or three-dimensional correlations. Solitons require then a conservative dynamic regime, whereas the collective oscillations described in previous sections implies a dissipative regime. ATP hydrolysis has been proposed as the energy supplier for solitons; it produces a quantum of about 0.25 eV of energy.

3. It has been proven that soliton lifetime is infinite in the limit of an infinite length of the chain. The actual finite length of the chain implies long lifetimes (order of magnitude of seconds).

4. Davydov has shown that a vibrational soliton is able to trap an electric charge and carry it along the chain (Davydov 1982). This charge transport does not require additional energy apart from the amount required for soliton formation. Consequently, a sort of "supercurrent" (i.e. an electric current flowing without losses) is present when solitons propagate on a molecular chain.

From the above points a picture emerges. Metabolic reactions occurring at definite sites of molecular chains can trigger solitons, which in turn produce a long-lasting electric current whose associated electromagnetic fields can orientate molecular dipoles of surrounding water. Once the electret is produced and, according to the discussion of Sect. 3, long-range three-dimensional correlations are established, the Davydov conservative regime is switched off and a Fröhlich regime enters (Del Giudice et al. 1982c), where the chain dipoles and the water dipoles cooperate by oscillating coherently. Therefore the appearance of a non-vanishing electric polarization in the ground state and the consequent breaking of the dipole rotational invariance are the effect of the soliton dynamics induced by metabolic activity on the chains uncorrelated yet.

Soliton and collective regimes cannot be simultaneous. As discussed in Sect. 5, molecular structures are assembled around the filaments where EM, fields are kept trapped by the collective dynamics of the system. These structures, however, can survive for a while after this dynamics is shut off because the hysteresis processes then provide the necessary conditions for the emergence of solitons.

References

Askar'yan GA (1974) The self-focusing effect. Sov Phys Usp 16:680

Barone A, Paternò G (1982) Physics and applications of the Josephson effect. Wiley, New York

Bonifacio R, Casagrande F, Milani M (1984) Superradiance and superfluorescence in Josephson junction arrays. Phys Lett 101 A:427

Celaschi S, Mascarenhas S (1977) Thermal-stimulated pressure and current studies of bound water in lysozime. Biophys J 20:273–278

Chu S, Bjorkholm JE, Ashkin A, Cable A (1986) Experimental observation of optically trapped atoms. Phys Rev Lett 57:314

Clark TD (1973) Electromagnetic properties of pointcontact Josephson-junction arrays. Phys Rev B8:137

Clegg JS (1981) Intracellular water, metabolism and cell architecture. Part I. Collect Phenom 3: 289

Clegg JS (1983) Intracellular water, metabolism and cell architecture. Part 2. In: Fröhlich H, Kremer F (eds) Coherent excitations of biological systems. Springer, Berlin Heidelberg New York

Davydov AS (1979) Solitons in molecular systems. Physica Scripta 20:1387

Davydov AS (1982) Biology and quantum mechanics. Pergamon, Oxford

Del Giudice E, Doglia S, Milani M (1982a) Self-focusing of Fröhlich waves and cytoskeleton dynamics. Phys Lett 90A:104

Del Giudice E, Doglia S, Milani M (1982b) Actin polymerization in cell cytoplasm. In: Earnshaw JC, Steer MW (eds) The application of laser light scattering to the study of biological motion. Plenum, London, New York, p 493

Del Giudice E, Doglia S, Milani M (1982c) A collective dynamics in metabolically active cells. Physica Scripta 26:232

Del Giudice E, Doglia S, Milani M (1983) Self-focusing and ponderomotive forces of coherent electric waves: a mechanism for cytoskeleton formation and dynamics. In: Fröhlich A, Kremer F (eds) Coherent excitations in biological systems. Springer, Berlin Heidelberg New York, p 123

Del Giudice E, Doglia S, Milani M (1984) Order and structures in living systems. In: Adey WE, Lawrence AF (eds) Nonlinear electrodynamics in biological systems. Plenum, New York, p 477

Del Giudice E, Doglia S, Milani M (1985a) Ordered structures as a result of the propagation of coherent electric waves in living systems. In: Chiabrera A, Nicolini C, Schwan H (eds) Interactions between electromagnetic fields and cells. Plenum, New York, p 157

Del Giudice E, Doglia S, Milani M, Vitiello G (1985b) A quantum field theoretical approach to the collective behaviour of biological systems. Nucl Phys B251 (FS 13):375

Del Giudice E, Doglia S, Milani M (1985c) Rouleau formation of erythrocytes: A dynamical model. J Biol Phys 13:57

Del Giudice E, Doglia S, Milani M, Vitiello G (1986) Electromagnetic field and spontaneous symmetry breaking in biological matter. Nucl Phys B275 (FS 17):185

Del Giudice E, Doglia S, Milani M, Vitiello G (1987) Physica scripta (submitted)

Frauenfelder H (1985) Dynamic aspects of protein reactions. In: Clementi E, Sarma RH (eds) Structure and dynamics: nucleic acids and proteins. Adenine Press, New York, p 369

Friedrichs KO (1953) Mathematical aspects of the quantum theory of fields. Interscience publ

Fröhlich H (1968) Long-range coherence and energy storage in biological systems. Int J Quantum Chem 2:641

Fröhlich H (1980) The biological effects of microwaves and related questions. In: Marton L, Marton C (eds) Adv Electronic Electron Phys 53:85

Itzykson C, Zuber JB (1980) Quantum field theory. McGraw-Hill, New York

Japhary-Asl, Smith CW (1983) Biological dielectrics in electric and magnetic fields. Ann Rep Insulation Dielectric Phenomena

Mańka R, Vitiello G (1987) Topological solitons and temperature effects in gauge theories, DFSA-10-87

Matsumoto H, Tachiki M, Umezawa H (1982) Thermofield dynamics and condensed states. North Holland, Amsterdam

Matsumoto H, Papastamatiou NJ, Umezawa H (1975) The boson transformation and the vortex solution. Nucl Phys 97B:90

Metcalf H (1987) Laser cooling and electromagnetic trapping of atoms. Optics News 13:6

Pavlov VI (1978) On discussions concerning the problem of ponderomotive forces. Sov Phys Usp 21:171

Popp FA (1986) In: Kilmister CW (ed) Disequilibrium and self-organization. Reidel, p 207

Prigogine I, Nicolis G (1977) Self-organization in non equilibrium systems, from dissipative structures to order through fluctuations. Wiley, New York

Reid BL (1987) The aetiological factors of cancer of the lower genital tract, a different view. Preprint 2006 Quen Elisabeth II Research Institute for Mothers and Infants University of Sydney (Australia)

Solymar L (1972) Superconductive tunnelling and applications. Chapman and Hall, London

Umezawa H, Matsumoto H, Tachiki M (1982) Thermofield dynamics and condensed states. North Holland, Amsterdam

Wolosewick I, Porter KR (1979) Microtrabecular lattice of the cytoplasmic ground substance. J Cell Biol 82:114

Resonant Cellular Effects of Low Intensity Microwaves

W. GRUNDLER [1], U. JENTZSCH[1], F. KEILMANN [2] and V. PUTTERLIK [1]

1 Introduction

Biological effects of low intensity microwaves have been the subject of a great number of studies and many different reactions were reported ranging from molecular to animal level. The acceptance level is low, however, due to difficulties both in reproducing the effects by others and in demonstrating their real athermal nature. In some cases great efforts were put into repeating athermal experiments, without clear results. In these studies the interpretation of data assumes a statistical behaviour which possibly is not adequate to the biological system (Kaiser 1984). These problems will probably remain as long as the mechanisms of these subtle effects are unknown. It seems worth thinking of new concepts for experimental investigations in athermal bioelectromagnetics. To focus on fundamental mechanisms, experiments must be oriented strongly by theoretical concepts rather than by simply investigating, for instance, biological reactions induced by a fixed frequency.

Trying to explore the relevance of a hypothesis eliminate the exploratory character of an experiment and can help to avoid redundant and time-consuming trials. In this context we see the phenomenon of resonant biological responses as a chance to discover much more interesting fundamental properties of nature.

The aim of this publication is to present work on resonant microwave effects from the recent literature, and to report new measurements confirming resonant effects on synchronized yeast cultures caused by low intensity millimeter waves. In the latter study the development of individual microcolonies is analyzed by microscopic observation, and the cell cycle times of each cell are determined and correlated in a computer.

2 Reported Frequency-Dependent Microwave Effects on Mamalian Cells and Bacteria

More than 10 years ago, a great effort in the Soviet Union determined for the first time frequency-dependent microwave biological effects in the frequency range from 39 GHz to 60 GHz (Devyatkov et al. 1974). Results were reported on a dozen different experiments in different laboratories. A greater part of these studies agreed on

1 Gesellschaft für Strahlen- und Umweltforschung mbH, Ingolstädter Landstr. 1, D-8042 Neuherberg, FRG

2 Max-Planck-Institut für Festkörperforschung, D-7000 Stuttgart 80, FRG

– the existence of irradiation effects which depend strongly on frequency, sometimes in a resonant manner,
– the existence of a threshold intensity necessary for effect induction, and of a large intensity range of several orders of magnitude above threshold where the induced effects do not vary with intensity,
– a significant dependence on irradiation time of the effects.

After the pioneering work of Webb on the frequency-dependent growth of *E. coli* (Webb and Dodds 1968), the Soviet experiments offered the first comprehensive information on a frequency-specific biological reaction of low intensity microwaves. Long before, theoretical considerations of Fröhlich had suggested that living objects could contain systems exhibiting coherence properties, e.g. large-amplitude oscillations at high frequencies (Fröhlich 1968, 1970). A trigger action of resonantly applied external radiation on these internal systems was anticipated to invoke a frequency- and intensity-dependent response as, in fact, was experimentally shown by the Russian results.

To obtain further understanding, several groups in the United States and Europe began to repeat some of these experiments and start now ones. These studies were, however, successful only in part (Grundler 1985). Some groups were able to induce effects and to reproduce them, others failed in induction or reproduction. In our opinion it has become clear that the reproducibility of biological effects of electromagnetic radiation presents a special problem (Pickard 1986), which may in itself lead to possibly interesting insights. Hence it becomes more important to consider stringent criteria how to design experimental variables from the outset. Some suggestions to this problem will be made later.

In this section, however, we discuss the observation, repeated by the original authors, of resonant microwave effects on mammalian cells, using an improved frequency resolution.

During the last years, information on further frequency-dependent microwave studies in the USSR has been scarce; therefore we will report also on a frequency-dependent change of metabolic activities in bacteria.

A protection effect of millimeter microwaves on mouse bone marrow cells exposed to damaging X-rays was published by Sevastyanova and Villenskaya (in: Devyatkov et al. 1974) for the first time. The effect is shown in Fig. 1.

It was found that the protective effect of microwaves before X-ray exposure of the animals is frequency-selective in nature. At certain wavelengths within the range of 6.6 and 7.7 mm and at certain intensities (> 10 mW/cm^2), the X-ray damage ($N/N_0 = 0.5$) is dramatically reduced ($N/N_0 \approx 0.85$).

Some years ago, this study was repeated with a much better frequency fine tuning for the wavelength range of 7.07 to 7.27 mm (marked on the abscissa of Fig. 1) (Sevastyanova 1981).

For each of the 21 different wavelengths equally spaced by 0.01 mm, the experiment was performed using four groups of mice: one as a control and the others for the three different treatments: (i) mm-wave irradiation, (ii) X-ray irradiation, and (iii) combined treatment of both. In these studies, the X-ray irradiation (dose 900 rad) as well as the mm-wave irradiation (power flux density > 10 mW/cm^2, irradiating horn area 1.8 cm^2) were applied to the hip (sometimes neck) of the animals.

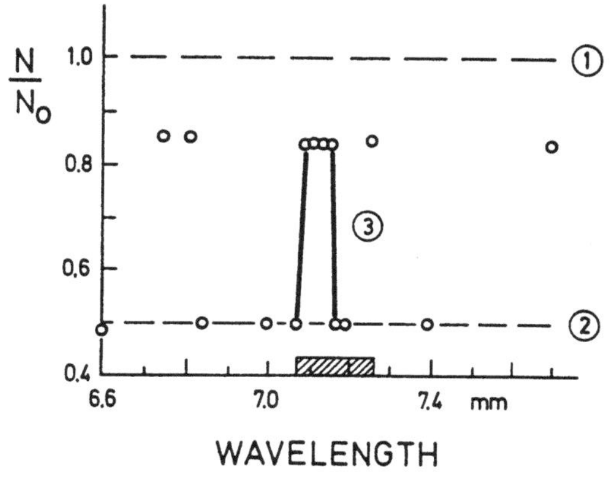

Fig. 1. Dependence of the relative number of bone marrow cells N/No on the wavelength of microwave radiation. *N* number of surviving cells; *No* number of cells without irradiation; *1* control; *2* X-ray irradiated; *3* millimeter wave and X-ray irradiated. (Sevastyanova and Vilenskaya 1974)

To test the concentration of bone marrow cells in animals with and without treatment, an amount of 5 to 8 x 10^6 cells was harvested 1 day after irradiation. To obtain high statistical accuracy, these tests were carried out with a great number of mice (more than 1000). The results are shown in Fig. 2.

The result is very interesting: it is apparent that microwave irradiation alone has no influence on cell viability, whereas the probability of survival is increased from about one half to about 80% by prior microwave irradiation at certain selected frequencies. No effect exists, however, at certain other frequencies. In Fig. 2 a curve is shown which connects data points taken at a fixed difference in frequency of about 60 MHz. While we cannot see the reason to draw this curve as it is shown, the data do in fact prove a very highly frequency-selective response which may very well indicate a multiply-resonant behaviour.

To explore a possibly narrow resonant response, a further study was performed (Sevastyanova et al. 1983). Within the wavelength region of 7.085 mm to 7.14 mm (marked on the abscissa of Fig. 2) corresponding to 42,343 MHz and 42,017 MHz

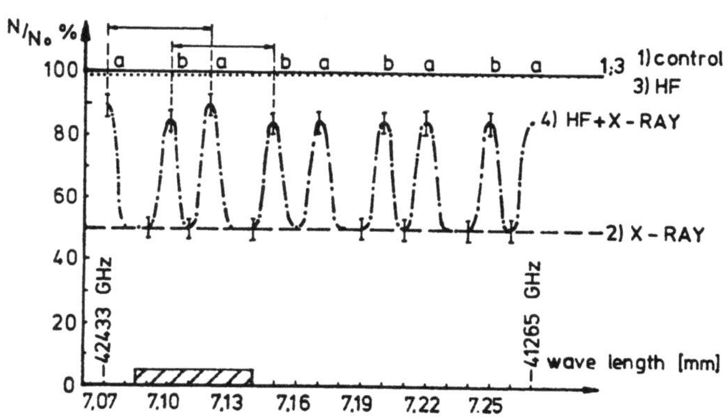

Fig. 2. Dependence of the relative number of bone marrow cells *N/No* on the wavelength of millimeter wave radiation. *N* number of surviving cells; *No* number of cells without irradiation; *1* control; *2* X-ray irradiated; *3* millimeter wave irradiated; *4* millimeter wave and X-ray irradiated. (Sevastyanova 1981)

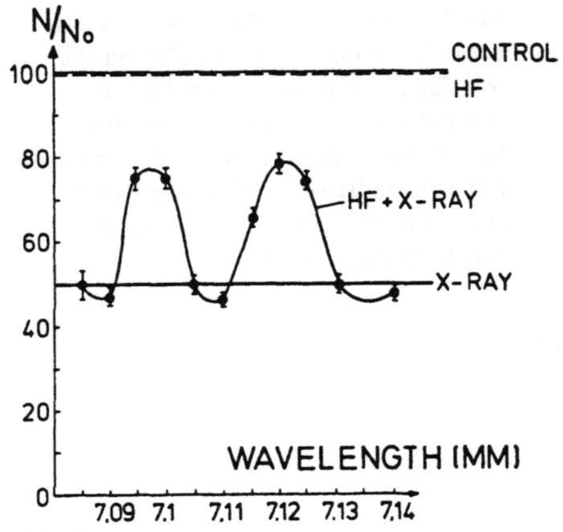

Fig. 3

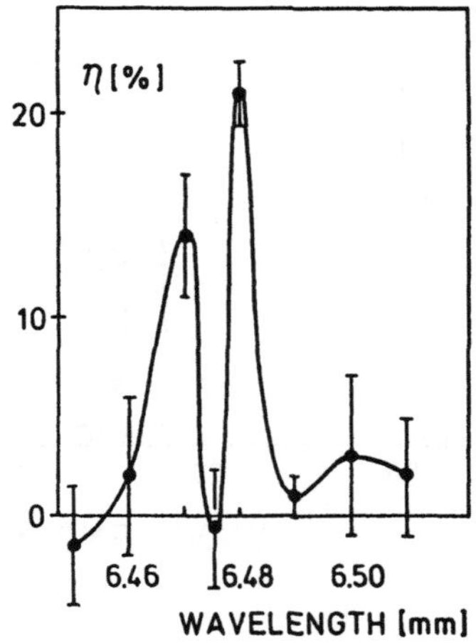

Fig. 4

Fig. 3. Relative number of bone marrow cells *N/No* in dependence on the wavelength of millimeter wave radiation. *HF* millimeter wave irradiation; *X-RAY* X-ray irradiation; *HF + X-RAY* combined radiation. (Sevastyanova 1983)

Fig. 4. Decrease of the synthesis of β-lactamase in *Staphylococcus aureus* in dependence on different microwave wavelengths. (Smolyanskaya 1981)

the tests described above were carried out for a stepwidth in wavelength of 0.005 mm (about 30 MHz).

Figure 3 shows now the remarkably frequency-dependent response of the relative cell survival after combined treatment with microwave and X-ray radiation. As long as there are no tests with even higher frequency resolution the response curve demonstrates resonances with a width of about 60 MHz at half maximum. It is quite reassuring that the strong frequency dependence has been confirmed twice by the original authors.

A further study, described in the 1981 publication of the Russian Academy of Science (Devyatkov 1981), shows also a strong correlation to the frequency of irradiation. Here the synthesis of β-lactamase in bacteria is measured under microwave irradiation (Smolyanskaya 1981).

A series of experiments was done at wavelengths in the interval 6.45 and 6.51 mm, using penicillin-resistant strains of *E. coli* and *Staphylococcus aureus*. The results obtained with the latter strain are shown in Fig. 4 and demonstrate a decrease of the enzyme synthesis by 15% to 20% at two wavelengths, whereas an intermediate wavelength between these two causes no effect.

It is important to notice that the active frequencies found here do not have any effect in *E. coli*. This indicates a specifity of the influence of microwave radiation.

3 Resonant Growth Rate Response of Yeast to 42 GHz and 84 GHz Radiation: Microscope Studies on the Generation Time of Single Cells

Microwaves near 42 GHz were found to influence the growth of yeast cells *(Rhodotorula ruba)* in a frequency-selective way, as reported by Devyatkov et al. (1974). Suggested by this study, we have analyzed the growth behaviour of yeast cultures *(Saccharomyces cerevisiae)* in aqueous suspension, by monitoring the visible light extinction. We found an exponential growth rate reproducible within ±4% limits. When the cultures were irradiated by continuous wave microwave fields of a few mW/cm^2, the growth rate either stayed constant or was considerably enhanced or reduced depending on frequency around 42 GHz. A spectral fine structure with a width of the order of 10 MHz was observed. Careful temperature monitoring excludes a trivial thermal origin of this effect (Grundler et al. 1977; Grundler and Keilmann 1978). Repetition of this experiment has confirmed that the growth of aqueous yeast cultures is indeed affected by weak microwave radiation in a frequency-selective manner (Grundler et al. 1983; Grundler and Keilmann 1983). To exclude a possibly influence of uncontrolled experimental parameters, many improvements were added in these studies. These include a refined frequency stabilization, refined power recording, impedance matching elements, two geometrically different antenna structures and two recording photometers. Laser thermometry was employed to locate any hot spots (Keilmann 1983). Two types of statistical calculations were applied to the experimental data, both aimed at assessing the significance of the important conclusions, (i) that yeast growth rates depend with a high probability on the frequency, thereby excluding a mere chance sequence of a frequency-independent microwave effect, and (ii) that the resonances found in two independent experimental series occur at the same frequencies.

These results, however, are obtained as an average reaction of nearly 10^6 cells, measured in a stirred suspension. Furthermore, in these experiments no exact measurement can be made of the power actually applied to an individual cell, or of the time during which a cell is affected by irradiation.

To overcome these disadvantages, inherent in the photometer experiment, and to obtain a more differentiated knowledge of the reaction at the single cell level, a new method was developed for studying the kinetics of single cell growth during the influence of microwave radiation.

3.1 General Description

The growing cells are locally fixed on the lower surface of a thin agar layer placed on a sapphire plate which is inserted in a metal-chamber holder. This cell chamber is mounted on a scanning microscope for observation. A flexible waveguide connects the chamber to a stabilized millimeter-wave source. To apply the radiation the microwaves are guided through the sapphire plate on which the cells are spread out in a monolayer. An automatic microphotographic method is used for continuous observation of the fate of these single cells during a growth period of three generations (one single cell develops a colony of eight cells).

At constant time intervals (e.g. 5 min), a set of six pictures is taken by moving the stage within an area of 0.2 mm^2, which constitutes a small fraction only of the total agar area of 78 mm^2. The pictures recorded on film are then digitized by a TV camera. In the resulting images the cells appear as white rings with high background contrast, which is optimal for further data analysis by a computer. Using pattern-recognizing programmes we can compute several important informations on the kinetics of cell growth. A detailed description of the cinematographic registration system and of the programmes for pedigree analysis can be found in (Grundler and Abmayr 1983).

3.2 Biological Material and Preparation Technique

Diploid yeast cells *(Saccharomyces cerevisiae)* were obtained from a stock culture of stationary state cells on solid nutrient agar (20 g/l glucose, 20 g/l Bacto agar, 5 g/l yeast extract; 4 °C) and suspended in phosphate buffer. A synchronized subpopulation (early G_1-phase) was selected by volume sedimentation within a linear density gradient. These cells are characterized by a mean volume of 30 μm^3, a mean duration of the first G_1-phase of about 3 h and a mean cell cycle time of 1.35 ± 0.03 h. For microscopic observations 10 μl of a buffer suspension containing 4×10^6 cells/ml are filled in the cell chamber shown in Fig. 5.

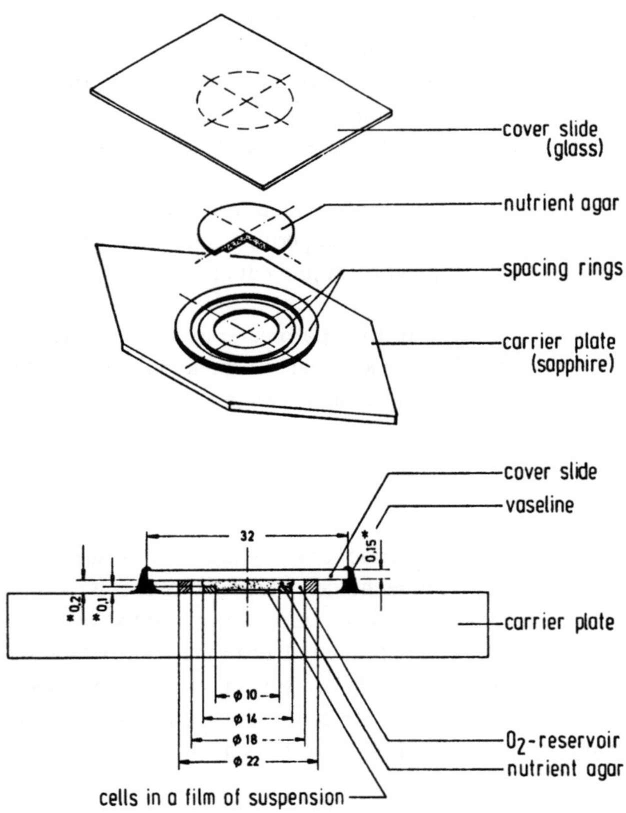

Fig. 5. Microscope cell chamber for yeast cell observation (*top*) and sectional side view. Figured dimension in mm; * indicates drawing enlarged by a factor of 10

The carrier slide for the cells is a thermally well-conducting tapered sapphire plate. Two plastic rings of different thicknesses (inner ring: 0.1 mm; outher ring: 0.2 mm) are fixed concentrically on it. After the inner ring is filled with warm fluid nutrient agar, a cover slide (0.15 mm thick) is put on the outer ring and the agar surface; afterwards the chamber is stored in a refrigator (4 °C) for some days. Before usage the cover slide is removed together with the agar layer and the inner ring is now partially filled with 10 μl cell suspension (10^4 till 10^5 cells) and afterwards agar and slide are carefully replaced. Finally, the cover slide is fixed elastically to the carrier slide by means of Vaseline.

The chamber is well suited to search for a possibly microwave-influenced cell growth kinetics. The cells lie in a plane directly on the holder plate and are surrounded by a thin fluid film. The slight pressure of the agar layer on the cells prevents them from migrating and rotating, and thus permits a local fixation within a range of ± 2 μm. Furthermore, the newborn cells are also forced to arrange in a monocellular layer without growing over each other. The chamber offers, therefore, also a good thermal coupling of the cells to the sapphire plate and allows them to be irradiated with a constant intensity at the same time.

To exclude a disturbed growth behaviour by the use of this artifical cell chamber, characteristic growth parameters such as lag time and generation time were studied first. Two different methods, extinction measurement on cells in suspension and microscopic cell count, were applied. The results were compared and found identical within the experimental error (Grundler and Abmayr 1983).

3.3 Cell Chamber and Irradiation Procedure

For microwave irradiation experiments the cell chamber was inserted into a metal chamber holder (Fig. 6) connected by standardized wave guides to a microwave system. This system was equipped with a backward wave oscillator as the radiation source, stabilized absolutely to ±1 MHz. The technique of stabilization and the components used were described in Grundler et al. (1983).

The incoming radiation passes through a flat horn in the inner part of the holder and enters the tapered front side of the sapphire plate. Those parts of the holder where the tapered parts of the cell-chamber are located were covered by spring-loaded Al plates to assure good coupling. At the other sides the sapphire plate is coated with Au except in circular areas of 10 mm diameter on the top and bottom surfaces. In the bottom surface, the open area is covered by a metal mesh (20 μm hole diameter) to permit the transmission of the visible light for cell observation and, at the same time, to avoid leakage of microwave radiation. Apart from the other tapered front end (where transmitted power is measured) the only way the microwaves can leave the crystal is through the central circular area on the top on which the cells are placed for irradiation.

The chamber holder thus constitutes a broadened wave guide which supports probably a multimode field distribution which is difficult of access. However, measurements could be taken for estimating both the power radiating out through the unloaded circular area (P_{rad}) and the power absorbed by the thin liquid layer surrounding the cells (P_{abs}). Both measurements and calculations were performed by Chenghe

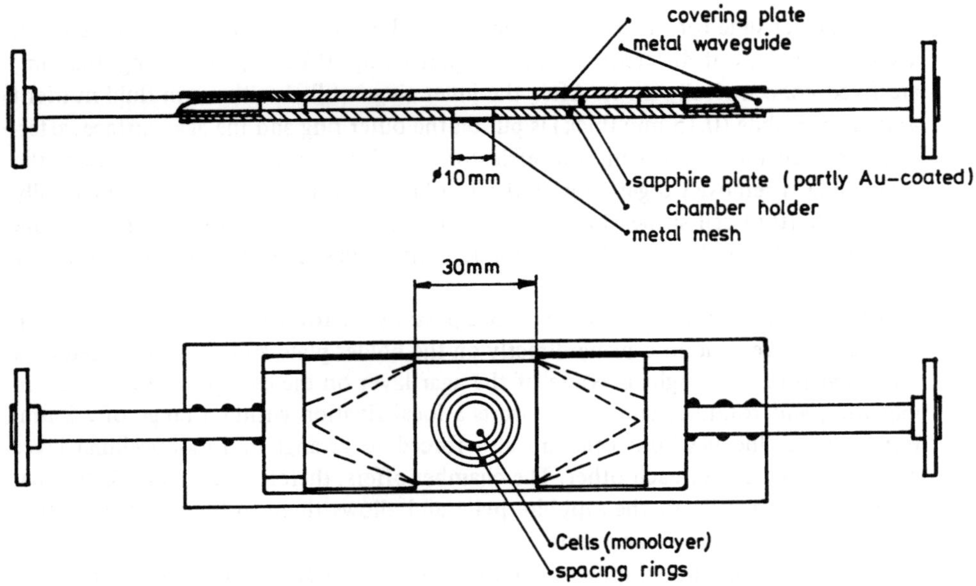

Fig. 6. Cell chamber holder in side (*above*) and top (*below*) view. The cell chamber (Fig. 5) is inserted

Xu (1986, pers. commun). Using a fixed input power P_0 of 20 mW and typical frequencies (e.g. 41,780 MHz) the reflected power P_R and the transmitted power P_T were measured on the chamber holder for three cases: The central circular area of the cell chamber

i. is uncovered, radiation emerges: $P_T = 0.09$ mW, $P_R = 0.15$ mW;
ii. is covered by a metal plate, radiation is transmitted: $P_T = 0.24$ mW, $P_R = 0.21$ mW;
iii. is covered by a 0.2 mm thick water layer, radiation is absorbed: $P_T = 0.12$ mW, $P_R = 0.26$ mW.

Using these data and simple principles of microwave transmission theory, an overall loss factor of the whole chamber system is calculated from which the power radiated by the uncovered circular area (1.13 cm^2) is $P_{rad} = 1.5$ mW and the power absorbed by a water layer is $P_{abs} = 0.6$ mW.

Taking all uncertainties into account, we assume a power flux density of the order of 1 mW/cm^2 on the sapphire surface. To estimate the SAR value a reflection of the radiation by the slide covering the cells and the agar layer (Fig. 5) is taken into account. The result is that about 5% of the radiated power flux density of 1 mW/cm^2 is absorbed in a 10 μm thick water layer, where the cells are located and that a SAR (specific absorption rate) of 26 mW/g must be taken in account. This SAR corresponds to an input power $P_0 = 20$ mW.

The relatively high specific absorption rate does not, however, cause a significant temperature increase of the cells because of the strong thermal coupling to the sapphire crystal, which in turn is an excellent heat conductor. The temperature of the sapphire is stabilized through contact of the holder with Cu tubes to a water bath, to

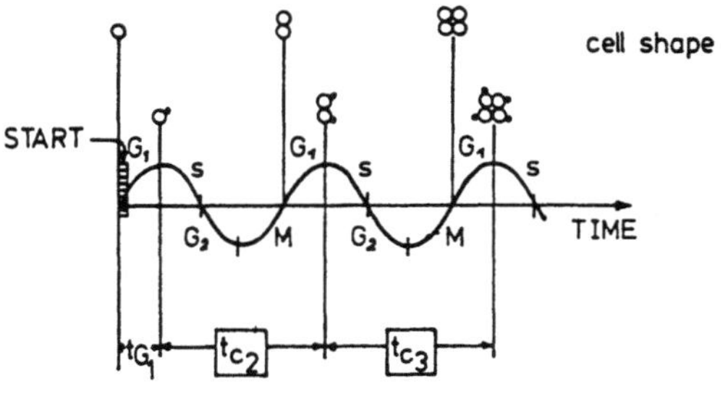

Fig. 7. Schematic diagram of the temporal correlation of yeast cell shape to cell cycle phases

31 ± 0.1 °C. Temperature measurements on the holder and on the sapphire proved that the maximum possible temperature increase is smaller then 0.1 °C when 20 mW of microwave power are applied.

3.4 Registration and Definition of Growth Parameters

In our experiments, the starting point of growth occurs when early G_1-phase cells come into contact with the agar of the cell chamber (see Fig. 7). In general, fixation of the chamber in the holder and start of the irradiation follow immediately. During the observation time of 6 to 10 h, up to 50 cells are registered at fixed time intervals between 5 and 10 min, by taking each time six photographs covering an area of 0.54 x 0.36 mm^2. The small extension of the observation area ensures a constant intensity.

These pictures were stored on film. The yeast cells appear as white rings on black background, very suitable for digitalization. This is achieved by recording with a TV camera and further treatment by pattern-recognizing programmes (Grundler and Abmayr 1983).

For detecting possible millimeter wave effects and for comparing them with the earlier results on cell growth in suspension, normalized growth rates were calculated, to which end cell cycle times had to be defined. Yeast is budding nearly at the boundary between G_1- and S-phase (Fig. 7).

We can thus take cell cycle times, by observing budding, as the time interval needed for a cell to start the S-phase, run through the G_2-, M-, and G_1-phases and arrive once again in the S-phase. After passing the first G_1-phase and two cycles (Fig. 7) for every cell, cycle times are known both t_{c2} for the first ($F_1 \rightarrow F_2$) and t_{c3} for the second ($F_2 \rightarrow F_3$) division. From the programme we obtain for the observed cell assembly both the cycle time distribution of all cells and the distributions of t_{c2} or t_{c3} alone. We observe that unirradiated cells yield logarithmic-normal distributions of cell cycle time. From these the mean cell cycle times and standard deviations can be determined.

3.5 Results and Discussion

We have developed this new observation method, using a scanning microscope technique, in order to answer the question whether frequency-dependent growth effects measured as mean values averaged over many cells in suspension can also be seen on locally fixed single cells. This aim required a strongly different irradiation geometry.

In order to compare the new results with the former ones, a frequency interval 41,760 MHz ⩽ f ⩽ 41,800 MHz was chosen, which covers a resonance region of the photometrically measured reaction spectrum (Grundler and Keilmann 1983).

We have performed a series of 17 control and 25 irradiation experiments. The input power P_0 ranges from 13 to 23 mW, equivalent to a specific absorption rate SAR from 17 to 30 mW/g. The transmitted power P_T was always smaller than 1 mW, and the reflected power was adjusted to zero by a matching unit.

Out of the series of control experiments three results of a normalized cell cycle time distribution are shown in Fig. 8 (upper left diagram), for the times of both divisions observed. We find the known shapes of logarithmic-normal distributions. After computing the mean values of all 17 control distributions, the arithmetic mean of

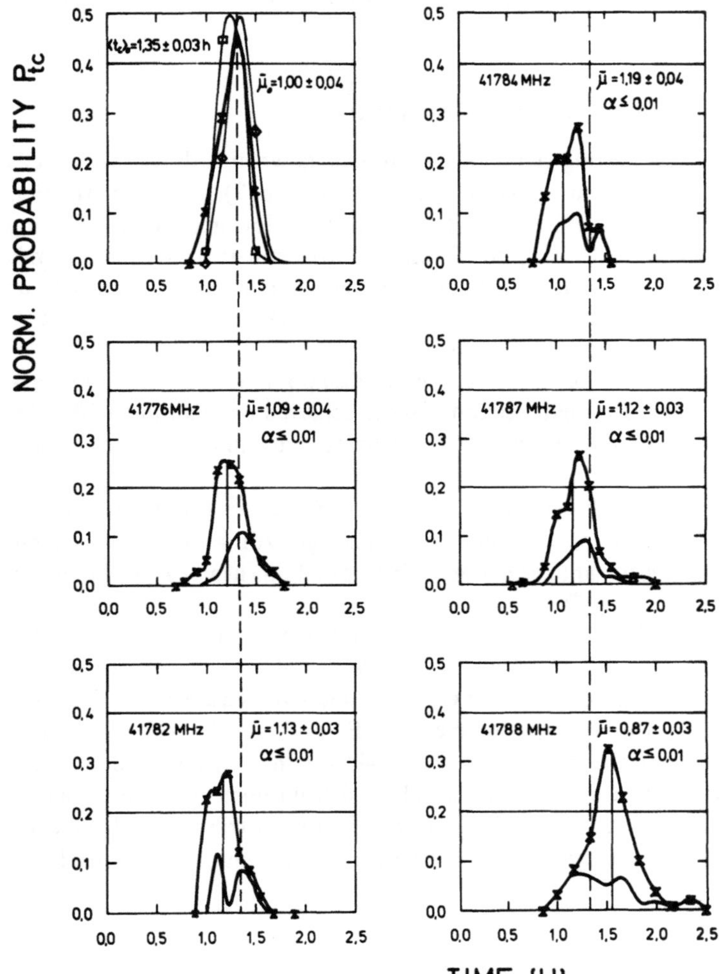

Fig. 8. Dependence of normalized cell cycle time distribution of single yeast cells on frequencies of irradiating energy. (Distributions marked by *symbols* involve cell cycle times for two divisions, the *lower smooth curves* show distributions for the first division)

these values is found to be $\langle t_c \rangle_0 = 1.35 \pm 0.03$ h, shown as dashed line in Fig. 8. The other five diagrams of Fig. 8 represent five irradiation experiments at different frequencies in increasing order. To differentiate between the two division steps, two distributions are drawn in each of these five graphs: the lower curves are related to the measured time intervals for the first division and the upper curves marked by symbols are the normalized distributions of the cycle times registered during both division steps. For visualizing a dependence on irradiation, the mean values $\langle t_c \rangle$ of these distributions are inserted as vertical lines. In order to estimate normalized growth rates for comparison with former results the ratio $\mu = \langle t_c \rangle_0 / \langle t_c \rangle$ is calculated in each diagram.

For estimating the significance levels of results in comparison with controls, a statistical analysis of the experimental data was performed. We start with, the mean values of the distributions and their experimentally defined standard deviations, in spite of the fact that some distributions were asymmetrical and multiply-peaked, which of course gives some uncertainty to the method. By using the calculated errors of the mean cell cycle times $\langle t_c \rangle$ and further error propagation calculations, the final errors were determined for the arithmetic mean $\langle t_c \rangle_0$ of all control distributions (see Figs. 8 and 11).

The same applies to the normalizd growth rates $\bar{\mu} = \langle t_c \rangle_0 / \langle t_c \rangle$ for both control and irradiation data. These errors were inserted as uncertainty intervals in the diagrams of Figs. 8 and 11 and drawn as error bars in Figs. 9, 10, and 12. Finally, by the use of the one-sided Student t-test, a significance coefficient α was determined testing the coincidence of the $\bar{\mu}$-values in reference to $\bar{\mu} = 1$.

The evidence of a marked microwave influence is clearly demonstrated in Fig. 8 in a twofold way:

i. In comparison to controls, the radiation is seen to induce asymmetries and double peaks in both cell cycle time distributions. These changed shapes indicate the existence of subgroups of time intervals related to cells which are decreased or increased in growth. In the case of double peaks it seems that one of them is close to the dashed line of the mean control value. This fact allows the astonishing conclusion that although all cells have been irradiated with the same intensity (note they are all located within an area of 0.2 mm^2) only a part of them reacts like the control, while the others have changed cell cycle times.
ii. The radiation shifts the mean values of the distribution $\langle t_c \rangle$ in comparison to the control line $\langle t_c \rangle_0$. More specifically, the normalized growth rate values $\bar{\mu} = \langle t_c \rangle_0 / \langle t_c \rangle$ become highly significant for all of the five diagrams shown.

To test the existence of a systematic dependence on frequency these μ-values are shown in Fig. 9 (marked by large circles) together with the results of the further 20 irradiation experiments and of the 17 control runs. The growth rates of the latter are shown relative to the mean value of $\langle t_c \rangle_0 = 1.35$ h.

For comparison, the formerly reported resonant microwave effects resulting from photometrical observation of cell suspensions (Grundler et al. 1983) are reproduced in the lower part of Fig. 9.

Comparing now the results obtained with the two widely differing experimental methods – single cells (up to 50) versus 10^6 cells in suspension – we come to the following conclusions:

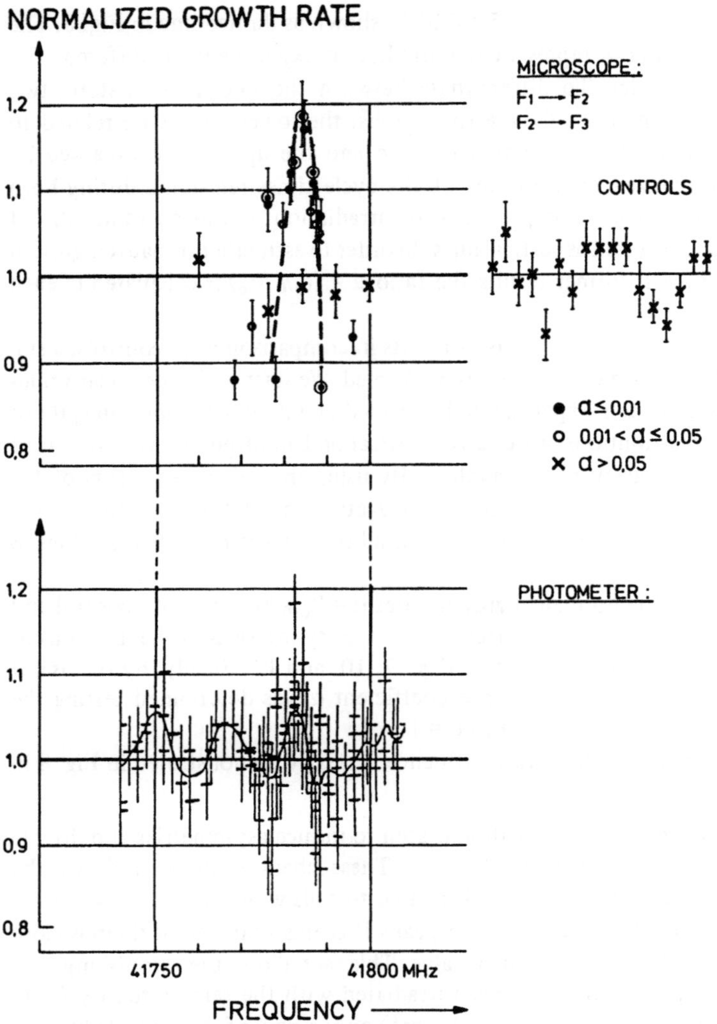

Fig. 9. Dependence of normalized growth rates of yeast cells on frequency of U-band microwaves. Comparison of the growth rates of microscopically observed single cells (*above*) with a photometrically registrated colony growth rate (Grundler and Keilmann 1983) (*below*). The *lower values* represent results and error bars of single measurements

– The existence of low-level microwave effects on cell growth is documented once more. Significant and in some cases highly significant growth rate values are found well outside the scatter of control runs.

– These effects show a strong dependence on frequency in a resonant manner, around 41,782 MHz. Comparison of both data sets of Fig. 9, furthermore, they reveal a great resemblance between both spectra, both qualitatively and quantitatively. The frequency of maximum effect, 41,784 ± 1 MHz measured microscopically agrees well with the corresponding value found in the photometer experiment, 41,782 ± 1 MHz. Even the same resonance width of about 8 MHz is obtained. Hence the resonance effect on yeast growth is manifested once more.

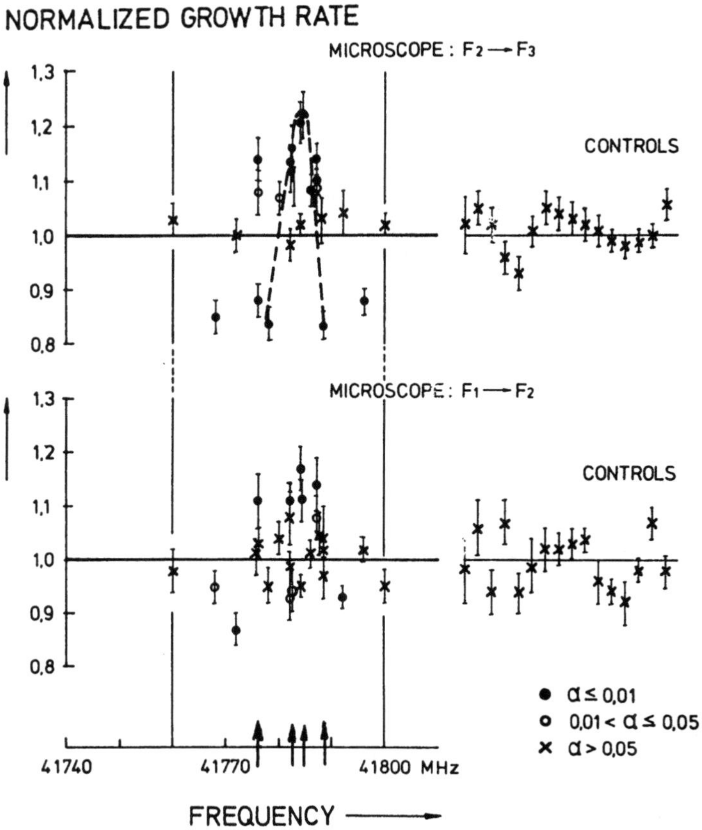

Fig. 10. Dependence of the normalized growth rates of single yeast cells on frequency of U-band microwaves. The data of irradiated and control cells are separately shown for the first and the second division. Frequencies are marked by *arrows* where ambiguous effects exist

As a matter of concern, the fact must be pointed out that at some frequencies both significant and non-significant effects were observed, and at others, significant effects of even opposite signs. This situation is documented quite clearly in Fig. 10, where the positions of the frequencies mentioned are marked by arrows. Although the problem of unreproducibility at certain frequencies was observed earlier (Fig. 9, below), the question is now more serious because of a better control of irradiation parameters and an improved registration method.

The question arises whether a better control of experimental parameters would help at all, or, whether irreproducibility might be an inherent feature of biological response experiments with electromagnetic radiation (Sect. 4).

Figure 10 differentiates between the influence on the first ($F_1 \rightarrow F_2$) and on the second ($F_2 \rightarrow F_3$) division in effects of frequency-dependent growth rates.

The effects described (Fig. 9) obviously correspond mainly to radiation-induced changes of the second division. This means that a microcolony of two budding cells produces four budding cells. In contrast, the first division step is less disturbed, in respect to both the amplitude of the response and a marked frequency dependence. Note that highly significant reactions are still observed.

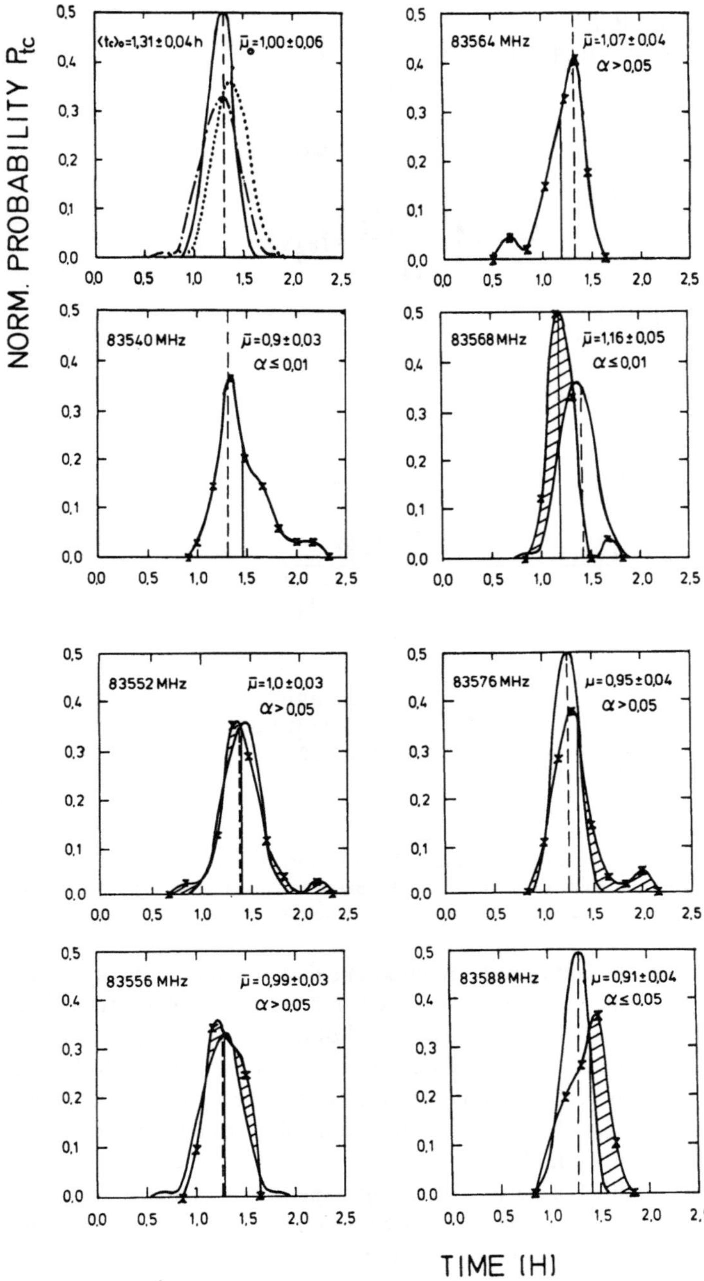

Fig. 11. Dependence of normalized cell cycle distributions of single yeast cells on frequency of E-band microwaves (first and second division)

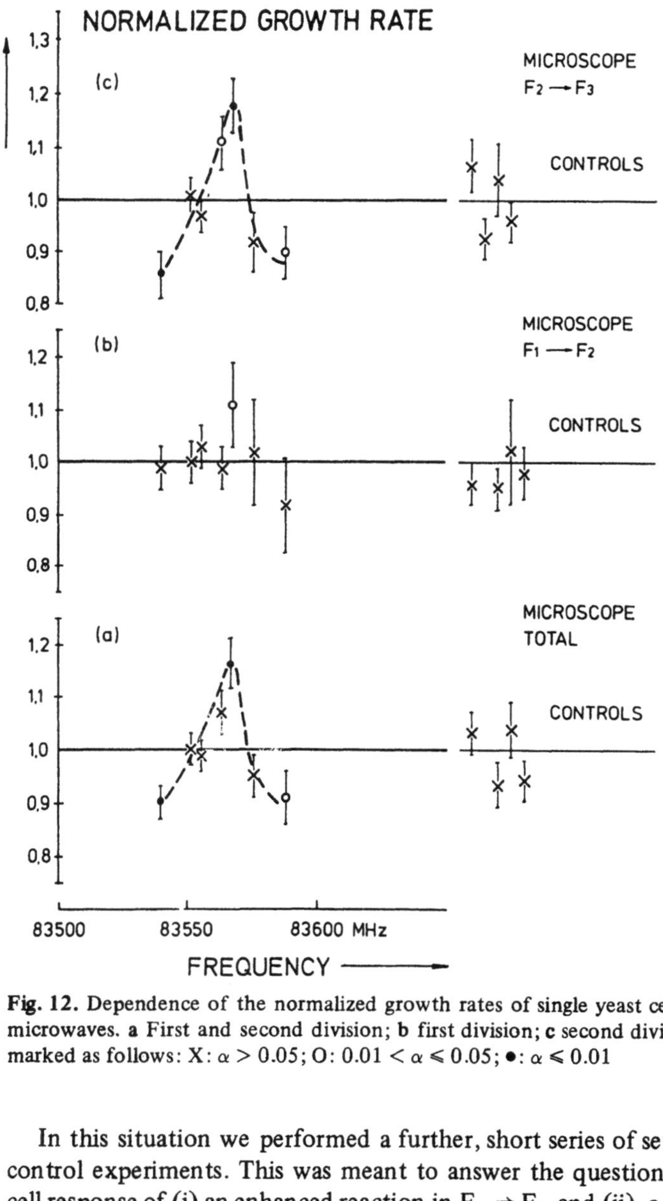

Fig. 12. Dependence of the normalized growth rates of single yeast cells on frequency of E-band microwaves. **a** First and second division; **b** first division; **c** second division. Significance levels are marked as follows: X: $\alpha > 0.05$; O: $0.01 < \alpha \leqslant 0.05$; •: $\alpha \leqslant 0.01$

In this situation we performed a further, short series of seven irradiation and four control experiments. This was meant to answer the questions whether the observed cell response of (i) an enhanced reaction in $F_2 \rightarrow F_3$ and (ii), a response by only a part of the population, was restricted to the frequencies used or whether effects could also be induced by E-band frequencies as applied already by others (Webb and Booth 1969; Berteaud et al. 1975). The experimental procedure used was the same as described in the above mention U-band experiments. The irradiation parameters were briefly as follows: frequency range from 83,540 MHz to 83,588 MHz, frequency uncertainty ± 2 MHz, incident power between 1 mW and 3 mW, corresponding to roughly estimated SAR values between 1.3 mW/g and 4 mW/g; the power transmitted $P_T < 0.1$ mW. The results of this pilot series at E-band frequencies are documented by Figs. 11 and 12.

Since only four controls were performed, the normalized growth rate $\bar{\mu}_0$ defined by the arithmetic mean of $\langle t_c \rangle_0$ = 1.31 ± 0.04 h scatters by ± 0.06 (Fig. 11, upper left). At two frequencies, 83,540 MHz and 83,564 MHz, the arithmetic mean of control runs was used as a reference to obtain $\bar{\mu}$, due to a lack of direct controls for these runs. For the other irradiation experiments, a direct control was available, as shown separately in every corresponding graph where it is characterized by a broken line, indicating the mean of the control distribution. In order to obtain such direct controls, synchronized cells were subdivided twice. One part was used in an irradiation experiment, while the other served in a control experiment performed sequentially 1 day later.

Surprisingly, when using much higher frequencies compared to the former experiments, qualitatively similar asymmetric changes of the shapes of the distributions occur (Fig. 8). Even in the cases where the effect on the average growth is strongest, be it positive, e.g. at 83,568 MHz, or negative, e.g. at 83,588 MHz, it can clearly be seen that one part only of the cells reacts on the radiation (roughly characterized by the shaded areas which represent the cycle times outside of the control distributions), while the other part behaves undisturbed, as already observed in Fig. 8. This observation may, on the one hand, explain the small size of averaged effects, and on the other hand, may suggest the influence of a still uncontrolled biological or other parameter synergistically acting with the radiation.

A further similarity between the results at 84 GHz and those obtained at U-band frequencies earlier is the different strength of the effect at different division steps. As presented in Fig. 12, significant irradiation effects are found mainly in the times for the second division. The effects are up to 20% and show also a frequency dependence with the highest value occurring at 83,564 ± 2 MHz. This frequency, surprisingly, is double frequency of 41,782 ± 1 MHz, where in the earlier photometer experiment yeast cells were found to respond with maximum positive growth rate effect. While the number of data points is still too small for a further evaluation of this "coincidence", we note that this observation may point in a very interesting direction, and should thus be investigated in further experiments. Finally we mention the small response of irradiated cells found during the first division, on observation already observed in the U-band experiment. As is well known, cells are able to metabolize many important macromolecules, such as proteins and RNA and to stock them for use in the later cycles. Therefore the conclusion may be drawn that microwaves act on physiological processes rather than directly on genetical ones. Due to the cumulative effect of metabolic production, the microwave effect is expressed more strongly in the second than in the first division cycle.

4 On the Induction and Reproduction of Frequency-Dependent Low Level Microwave Effects

There is no doubt for the importance of a convincing verification of frequency-dependent low level microwave biological effects. Of course, agreement on the existence of these effects is a prerequisite to any further judgement on their benefit or harmfulness.

Since the first publications in this field, however, the experimental proof of the biological responses seems to have been characterized by two difficulties,

i. the effects measured are mostly small,
ii. their reproducibility is frequently questionable.

Thus significant measurements of one laboratory could be reproduced by other laboratories with marginal success only or in other cases failed to be repeated.

Possibly these difficulties are of a fundamental nature, and may even be considered as an inherent feature of microwave biological effects. While mentioning interesting theoretical conjectures in this direction (Kaiser 1984), we would like to discuss in this chapter several lines of thinking where, from our experience, a key to solving reproducibility problems may be found.

– A stressed biological system is more sensitive to microwaves (synergistic effects)

The results of the bone marrow experiment (Sevastyanova and Vilenskaya 1974; Sevastyanova 1981; Sevastyanova et al. 1983), as shown in Figs. 1 and 2, demonstrate a lack of effects on viability when irradiated with microwaves alone. Only when combining microwaves with X-ray radiation did a marked microwave influence on the cell system become visible. Generalizing this result would mean that some stress is necessary in order for microwave effects to become visible. A similar result is available with yeast cells: with microwaves only (17 GHz, SAR = 5 mW/g) or in combination with X-rays, the frequency of the induction of genetically aberrant colonies was investigated (Dardalhon et al. 1986). Microwaves alone were unable to produce genetic effects, but microwave exposure, in addition to X-rays, resulted in a significant increase in frequency of X-ray-induced aberrant colonies including mitotic intergenic recombination.

– Frequency-specific biological reactions may depend on the biological system

Different results were obtained when the same specific enzymatic reaction, the synthesis of β-lactamase, was tested in different cell systems: a decrease in reaction rate caused by irradiation at two of several frequencies was observed in *Staphylococcus aureus* only (Fig. 4; Smolyanskaya 1981). No effect was observed by the same authors with a second cell system *(E. coli)*.

– Biological parameters to be tested should not be of too specific a nature

Studies performed on the genetic activity of microwaves at low power levels and different frequencies revealed no lethal or mutagenic effects in microorganisms (Blackman et al. 1975; Dutta et al. 1979; Dardalhon et al. 1981) and no chromosomal changes (Huang et al. 1977; McRee et al. 1981). Microwaves were found, however, to affect cellular growth in a frequency-specific manner (Webb and Dodds 1968; Berteaud et al. 1975; Grundler et al. 1983; Grundler 1985) or in one case, where frequency was not varied systematically, to enhance growth (Hamnerius et al. 1985). From these studies microwaves seem to cause physiological rather than genetical reactions. Altogether, growth is a test parameter general enough in detecting a possible frequency-dependent microwave response. Since growth may be affected by many physiological parameters, the chance of selecting the one responding to the microwave irradiation in a more specific experiment is rather small.

– Microwave influence can be restricted to certain cell cycle phases

To specify microwave effects on yeast growth we have for the first time tested the restriction of the influence to both to (i) certain cell cycle phases and (ii) certain metabolic pathways (respiration). For this purpose we used a respiration-defective mutant (strain 211 pμ, Laskowski 1960) which we synchronized by two subsequent methods, the sedimentation gradient and isokinetic volume selection, to the early G_1-phase. The narrow volume distribution for t = 0h (Fig. 13) demonstrates the quality of this procedure and characterizes the cells used at the beginning of the experiment. The volume of stationary yeast cells is known to be strongly correlated to the position within a cycle.

By means of a computer-controlled coulter counter, the time-dependent shifting of this distribution is observed over one cell cycle and used as test parameter (Fig. 13). To use an elongated cell cycle time of 5 h we lowered the temperature from the usual 30 °C to 25 °C. Inspection of the results in Fig. 13 reveals that the mean volume of cells developed for 4 h is greater than that for 5 h which is explained by assuming that after 5 h single cells have already developed due to cytokinesis. The reproducibility of the method is very good, since the results of repeated growth experiments are identical within the width of the symbols drawn (Fig. 8).

To analyze the influence of microwaves using this method we performed a dozen experiments applying the frequency 41,778 ± 0.5 MHz at 1 mW/cm², which earlier (Fig. 9) had effected a growth reduction. The irradiation chamber was equipped with a circular horn antenna. An identical but inactive horn was positioned in the control chamber.

The results of a sham control and two irradiation experiments after 5 h growth are shown in Fig. 14. They demonstrate a cell cycle-specific influence. Irradiation during the lag and G_1-phases (0 to 2.5 h) causes a reduction in volume growth velocity. This can be seen by the quite good coincidence of the measured distribution (Fig. 14 "...") with the 4-h distribution of Fig. 13, in comparison to the sham control (Fig. 14). In contrast, irradiation during the S, G_2, and M-phases (2.5 to 5 h) causes no shift of the volume distribution in respect to the control. In conclusion, the microwave influence is restricted to the G_1-phase at the frequency used.

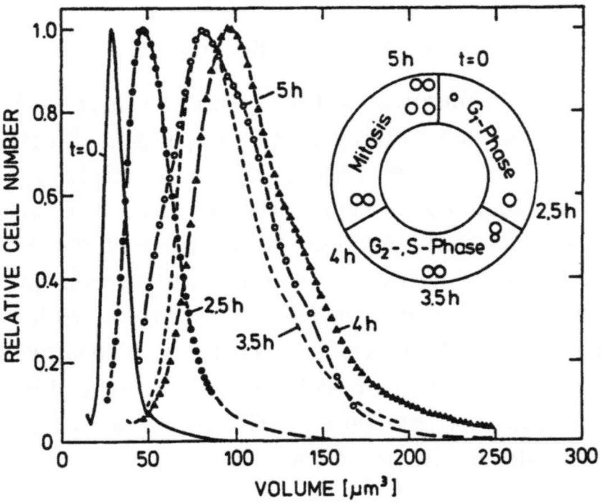

Fig. 13. Normalized volume distributions of yeast cells grown during one cell cycle at 25 °C. Schematically, the correlation of the different distributions to the position within the cell cycle is shown

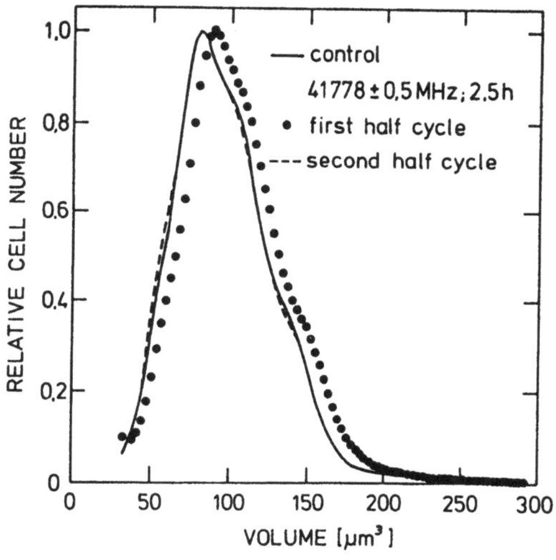

Fig. 14. Normalized volume distributions of yeast cells measured at the end of mitosis for control (——) and for cells irradiated with 41,778 MHz during the first (·····) and the second (– – – – –) half of the cycle

This result has been confirmed by a series of 12 similar experiments.

The respiration-defective mutant reduces its growth velocity, as does the wild-type strain when irradiated at the some frequency. Therefore, it is probably not the respiration metabolism which primarily becomes disturbed and leads to the reduction of yeast cell growth.

– The temperature of cells in pre-experiment storage may affect the interaction with microwaves

Using the method of measuring volume distributions as described before, wild-type yeast cells were synchronized within the early G_1-phase and subsequently stored in buffer at 4 °C and 30 °C. For testing the duration of the G_1-phase of these cultures in dependence of microwave radiation (41,652 MHz), the volume distribution was measured at the time of contact to the medium (30 °C) and 70 min later, when the cells were about to move into S-phase. Comparison was made with the corresponding control distributions. As a preliminary result we observed an identical behaviour of sham controls and of irradiated cultures, i.e. no microwave effects when the storage temperature was 4 °C. When irradiating cells stored at 30 °C, however, an increased volume growth velocity was mostly observed.

5 Conclusion

The main and most important conclusion to be drawn from this survey is that resonant biological effects of microwaves are now well documented to exist. Active frequencies have been confirmed with independent methods. Active frequencies differing by a factor of 2 have been found in one case, while in general active frequencies have been systematically sought in very small relative frequency ranges (< 1%) only. Thus many more frequencies might be observable in future.

The experiments are not easy, since the effects are small and depend on biological variables in a way which has only started to become visible. One example is the temperature at which yeast cells are stored before the actual growth experiments begin. While strict control of this temperature has been seen to increase the reproducibility of the microwave effect, a full reproducibility has not been achieved, thus pointing to further unidentified parameters.

There is reason to expect a certain amount of non-reproducibility in resonant microwave biological effects, in a quite fundamental way. First, there are theoretical concepts dealing with the occurrence of coherence and deterministic chaos in biological systems. Secondly, experiments with *individual* cells directly prove that cells react in several distinct subgroups, although they all come from the same batch and growth procedure and are exposed simultaneously in close proximity. This new observation points to a very interesting new direction. On the other hand, studies which have failed to prove effects in the past or have been performed with marginal success only (e.g. Furia et al. 1986) should be reanalyzed, since cellular effects may well have been washed out by improper averaging.

Acknowledgements. This study was supported by the Deutsche Forschungsgemeinschaft and the Stifterverband für die Deutsche Wissenschaft. We thank H. Fröhlich (Liverpool), L. Genzel (Stuttgart) and W. Pohlit (Frankfurt) for their support and collaboration. The authors are indebted to Chenghe Xu (Beijing) for analyzing microwave structures. The technical assistance of I. Zimmermann, O. Lock, F. Müller and L. Mindl is gratefully acknowledged.

References

Berteaud AJ, Dardalhon M, Rebeyrotte N, Averbeck D (1975) Action d'un rayonnement électromagnétique à longueur d'onde millimétrique sur la croissance bactérienne. CR Acad Sci (D) (Paris) 281:843–846

Blackman CF, Benane SG, Weil CM, Ali JS (1975) Effects of nonionizing electromagnetic radiation on single cell biologic systems. Ann NY Acad Sci 247:352–366

Dardalhon M, Averbeck D, Berteaud AJ (1981) Studies on possible genetic effects of microwaves in procaryotic and eucaryotic cells. Radiat Environ Biophys 20:37–51

Dardalhon M, Averbeck D, Berteaud AJ, Rayary V (1986) Action of 17 GHz microwaves combined with ultraviolet or x-irradiation on *Saccharomyces cerevisiae*. Thesis (D.M.), Univ Paris, pp 132–156

Devyatkov ND (ed) (1981) Nonthermal effects of millimeter wave irradiation. Acad Sci USSR, Inst Radiotech Electrotech Moscow (in Russian)

Devyatkov ND (ed) (1983) Nonthermal effects of millimeter wave irradiation on biological objects. Acad Sci USSR, Inst Radiotech Electrotech Moscow (in Russian)

Devyatkov ND, Sevastyanova LA, Vilenskaya RL, Smolyanskaya AZ, Kondrateva YF, Chistyakova EN, Shmakova IF, Ivanova NB, Treskunov AA, Manoilov SE, Zalyubovskaya VA, Koselev RJ, Gaiduk VI, Khurgin YI, Kudryashova VA (1974) Sov Phys-Usp 16, 4:568–579

Dutta SK, Nelson WH, Blackman CF, Brusick DJ (1979) Lack of microbial genetic response to 2.45 GHz (CW) and 8.5 to 9.6 GHz pulsed microwaves. J Microwave Power 14:275–280

Fröhlich H (1968) Long range coherence and energy storage in biological systems. Int J Quantum Chem 2:641

Fröhlich H (1970) Long range coherence and the action of enzymes. Nature 228:1093

Fröhlich H (1980) In: Marton L (ed) Advances in electronics and electron physics. Academic Press, New York, 53:85–152

Fröhlich H (1986) Coherent excitation in active biological systems. In: Gutmann F, Keyzer H (eds) Modern bioelectrochemistry. Plenum, New York, pp 241–261

Fröhlich H, Kremer F (1983) Coherent excitations in biological systems. Springer, Berlin Heidelberg New York

Furia L, Hill DW, Gandhi OP (1986) Effect of millimeter-wave irradiation on growth of *Caccharomyces cerevisiae*. IEEE Trans Biomed Eng Vol BME-33, 11:993–999
Grundler W (1983) Biological effects of RF and MW energy at molecular and cellular level. In: Grandolfo M, Michaelson SM, Rindi A (eds) Biological effects and dosimetry of nonionizing radiation. Plenum, New York, pp 299–318
Grundler W (1985) Frequency-dependent biological effects of low intensity microwaves. In: Chiabrera A, Nicolini C, Schwan HP (eds) Interactions between electromagnetic fields and cells. Plenum, New York, pp 459–481
Grundler W, Abmayr W (1983) Differential inactivation analysis of diploid yeast exposed to radiation of various LET. Res 94:464–479
Grundler W, Keilmann F (1978) Nonthermal effects of millimeter microwaves on yeast growth. Z Naturforsch 33c:15–22
Grundler W, Keilmann F (1983) Sharp resonances in yeast growth prove nonthermal sensitivity to microwaves. Phys Rev Lett 51, No 13:1214–1216
Grundler W, Keilmann F, Fröhlich H (1977) Resonant growth rate response of yeast cells irradiated by weak microwaves. Phys Lett 62A:463
Grundler W, Keilmann F, Putterlik V, Santo L, Strube D, Zimmermann I (1983) Nonthermal resonant effects of 42 GHz microwaves on the growth of yeast cultures. In: Fröhlich H, Kremer F (eds) Coherent excitations in biological systems. Springer, Berlin Heidelberg New York, pp 21–37
Hamnerius Y, Rasmuson A, Rasmuson B (1985) Biological effects of high-frequency electromagnetic fields on *Salmonella typhimurium* and *Drosophila melanogaster*. Bioelectromagnetics 6: 405–414
Huang AT, Engle ME, Elder JA, Kinn JB, Ward TR (1977) The effect of microwave radiation (2450 MHz) on the morphology and chromosomes of lymphocytes. Radio Sci 12:173–177
Kaiser F (1984) Entrainment-quasiperiodicity-chaos-collapse, bifurcation routes of externally driven self-sustained oscillating systems. In: Adey WR, Lawrence AF (eds) Nonlinear electrodynamics in biological systems. Plenum, New York, pp 393–412
Keilmann F (1983) Experimental RF and MW resonant nonthermal effects. In: Grandolfo M, Michaelson SM, Rindi A (eds) Biological effects and dosimetry of nonionizing radiation. Plenum, New York, pp 283–297
Keilmann F (1985) Biologische Resonanzwirkungen von Mikrowellen. Physik in unserer Zeit 16:33
Keilmann F (1986) Triplet-selective chemistry: a possible cause of biological microwave sensitivity. Z Naturforsch 41c:795–798
Laskowski W (1960) Inaktivierungsversuche mit homozygoten Hefestämmen verschiedenen Ploidiegrades. Z Naturforsch Teil B 15:495
McRee DI, Macnichols G, Livingston GK (1981) Incidence of sister chromatic exchange in bone marrow cells of the mouse following microwave exposure. Radiat Res 85:340–348
Pickard WF (1986) Criteria for the design or selection of experiments in bioelectromagnetics. Bioelectromagnetics Society Newsletter Jan., Febr., March
Sevastyanova LA (1981) Specific influence of millimeter waves on biological objects. In: Devyatkov ND (ed) Nonthermal effects of millimeter wave irradiation. Acad Sci USSR, Inst Radiotech Electrotech Moscow, pp 86–113 (in Russian)
Sevastyanova LA, Vilenskaya RL (1974) A study of the effects of millimeter-band microwaves on the bone marrow of mice. Sov Phys Usp (Engl Transl) 16:570
Sevastyanova LA, Gorodnina ES, Zubenkova MB, Golant TB, Rebrova VL, Iskrickij VL (1983) Resonant influence of millimeter waves on biological systems. In: Devyathov ND (ed) Nonthermal effects of millimeter wave irradiation on biological objects. Acad Sci USSR, Inst Radiotech Electrotech Moscow, pp 34–47 (in Russian)
Smolyanskaya AZ (1981) Influence of electromagnetic waves on microorganisms. In: Devyatkov ND (ed) Nonthermal effects of millimeter wave irradiation. Acad Sci USSR, Inst Radiotech Electrotech Moscow, pp 132–146 (in Russian)
Webb SJ, Dodds DD (1968) Inhibition of bacterial cell growth by 136 gc microwaves. Nature (Lond) 218:374–375
Webb SJ, Booth AD (1969) Absorption of microwaves by microorganisms. Nature (Lond) 222: 1199

The Influence of Low Intensity Millimetre Waves on Biological Systems

F. KREMER[1], L. SANTO[2], A. POGLITSCH[3], C. KOSCHNITZKE[4], H. BEHRENS[5], and L. GENZEL[2]

1 Introduction

It is a well-known fact that millimeter waves, when absorbed in biological material, generate heat. The question whether millimeter waves cause further effects in living systems, so-called non-thermal effects, which cannot be explained on a thermal basis, is still a controversial topic in scientific discussion (Nato Advanced Research Workshop, Erice, 1984). In the course of this chapter three experiments will be presented which give insight into the variety of biological systems sensitive to low intensity millimetre wave radition. Examples will be given (i) for a thermal effect of low intensity millimetre waves on the growth of cress roots (Kremer et al. 1985), (ii) for a low intensity cytologically manifested influence on root tips of cress roots (H.M. Behrens et al., in prep.) and (iii) for a non-thermal effect on the puffing of giant chromosomes from salivary glands of larvae of the midge *Acricotopus lucidus* (Koschnitzke et al. 1983, 1986; Kremer et al. 1983). The experiments will be described in detail and, as far as the non-thermal nature of the effect is confirmed, mechanisms of action of the electromagnetic waves with the biological system will be discussed.

2 The Root Growth Experiment

Growing roots of cress seedlings are widely used as a model system for the analysis of growth processes and for the investigation of gravitropism (Iversen 1969; Behrens 1982; Volkmann 1979). Two different processes contribute to the growth of cress roots; growth by elongation and growth by cell division. The latter is localized in the meristematic zone of the root tip and contributes approximately 10% of the total elongation of the root (Behrens 1982). The fact that the seedlings grow well in an atmosphere of high relative humidity makes them especially suitable for irradiation experiments with millimetre waves, which would have a penetration depth of only about 0.5 mm in an aqueous medium (Szwarnowski and Sheppard 1977).

The cress seeds commercially available (Garten-Kresse) were submerged in distilled water for 1 h at 24 ± 0.5 °C. The seedlings were then placed on a moist filter pad in

---

1 Max-Planck-Institut für Polymerforschung, Postfach 3148, 6500 Mainz, FRG
2 Max-Planck-Institut für Festkörperforschung, Heisenbergstraße 1, 7000 Stuttgart 80, FRG
3 Max-Planck-Institut für Physik und Astrophysik, 8000 München 40, FRG
4 Institut für Allgemeine Genetik Universität Hohenheim, Garbenstr. 30, 7000 Stuttgart 70, FRG
5 Institut für Botanik Universität Bonn, Venusbergweg 22, 5300 Bonn-Bad Godesberg, FRG

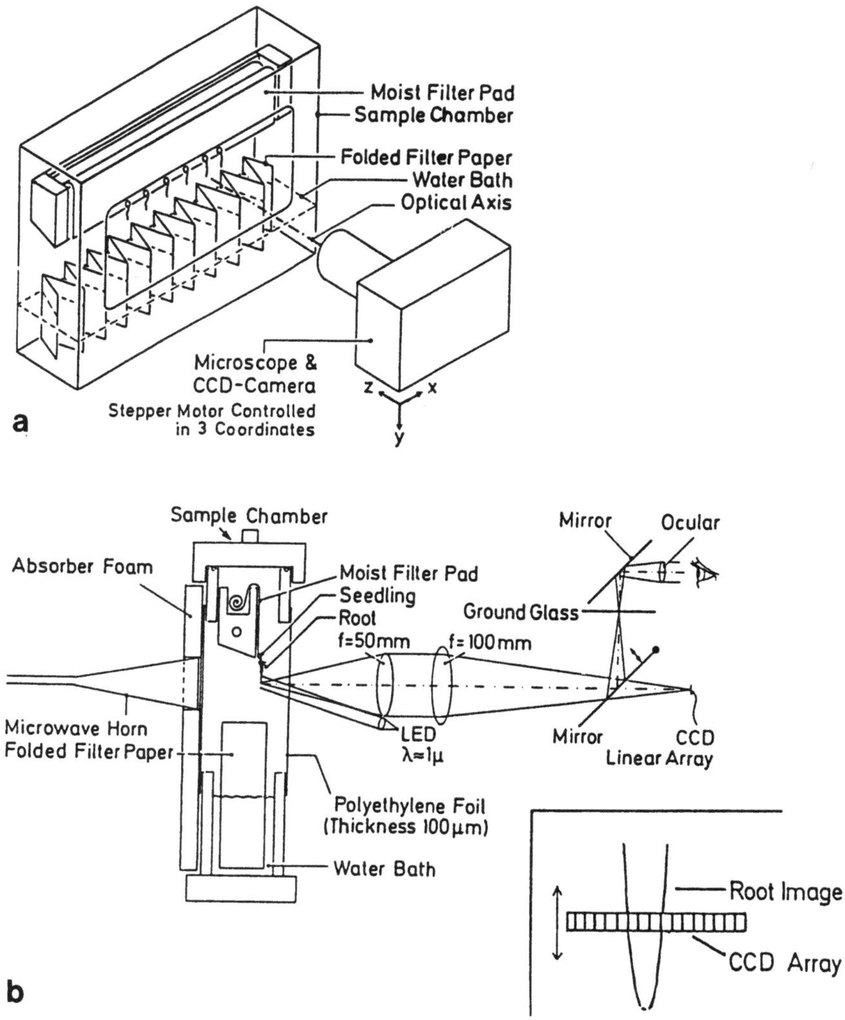

Fig. 1. **a** Schematic description of the sample chamber and the microscope. **b** Scheme of the optical path of the microscope and the millimetre-wave horn antenna (not so *scale*). *Inset* image of the root on the CCD array

the sample chamber (Figs. 1a,b) and incubated for 20 h at 24 ± 0.5 °C before irradiation (root length at the time: about 5 mm). The sample chamber was a sealed Plexiglas container with windows of polyethylene foil (area: 14 x 6 cm^2, thickness: 100 μm), which is almost transparent to millimetre waves. Inside the chamber, the atmosphere was saturated with humidity, using a water bath at the bottom on which a piece of folded filter paper was laid in order to increase the evaporating surface area.

The chamber and the optical system were kept in a dark, temperature-controlled (24 ± 0.5 °C) wooden box. To determine the growth rates of individual roots, a computer-controlled optical system (Fig. 1a,b) was employed. It included a microscope with twofold magnification. Four LED's ($\lambda \sim 1$ μm) for weak illumination of the seedlings and a linear array of 128 photodetectors (CCD array: area of each detector element was $26 \times 26\ \mu m^2$). The whole optical system was mounted on a stepper motor-controlled platform that could be moved in three dimensions. Use of image processing techniques in the computer (HP 9826) allowed quasi-continuous measurements of the positions of one or more root tips within an accuracy of $\pm 2\ \mu m$. This was achieved by automatically focussing the image of the root on the CCD array as part of the image processing. To measure the position of an individual root tip a time of about 50 s was necessary. Usually four roots were observed, two as controls and two for irradiation. To demonstrate that the measurement system including the illumination with light of $\lambda \sim 1\ \mu m$ had no effect on the root growth, tests were carried out in which a group of roots whose growth rate was measured with the optical system was compared with a control group in the same container. No significant difference in root lengths of these two groups was found.

The millimetre-wave irradiation system, schematically shown in Fig. 2, included a backward-wave oscillator (Siemens RWO 60), an isolator, a precision attenuator, a frequency metre and a calibrated thermistor mount (Hughes) in connection with a power metre (HP 432 A). The frequencies were arbitrarily chosen between 42 GHz and 58 GHz, and measured to an accuracy of ± 25 MHz. The forward power in the output waveguide ranged between 0.7 mW and 21 mW, corresponding to a power density of 0.2 to 6 mW/cm^2 respectively (horn area: 3.6 cm^2, gain of the horn: 20 dB). To protect the control roots in the sample chamber from the millimetre-wave radiation, a mask made from absorber foam (Eccosorb AN 72) was placed across the front of the sample chamber (see Fig. 1b). The microwave-induced temperature increase was measured by two different methods: (i) an IR-radiometric system and (ii) a microminiature thermal probe. The former contained a KRS5 lens, a black polyethylene filter (thickness: 0.1 mm), a germanium lens and a metal mesh (spacing: 0.4 mm) and allowed the root tip to be focussed on a Golay cell so that its thermal emission could be measured (see Fig. 2). For calibration a free-standing, temperature-controlled water jet was used.

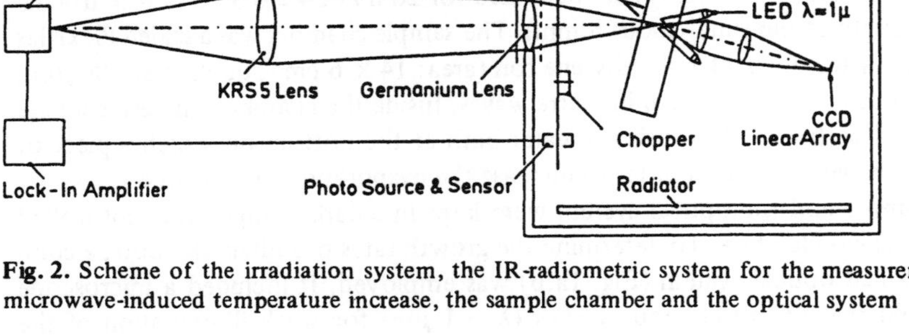

Fig. 2. Scheme of the irradiation system, the IR-radiometric system for the measurement of the microwave-induced temperature increase, the sample chamber and the optical system

In the second method the microwave-induced temperature increase was measured with a micro-miniature thermocouple (ϕ: 25 μm). This thermocouple was brought into contact with the root tip on the side opposite to the millimetre-wave horn. It was aligned perpendicular to both the electric and the magnetic millimetre-wave field. Thus the perturbation of the field was about the same for both polarizations. Microwave-induced temperature increases measured with both methods were equal within limits of experimental accuracy ($\pm$0.1 °C for the IR-radiometric system and $\pm$0.02 °C for the micro-miniature thermal probe).

2.1 Results

The temperature dependence of the growth rate of cress seedlings under our conditions is shown in Fig. 3. It shows a large variability in the growth rate of individual seedlings. This shows the necessity to study a possible microwave effect with single roots of which the growth can be measured quasi-continuously with and without irradiation. At around 25 °C a maximum in the growth rate is found. Hence in the experiments an ambient temperature of 24 °C was chosen to make sure that an overall irradiation-induced temperature rise of even 0.1 °C could not cause a decline in growth rate.

A typical irradiation experiment is shown in Fig. 4a. The growth rate was observed with the computer-controlled optical system for several hours and a value of about

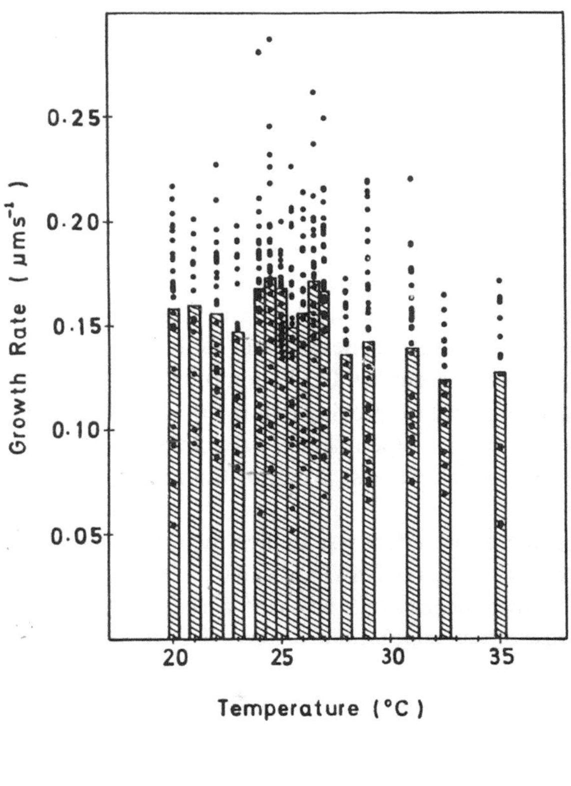

Fig. 3. Temperature dependence of the growth rate determined by measuring the length of a root after about 45 h of growth. Each *dot* represents the growth rate of a single root kept for 45 h in a chamber of corresponding temperature. *Shaded bars* indicate the mean values of the growth rate

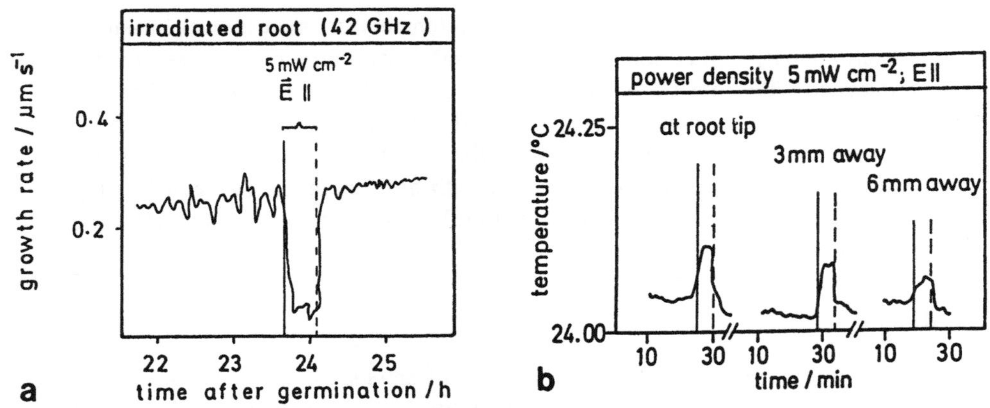

Fig. 4. a Growth rate vs. time after germination of a root irradiated with 42 GHz ± 0.025 GHz microwaves. The irradiation started (*solid line*) at 23.7 h and stopped (*broken line*) at 24.1 h after germination. The power density was 5 mW/cm², the E-vector of the microwaves was parallel to the central axis of the root. **b** Temperature conditions of the experiment reported in **a**. The microwave-induced temperature increase was measured with the micro-miniature thermocouple (left to right) (i) at the root tip, (ii) 3 mm away from the root tip and (iii) 6 mm away from the root tip

0.2 µm/s was measured. Irradiating with millimetre waves at a frequency of 42 GHz and a power density of 5 mW/cm² with the E-vector polarized parallel to the central root axis resulted in a strong reduction of the growth. After switching off the irradiation, the root recovered and reached a level of growth comparable to the value before the irradiation. On measuring the time constant for the occurrence of the irradiation effect, one finds a value of 100 s ± 50 s. This indicates that the irradiation primarily influences the growth by elongation rather than the growth by cell division. The microwaves-induced temperature rise of the root tip was measured with the thermocouple to be less than 0.1 °C. To measure the temperature gradient in the direct neighbourhood of the root, the microwave-induced temperature rise was measured 3 mm and 6 mm away from the root tip as well (Fig. 4b).

To study whether the microwave effect is localized in the root tip or in another part of the root, irradiation experiments were carried out in which, by use of an aluminium mask, different parts of the roots were shielded (Fig. 5a,b). The power deposition (measured as the microwave-induced temperature increase) for both experiments was found to be equal within experimental accuracy. Irradiation of the root hair zone did not show an effect on the growth rate, thus proving that the microwave-sensitive process is located in the root tip. For further experiments an aluminium mask was always used, which shielded the roots so that the last 2 mm of the root tip were preferentially irradiated.

The power dependence of the microwave effect was studied by irradiating a single root sequentially with power densities between 0.2 mW/cm² and 6 mW/cm² (Fig. 6a,b). Figure 6a shows that even with a power density of 2 mW/cm² a clear decrease of the growth rate resulted. For the dose-effect relationship a sigmoid function is found (Fig. 6b).

To study a possible frequency dependence of the irradiation effect, frequencies of 42 GHz, 55.8 GHz, 56 GHz, 56.2 GHz, 57.8 GHz and 58 GHz were arbitrarily chosen.

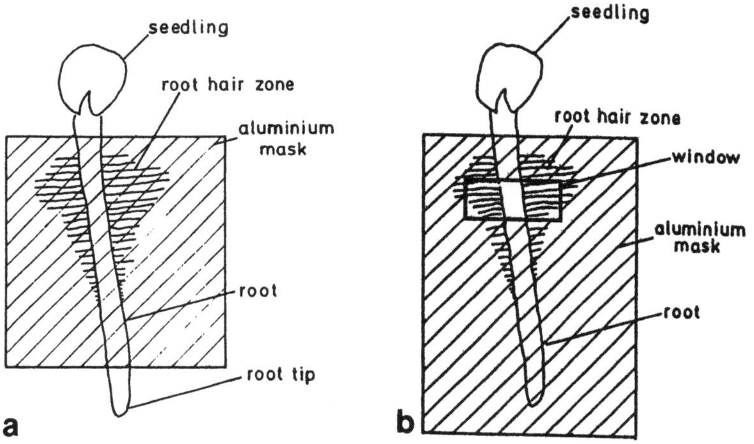

Fig. 5. a Scheme of a root shielded in such a way that mainly the root tip is irradiated. **b** Scheme of a root shielded so that mainly a part of the root hair zone is irradiated

In all cases a similar effect was found. A typical example is shown in Fig. 7a, in which one root was irradiated sequentially with 56.0 GHz and 58.0 GHz millimetre waves. The power density was 3 mW/cm². The temperature conditions in Fig. 7a are plotted in Fig. 7b.

For the study of the interaction mechanism between millimetre waves and the biological system, it could be of interest to determine whether the biological system would react to continuous wave (c.w.) irradiation in a different way compared to an amplitude modulated irradiation (a.m.). The amplitude modulation had a modulation frequency of 1 kHz. The power density was in both cases (c.w. and a.m.) adjusted to 2.5 mW/cm². As shown in Fig. 8, no difference is observable for both irradiation conditions. This excludes a non-linear effect of the electromagnetic field.

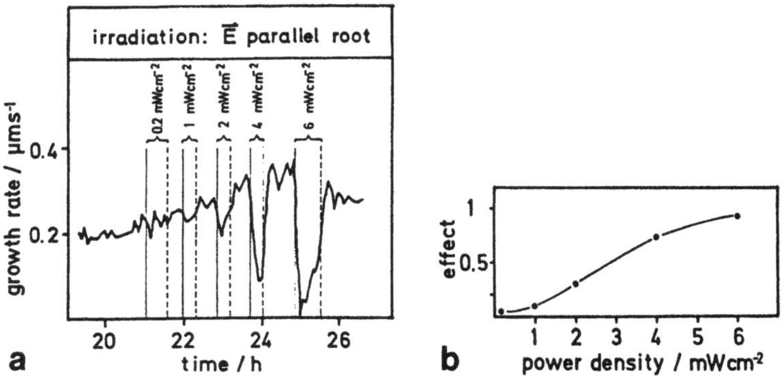

Fig. 6. a Power dependence of the effect. The root was successively irradiated with power densities between 0.2 mW/cm² and 6 mW/cm². The frequency was 56 GHz. The E-vector was parallel to the central axis of the root. **b** "Dose-effect-relationship" as determined from **a**. The value of an "effect" is given as the ratio between the growth rate before the irradiation with a certain power density started and the difference between this value and the growth rate after irradiation with this power density

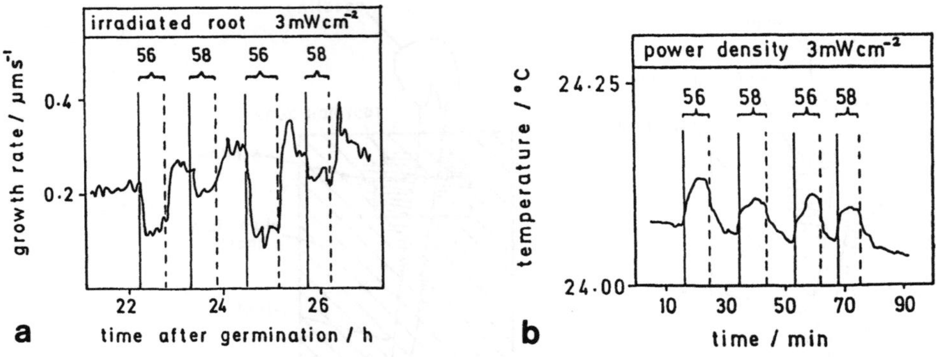

Fig. 7. a Growth rate vs. time after germination for irradiation of the roots with 56 ± 0.025 GHz and 58 ± 0.025 GHz millimetre waves in sequence. The power density was 3 mW/cm². **b** Microwave-induced temperature increase of the root tip for the irradiation described in **a**

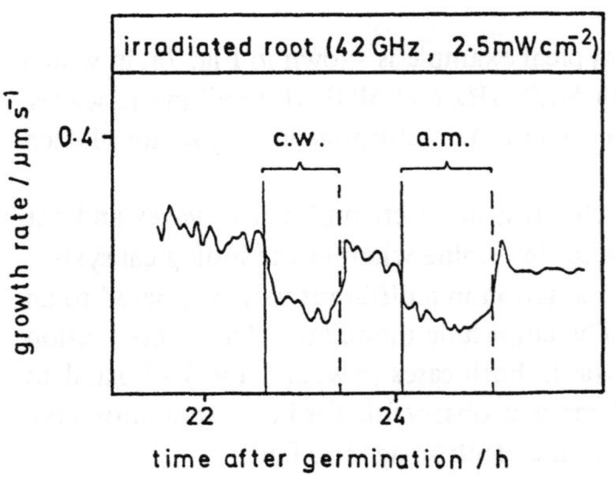

Fig. 8. Comparison of an irradiation with amplitude modulated (1 kHz) and continuous wave microwaves. Frequency: 42 ± 0.025 GHz. Power density: 2.5 mW/cm²

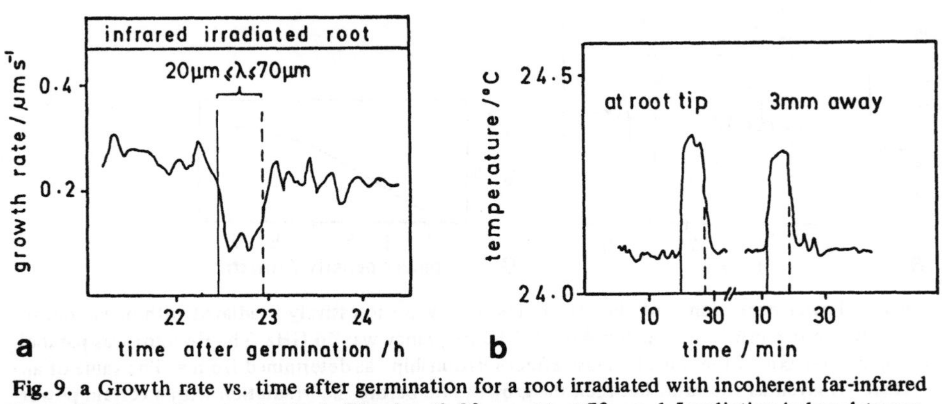

Fig. 9. a Growth rate vs. time after germination for a root irradiated with incoherent far-infrared light of a thermal source (globar). Wavelength 20 $\mu m \leqslant \lambda \leqslant 70$ μm. **b** Irradiation-induced temperature increase at the root tip and 3 mm away for the experiment described in **a**

The question arises if the small microwave-induced temperature gradients in the root tip exert an influence on the growth rate. In order to study this, the microwave-induced temperature rise at the location of the root tip was simulated with incoherent far-infrared light ($20\ \mu m \leqslant \lambda \leqslant 70\ \mu m$) of a thermal source. The Golay cell in Fig. 2 was replaced by a globar and additionally a PTFE filter (thickness 0.5 mm) was mounted in front of the germanium lens, so that only radiation of $20\ \mu m \leqslant \lambda \leqslant 70\ \mu m$ could reach the sample. Thus infrared light, which might influence the cytochrome C in the root tip (Mohr 1964) was filtered out. The radiation was focussed on the root tip and the effect was observed with the computer-controlled system. A typical result is shown in Fig. 9. The temperature conditions for the experiment are described in Fig. 9b. It shows that for temperature changes comparable to those observed in the millimetre wave irradiation experiment similar effects are found.

To summarize, the root growth of cress seedlings between 20 °C and 27 °C is only weakly dependent on the ambient temperature. However, small temperature gradients across the surface of the root tip, induced by microwave or far-infrared radiation, have proven to be of high influence on root growth. Thus the effect does not depend on temperature but rather on temperature gradient.

3 A Cytological Effect on Root Tips of Cress

After finding a strong influence of low intensity millimetre waves on the growth of cress roots, it seemed to be interesting to look for a structural radiation effect. The experiments were carried out under the same conditions as described in the growth experiment. Only roots with a strong irradiation-induced reduction in growth rate were used for the cytological examination. Directly after irradiation the fixation of the roots was caried out. Two methods were employed: (1) Fixation with 2% potassium permanganate in aqueous solution dehydrated through a graded aceton series and embedding in epoxy resin according to Spurr (Hensel 1984a,b; Spurr 1969).

(2) Fixation with glutaraldehyde (GA) with phosphate buffer and 2% Tanin at pH 7.0. The staining was done according to Reynolds (1963). From each preparation,

Table 1. Statistics of the cytological examination which was carried out in a blind experiment. "Success": Number of correctly determined control or irradiated samples. "Failure": Number of incorrectly determined control or irradiated samples. P: Probability in percent that the observed Relation "Successes" to "Failures" occurs by chance

Type of experiment	Number of roots	"Successes"	"Failures"	P [%]
Irradiation with 59 GHz, 5 mW/cm²	36	34	2	10^{-11}
Irradiation with 42 GHz, 3 mW/cm²	23	19	4	10^{-5}
Irradiation with 72 GHz, 5 cmW/cm²	15	12	3	1.4×10^{-4}

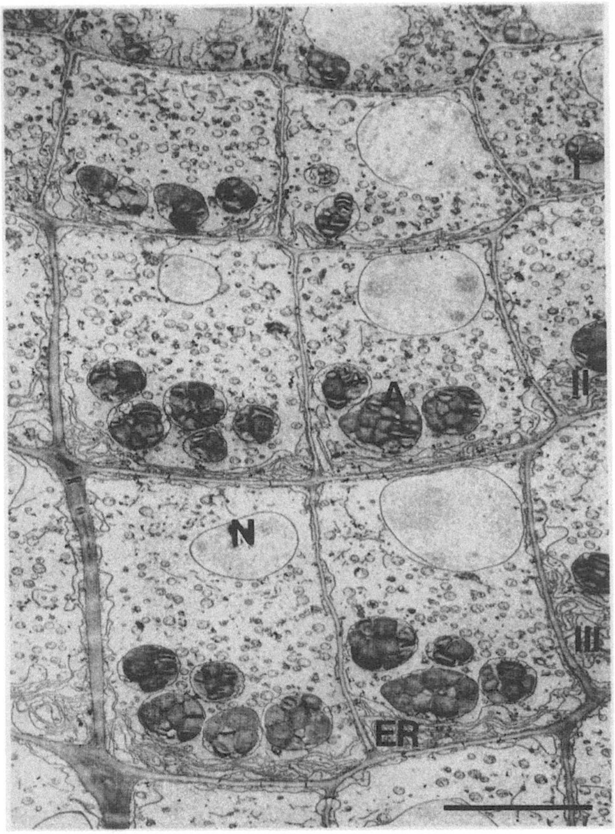

Fig. 10. Electron micrographs of longitudinal section through the statenchyma of the root tip from control experiments. Fixation: 2% $KMnO_4$. The statocytes (graviperceiving cells) of storeys I to III show, by the distal arrangement of the ER complex (*ER*) and the proximal lying nucleus (*N*), the characteristic cell polarity. *N* nucleus; *A* amyloplast; *ER* endoplasmic reticulum; bar = 10 μm

three to five longitudinal sections through the calyptra were investigated. Irradiation frequencies between 42 GHz and 72 GHz with power densities between 3 mW/cm^2 and 5 mW/cm^2 were arbitrarity chosen.

In pilot studies we examined cells from the root cap (secretion cells), the statenchyma (gravi-perception cells) and meristematic cells. Only in tissue from the statenchyma was an irradiation effect observed: lysosomes with multivesicular structures occurred with a probability which was atypical for cells in this stage of development (Figs. 10–14). This indicates a disturbance of the metabolism as well.

A cytological study cannot easily be quantitative. To exclude the possibility of a biased examination of the electron microscopic pictures, we carried out blind experiments. The biologist received only numbered samples and he had to find out which

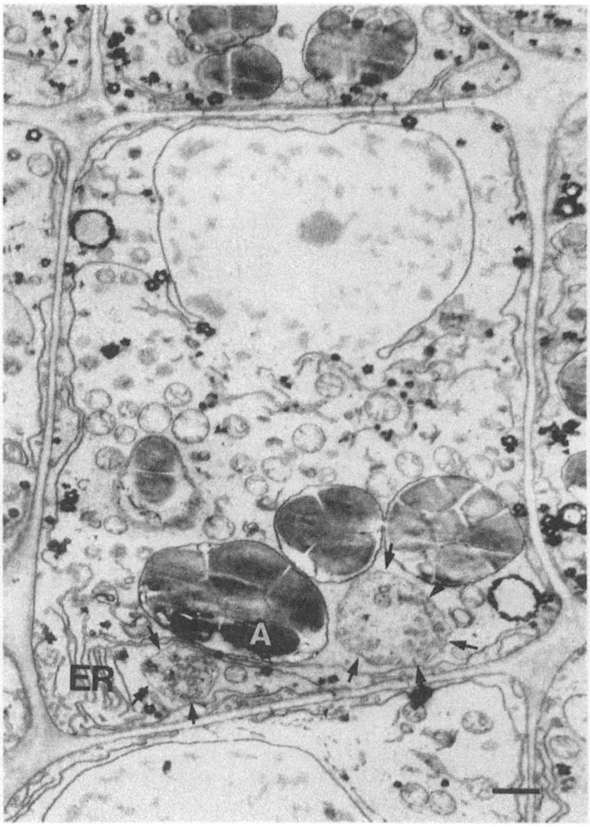

Fig. 11. Irradiated root. Statocyte from the median plane of the root tip after 55 min irradiation. 65 GHz, power density: 3 mW cm^{-2}; Fixation: 2% $KMnO_4$. Two lysosomes (*arrows*) can be seen in the proximate cell site comparable in size to the amyloplasts. This is atypical for cells in this stage of development and indicates a strong disturbance of its cell metabolism. *N* nucleus; *A* amyloplast; *ER* endoplasmic reticulum; bar = 1 μm

was irradiated and which had served as control. (The number of irradiated and control samples was equal). The result (Table 1) proves with high significance an irradiation effect. The question if this is of thermal or non-thermal origin has not yet been answered.

4 The Giant Chromosome Experiment

Giant chromosomes which are obtained in gland tissues (e.g. salivary glands) of insects, are a unique phenomenon in nature. In these chromosomes strands of identical DNA are many thousandfold aligned in parallel. They are condensed in their tertiary structure and stabilized by proteins, the histones. In order to transfer the DNA information into messenger RNA, the giant chromosomes must partly decondense. Such loci along the chromosome, the so-called Balbiani rings or puffs, are directly observable with the light microscope (magnification: 200). They form a pattern which is

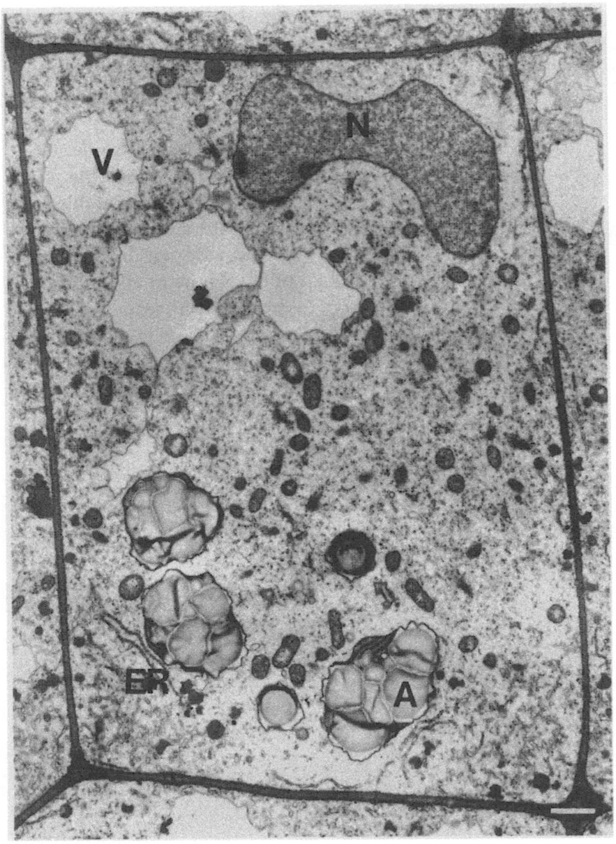

Fig. 12. A central statocyte from the median plane of the root tip from a control experiment. Fixation: glutaraldehyde. After fixation with GA-O_sO_4 the ER complex is only partially preserved. Close to the amyloplast (*A*) short ER-cisternae (*ER*) can be recognized. In the vicinity the ER-complex is artefactually disintegrated into vesicles. *N* nucleus; *V* vacuole; bar = 1 μm

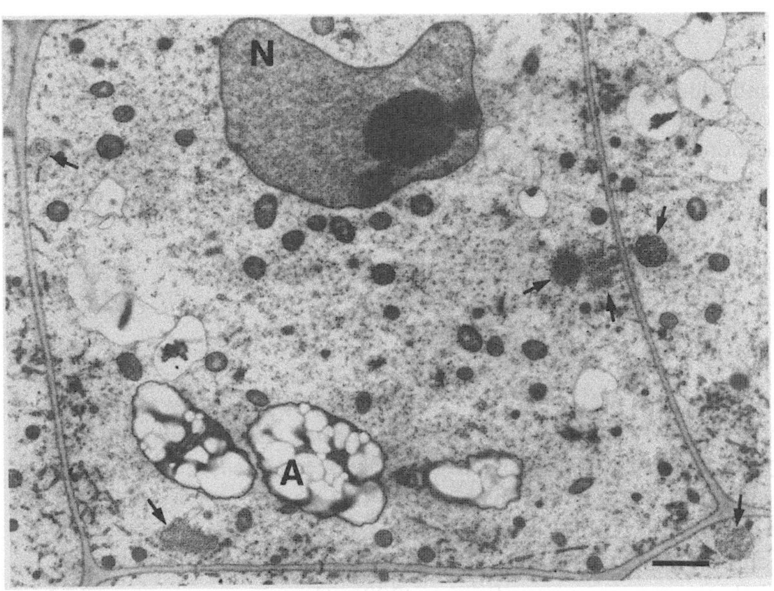

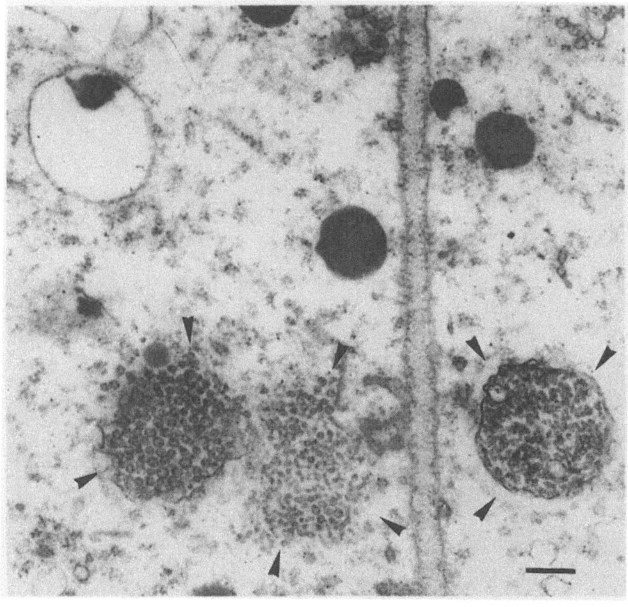

Fig. 14. Detail of the multivesicular arrays shown in Fig. 13. Bar = 1 μm

characteristic for the developmental stage of the insect (Panitz 1967, 1972; Bandisch and Panitz 1968; Bandisch 1977). Thus giant chromosomes offer the possibility of studying directly the influence of agents or radiation on the level of the genome (Fig. 15a,b).

4.1 Experimental

The irradiation set up included two electromagnetically isolated chambers: one for irradiation and one for control. As millimetre wave source a backward-wave oscillator (RWO 60, RWO 80, Siemens) were used (Fig. 16). Different microwave frequencies between 40 GHz and 80 GHz were arbitrarily chosen (41.2 GHz, 45.2 GHz, 80.2 GHz). By use of a phase-locked loop, the frequencies were stabilized with an accuracy of ± 0.5 MHz. The forward power was measured as 20 ± 0.5 mW and the power reflected by the sample container was 2 ± 0.5 mW. Thus a power of 18 ± 2 mW entered the sample container. After passing through the medium layer of at least 200 μm thickness, the power was reduced to 4.5 ± 0.5 mW [calculated by using the absorption coefficient $\alpha \approx 70\ cm^{-1}$ of water (Swarnowski and Sheppard 1977) at 70 GHz for the medium]. This resulted in a power density of less than 6 mW/cm^2 at the sample location (horn area 1.6 cm^2).

Fig. 13. Statocyte from the median plane of a root tip after irradiation with millimetre waves 72 GHz, 5 mW cm^{-2}, duration: 50 min. Fixation: Glutaraldehyde. Arrays (*arrows*) with tubular and vesicular structures near the anticlinical and the distal cell walls are seen. This is atypical for cells in this stage of development after GA-O_5O_4-fixation. *N* nucleus; *A* amyloplast; bar = 1 μm

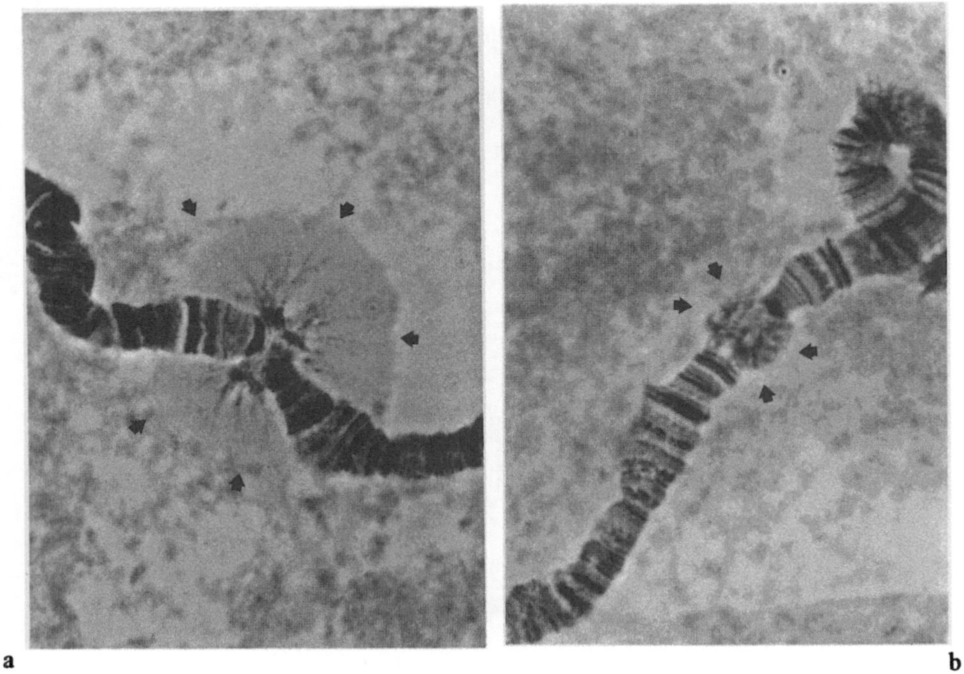

Fig. 15. a Balbiani ring BR 2 (control), fully decondensed (*arrows*). b The same chromosomal section with BR 2 locus after (2 h) irradiation with millimetre waves. The BR 2 has regressed, the chromatin fibrils are totally condensed and the surrounding puff material; (ribonucleoprotein) has disappeared. Magnification: 1500, power density ⩽ 6 mW cm^{-2}

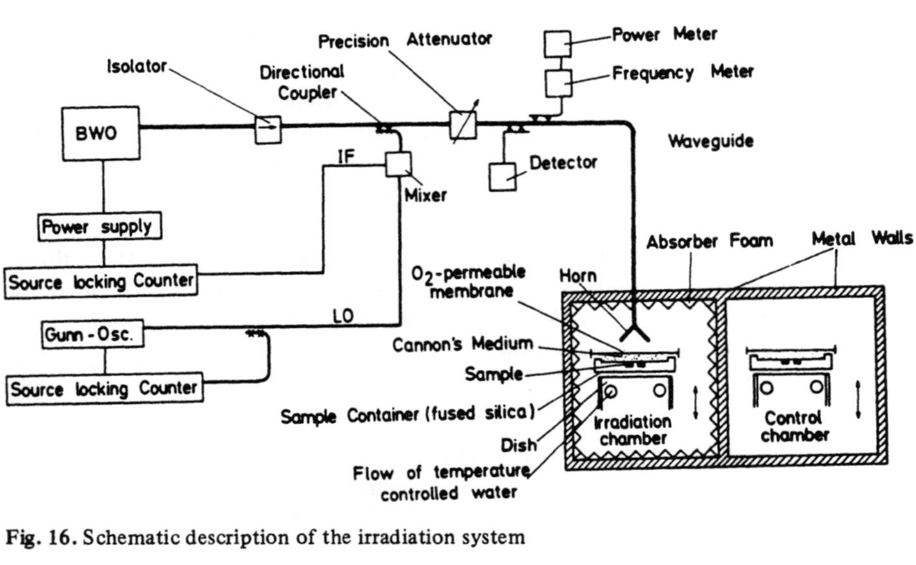

Fig. 16. Schematic description of the irradiation system

The sample container consisted of a fused silica plate with an indented circle of radius 40 mm and depth 0.2 ± 0.02 mm. In its centre, two further circular incisions of diameter 2 mm and depth of 0.3 ± 0.02 mm were prepared. One salivary gland was put in each of these last indentations. Cannon's medium (Ringborg et al. 1970) was added in order to supply the glands with nutrient and then the sample container was covered by an oxygen-permeable membrane to guarantee gas exchange. The glands are sac-like, having diameter of about 0.25–0.30 mm. The sample container was positioned on a fused silica, temperature-controlled dish (9.0 ± 0.1 °C). An identical dish was mounted in the control chamber. The microwave-induced temperature increase was found to be less than 0.3 °C for a forward power of 20 mW.

For the experiment the paired salivary glands from larvae of the fourth larval instar of *Acricotopus lucidus* were dissected. One gland was placed in the irradiation chamber and the other in the control chamber. Immediately after irradiating for 2 hours, each gland was fixed with ethanol-acetic acid (3:1). After staining the samples, squash preparations were made and the number q of strongly reduced Balbiani rings BR2 in one gland (with t examined chromosomes) was determined. Using q, the reduction probability $r = q/t$ was calculated for the gland in the (sham)-irradiation chamber ($r^{irr.}$) and in the control chamber ($r^{contr.}$).

All experiments were carried out blind, i.e. the examining biologist did not know which sample was irradiated and which served as control. Furthermore, without the biologist being informing, sham-exposure experiments and additional experiments were carried out to examine whether the irradiation effects are of thermal or non-thermal origin. To study the temperature sensitivity of the giant chromosomes, sham-exposed samples were warmed up by 2.5 °C (over the temperature of the controls), which is more than the eightfold microwave-induced temperature increase of 0.3 °C. To examine if the microwave-induced temperature gradients across the sample exert an influence on the puffing pattern of giant chromosomes, irradiation experiments with (incoherent) FIR radiation (20 μm $< \lambda <$ 70 μm) of a thermal source (globar) were performed (experimental series "IR"). By use of micro-miniature thermocouples it could be shown that – within the limits of accuracy – the microwave-induced temperature gradients were simulated by the FIR light.

The result is shown in Fig. 17 and Table 2. For the sham-irradiated samples no significant difference could be detected between samples placed in the (sham)-irradia-

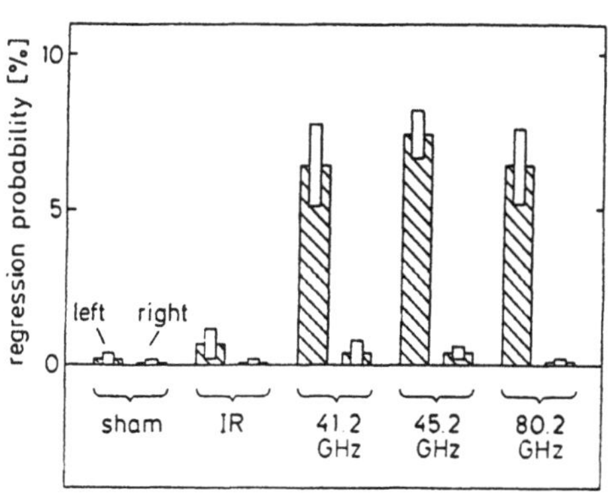

Fig. 17. Percentage regression probability $r^{irr.}$ for the *left* (irradiation)-chamber and $r^{contr.}$ for the *right* (control) chamber. The *shaded area* describes the mean, the *open area* the standard deviation of the mean. Because of graphical reasons not all data presented in Table 2 are plotted in this figure

Table 2. Regression probability after sham-, "IR"- and millimetre wave irradiation: n describes the number of gland pairs used in one type of experiment. The number in brackets indicates the number of chromosomes used for that experimental series. r is the mean of the percentage regression probability for the glands placed in the left or right chamber. The number in brackets indicates the standard error of the mean. P is the probability in percent that the samples in the left chamber belong to the same distribution as the samples in the right chamber according to the U-Test of Mann-Whitney (Siegel 1956)

Type of experiment	n	r [%] Left chamber	r [%] Right chamber	P [%]
Sham exposure	12 (908)	0.2 (±0.2)	0.0 (±0.0)	36.3
Sham exposure with additional heating of 3 °C	12 (891)	0.7 (±0.6)	0.0 (±0.0)	24.5
Infrared irradiated sample ($20 < \lambda < 70$ μm)	10 (724)	0.9 (±0.5)	0.5 (±0.3)	32.6
41.2 ± 0.0005 GHz	14 (912)	6.4 (±1.3)	0.4 (±0.4)	0.1
45.2 ± 0.0005 GHz	10 (759)	7.2 (±0.8)	0.4 (±0.3)	0.007
67.2 ± 0.0005 GHz	13 (707)	6.9 (±1.7)	0.8 (±0.8)	0.2
68.2 ± 0.0005 GHz	17 (931)	6.0 (±1.2)	0.7 (±0.3)	0.1
80.2 ± 0.0005 GHz	14 (947)	6.4 (±1.2)	0.0 (±0.0)	0.007

tion- and the control chamber. In contrast, the millimetre wave-irradiated samples showed an approximately tenfold increase of the regression probability of the BR 2, compared to controls. The effect is highly significant, as shown by the U-Test of Mann-Whitney (Siegel 1956) which is non-parametric (Table 1). Simulating the microwave-induced temperature gradient in the sample with far-infrared light or warming up the sample by 2.5 °C over the control did not result in a significant effect. Hence the non-thermal nature of the microwave irradiation is experimentally proven.

5 Discussion

Only for the giant chromosome experiment, was it possible to confirm experimentally the non-thermal nature of the irradiation effect. In view of the fact that the puffing phenomenon is not yet fully understood on a molecular basis, the explanation of the millimetre wave effect must be speculative. The observed effects could be understood following Fröhlich's (Fröhlich 1968, 1980) conjecture of coherent electric vibrations in biological systems. Their frequency – if based on membrane oscillations – is estimated to be of the order of 10^{10} Hz to 10^{11} Hz. Thus the observed effect could possibly be comprehended by assuming that the externally applied radiation field influences the excitations of the biological system.

From the viewpoint of an experimentalist, it is at the moment of primary interest to examine the nature (thermal or non-thermal) of known microwave irradiation effects.

Therefor it might be helpful to observe the following guidelines:

1. Carry out sham-irradiation experiments.
2. Examine the temperature sensitivity of the biological system.
3. Simulate the microwave-induced temperature increase.
4. Simulate the microwave-induced gradients in the sample.
5. Compare amplitude modulated and continuous wave irradiation.
6. Perform the experiments as blind experiments if the irradiation effect is small ($\leqslant \pm 20\%$) or cannot easily be quantified.

References

Baudisch W (1977) In: Beermann W (eds) Biochemical differentiation in insect glands. Springer, Berlin Heidelberg New York, p 197

Baudisch W, Panitz R (1968) Exp Cell Res 49:470

Behrens HM, Weisenseel MH, Sievers A (1982) Plant Physiol 70:1079

Buff E, Baake M, Sievers A (1987) Plant Physiol 83:685

Fröhlich H (1968) Int J Quantum Chem 2:64

Fröhlich H (1980) In: Marton L (ed) Advances in electronics and electron physics. Academic Press, New York, 53:85

Hensel W (1984a) Planta 162:404

Hensel W (1984b) Protoplasma 119:121

Koschnitzke Chr, Kremer F, Santo L, Quick P, Poglitsch A (1983) Z Naturforsch 38c:883

Koschnitzke Chr, Kremer F, Santo L, Poglitsch A, Genzel L (1986) Z Naturforsch 41c:321

Kremer F, Koschnitzke Chr, Santo L, Quick P, Poglitsch A (1983) In: Fröhlich H, Kremer F (eds) Coherent excitations in biological systems. Springer, Berlin Heidelberg New York, p 10

Kremer F, Poglitsch A, Santo L, Sperber D, Genzel L (1985) Z Naturforsch 40c:336

Mohr H (1964) Lehrbuch der Pflanzenphysiologie. Springer, Berlin Heidelberg New York

Nato Advanced Research Workshop 18.–21. Sept. 1984 on Interaction between electromagnetic fields and cells. Erice, Italy

Panitz R (1967) Biol Zentralbl 86 Suppl:147

Panitz R (1972) In: Beermann W (ed) Results and problems in cell differentiation, vol IV. Springer, Berlin Heidelberg New York, p 209

Reynolds ES (1963) J Cell Biol 17:208

Ringborg U, Daneholt B, Edstroem JE, Egyhazhi E, Rydlander LJ (1970) Mol Biol 51:679

Siegel S (1956) Non-parametric statistics for the behavioural sciences, Mc Graw-Hill, New York

Spurr DR (1969) J Ultrastruct Res 26:31

Szwarnowski S, Sheppard RJ (1977) J Phys E10:1163

Volkmann D, Sievers A (1979) Graviperception in multicellular organs. In: Haupt H, Feinleib M (eds) Encyclopedia of plant physiology, N.S., vol 7: Physiology of movements. Springer, Berlin Heidelberg New York, pp 573–600

Metastable States of Biopolymers

J. B. HASTED [1]

1 Introduction

Energy can be supplied to biopolymers in a number of different ways. The most important of these, the most common of natural occurrence, are collisional thermal energy or heat, electromagnetic energy or light, and metabolic or chemical energy.

Thermal energy, supplied by thermal conduction, is transmitted by movement of chemically bound atoms, usually in large numbers of vibrational modes, causing larger amplitudes of vibration and thereby faster movement of the atoms. This is the classical picture, to which the quantum picture corresponds, because of the contributions from numbers of modes in which the restoring forces are so weak that the quantum energy levels are very closely spaced; large quantum numbers must be present except at the lowest temperatures, and where the quantum numbers are sufficiently large, the correspondence principle ensures that the picture is classical.

However, the supply of thermal energy is usually balanced by the removal of thermal energy by conduction, which is similarly interpreted by classical theory. A spatially uniform temperature indicates that there is such a balance, but a spatial temperature gradient implies the local inequality of supply and removal of energy. It should be borne in mind that the extreme form of thermal gradient is the shock front. The important process which governs the passage of a shock through non-metallic matter is vibrational energy exchange. Hence the classical theory calculations that form the basis of coherent excitation theory could be central to energy transport in biomolecular systems.

Metabolic energy is being continuously exchanged in living biological or in biomolecular systems. It is normally chemical in nature. Molecular rearrangements occur both endothermically and exothermically. An exothermic process releases energy, which, apart from the electromagnetic form, can be either in the translational or vibrational modes. The former, if reabsorbed, adds to the Brownian motion and thereby to the thermal energy of the system. The latter, as has been seen, adds kinetic energy to the moving atoms. Metabolic energy, therefore, can generally be regarded as having the same consequences as thermal energy. There may be other consequences, such as the emission of light, or of electrical energy, or further chemical change.

The transmission of electromagnetic energy takes place by a different mechanism, and is dependent on the frequency at which the electromagnetic field alternates. In the radiofrequency region the field can be regarded as classically re-orienting the elec-

1 Department of Physics, Birkbeck College, Malet Street, London WC1E 7HX, U.K.

tric dipoles or polarizing the molecule by other means. Relaxation of the displaced electric charge takes place after a delay sufficient for energy to be absorbed and distributed around the molecule. It is usual for only a small fraction of the electromagnetic energy to be absorbed, since the wavelengths are much larger than the molecule; but as the frequency is increased into the infrared this ceases to be the case, and eventually a quantum picture becomes more appropriate. The very broad relaxation absorptions, more than a decade in breadth, are replaced by somewhat narrower molecular absorption bands; but because of the complicated nature of biopolymers, the absorption bands are still on the whole not sharp, because many overlap, and there is collisional broadening.

Biomolecular systems and biopolymers can nearly always be regarded as wet; that is, there is a large amount of water both bound to the biopolymers and also freely surrounding them. Both bound and free water absorb electrical energy, although not identically; and the presence of transportable ionic, protonic and possibly electronic charge also contributes to the absorption of electrical energy. By contrast, magnetic absorptions are on the whole very weak.

The dielectric processes which contribute to these absorptions of electromagnetic energy have been widely studied and are described in a number of publications (Fröhlich 1958; Hasted 1973; Pethig 1979). Here we shall not include a detailed discussion, because very few of the processes can be considered to be generating the proposed "metastable state" which is the subject of this chapter. Such a state must be considered to have the property of existing for a period of time which is longer than either (a) the characteristic period of oscillation of the electromagnetic field, or (b) the natural lifetime of a quantum energy level, or excited state. Such lifetimes are of the order of tens of nanoseconds for electric dipole transitions of atomic energy levels, but they can be much larger for large principle quantum number n; the lifetime can be taken to be proportional to n^3.

A metastable state does not, by definition, last forever. In quantum theory it might last for a very long period if completely isolated in a universe of its own. Consider the classical model of a sphere resting upon a convex spherical pimple in the centre of a concave spherical bowl in a vertical gravitational field. The lowest stable configuration is with the sphere entrenched on the lowest circle of the annular valley; but there is another temporary stable configuration with the sphere resting on the top of the pimple. This is a metastable configuration, which can be destroyed by perturbation, knocking the sphere off. But the question is, how large is the perturbation necessary? Atomic metastable states will, if they are not quenched collisionally or stimulated by radiation, last for a characteristic lifetime (sometimes seconds), after which they spontaneously decay by such processes as magnetic dipole, electric multipole or multiphoton transitions. But with the type of molecular metastable state envisaged in coherent excitation theory, quite a lot of energy may be necessary to destroy the state.

The absorption of thermal or electromagnetic energy can produce a metastable state only when the supply exceeds a certain threshold value. As the supply of energy to a strongly polar elastic biopolymer is gradually increased, the polarization energy increases monotonically, but eventually maximizes and subsequently minimizes; this minimum represents a metastable configuration, which will last until a much larger supply of energy destroys it. It is possible that several minima occur, that is, that several metstable states might exist.

It should be possible to detect or monitor the presence of the metastable state, which could have different electromagnetic, chemical or biological properties from those of the normal state. But the monitoring process must not perturb the metastable state sufficiently to destroy it. Monitoring by absorptive properties can be superior to monitoring by emission, but only if the radiation absorbed is sufficiently weak.

2 Dielectric Relaxation Phenomena

The electrical properties of proteins in aqueous solution and also of solid moist proteins precipitated as "cast film" have been intensively studied, and the dielectric relaxation processes they demonstrate follow the usual frequency dependence of complex permittivity $\hat{\epsilon}$:

$$\hat{\epsilon} = \epsilon_\infty + \frac{\epsilon_s - \epsilon_\infty}{1 + j\omega\tau} = \epsilon' + j\epsilon'' . \tag{1}$$

The real permittivity ϵ' falls from its static, or low frequency, value ϵ_s, to its high (or infinite) frequency value ϵ_∞, as the frequency of the applied electromagnetic stimulation is increased. This fall is centred about the radial frequency

$$\omega = 1/\tau ,$$

where τ is the relaxation time over which the orientational perturbation of the permanent molecular dipole relaxes exponentially to its normal orientation. The resulting polarization is of the type

$$P(t) = P\,[1 - \exp(-t/\tau)] . \tag{2}$$

As the real permittivity falls from ϵ_s to ϵ_∞, the dielectric loss displays an absorption band, whose shape is governed by the relation

$$\epsilon'' = \frac{\omega\tau\,(\epsilon_s - \epsilon_\infty)}{1 + \omega^2\tau^2} . \tag{3}$$

The width of this absorption band is more than a decade of frequency, enormously broad when compared to a resonance process. Moreover, many bands are broader still, because contributions are made from a number of relaxation times distributed over a range; the shapes of such broad bands can be described by equations such as the Cole-Cole Eq. (4):

$$\hat{\epsilon} = \frac{\epsilon_s - \epsilon_\infty}{1 + (j\omega\tau)^{1-\alpha}} \quad \text{(Cole and Cole 1941)} , \tag{4}$$

where α is a relaxation time spread parameter. The particular form of the relaxation function, somewhat analogous to line shape in spectroscopy, is capable of interpretation in terms of the molecular environment in which the dipole relaxation occurs; it should therefore be sensitive to molecular structure, and its measurement could constitute a method of monitoring the presence of a metastable state. The spread parameter, the relaxation time and the static permittivity should all be characteristic of the state.

The significant factor in the time-dependence of electrical polarization is the dipole correlation function $\mu(t) \cdot \mu(0)$ which is found at a time t after the application of a step perturbation at t = 0. When this varies exponentially with time, dielectric behaviour of the type of Eq. (1) is found. A transformational relationship (Klug et al. 1969) connects the time dependence of orientation of the dipole and the frequency dependence of permittivity:

$$\frac{\epsilon_s\,(\hat{\epsilon}(\omega)-\epsilon_\infty)\,(2\,\hat{\epsilon}(\omega)+\epsilon_\infty)}{\hat{\epsilon}(\omega)\,(\epsilon_s-\epsilon_\infty)\,(2\,\epsilon_s+\epsilon_\infty)}=\frac{4\,\pi\rho\mu^2}{3\,\mathrm{kT}}\int -\frac{\mathrm{d}\gamma}{\mathrm{dt}}\exp(\mathrm{j}\omega \mathrm{t})\cdot \mathrm{dt} \tag{5}$$

where

$$\gamma(\mathrm{t})=\mu(0)\cdot\mu(\mathrm{t})\,/\,\mu(0)\cdot\mu(0)\,. \tag{6}$$

When the dipole correlation function y(t) is known, or at least conjectured, the form of the dielectric relaxation function can be calculated. ρ is the density and k the Boltzmann constant.

The relaxation function is often represented as a locus of points, at different frequencies, in the complex $\epsilon''(\epsilon')$ plane; for a pure relaxation this locus takes the form of a semi-circular arc on the ϵ' axis; for a spread of relaxation times the arc is depressed onto an axis, making an angle arctan α below the ϵ' axis; but for a pure resonance process the locus is a full circle resting on the ϵ' axis. Combinations of these types of behaviour are usually found.

Many factors affect not only the principal relaxation time, τ, but also the parameter α. When energy is supplied to a biopolymer sample, whether the energy is thermal, biological (chemical) or electrical, there is often a change of temperature. The relaxation time is strongly dependent on temperature, and this dependence might demonstrate the appearance of a different state, such as a metastable state. The simplest temperature dependence or relaxation time is exponential in character:

$$\frac{1}{\tau}\propto\exp\left(-\,\Delta E/kT\right), \tag{7}$$

with Boltzmann constant k and the activation energy ΔE. The interpretation of this Arrhenius equation is in terms of an activated state through which the dipole and its environment must pass in order to achieve a reorientated state. In Boltzmann statistics the mean time that elapses before the activated state is achieved is exponentially proportional to the inverse of the temperature. The model is the familiar statistical one in which the system waits until thermal agitation produces activated states.

The exponential behaviour of relaxation time with temperature dominates dielectric relaxation behaviour, but there are many complications which indicate the limitations of the purely statistical model. It is often found that there is a range of temperature over which an exponential relationship holds well, and another range over which there is another exponential relationship, with different activation energy. This is usually taken as indicating an actual structural change, as for example the phase change between ice I and liquid water. The transition, however, need not be sudden, it may cover an intermediate range of temperature, with both processes participating.

The temperature dependence of the microwave dielectric loss in proteins and polypeptides affords a good example. Below about 50 K, the dielectric loss rises rather rapidly with increasing temperature, but at higher temperatures the rise is much slower. This has been interpreted (Kremer et al. 1984) in terms of two picosecond relaxation processes occurring in asymmetric double-well potentials. The activation energies required for this interpretation are, with one exception, all lower than 0.05 eV; the most probable interpretation is in terms of the NH...OC hydrogen bridges of the peptide backbone. However, this anomalous thermal dependence might possibly indicate the onset of a metastable state above 50 K, although such a speculation would have to be justified by the demonstration of anomalous absorption or emission resonance properties of the molecules with an onset at this temperature.

3 Dielectric Behaviour of Thin Biopolymer Films

The behaviour of relaxation spectra is rather different when considerable electric energy is supplied, by the application of a high electric field, so that materials with appreciable conductance absorb energy. The conductance can arise from both carrier transport and dielectric loss. Although most biological materials can be regarded as lossy dielectrics, with appreciable ionic and possibly electronic transport, dielectric studies on solid or liquid samples are usually made under sufficiently low-field conditions for no non-linear effects to be observed. But non-linear high field phenomena make their appearance when thin films of biological materials are used.

Thin films are one of the important phases in which certain biomolecules occur naturally; in particular, the cell membrane normally consists of lipid molecules stacked in two monolayers with their hydrophobic ends meeting in the membrane interior, and hydrophilic groups exposed both to the inside and outside of the cell. Within the natural membrane some protein molecules are embedded, but a planar lipid bilayer can be produced in the following way. Two aqueous solutions are separated by a small orifice drilled in a thick Teflon barrier; a fine capillary tube is passed within the thick barrier and its end exposed at a point on the circumference of the orifice. A small amount of liquid solution is injected so that it forms a drop within the orifice. The solvent evaporates through the aqueous electrolyte until a lipid film is formed across the orifice (T.J. Lewis, private communication, 1983). Alternatively, a lipid solution may be painted over the orifice before the electrolyte solutions are introduced (Rosen and Sutton 1968).

Optical interference shows that such a "black" membrane is only two molecular chains in thickness; the thickness of the bilayer lipid membrane (BLM), typically 5-10 nm, is such that very large electric fields can easily be developed across it. This is the characteristic of the situation for a biological cell in vivo.

When the potential across a BLM is raised very slowly (0.01 V/s), the voltage-current characteristic is at first linear; but eventually a more rapid rise of current is noticed, together with some instabilities, and after this "pre-breakdown region" there is actual electrical breakdown. If the breakdown is now inhibited, by reduction of the potential, the current is reduced, although not in a completely repeatable fashion. Continued electrical cycling shows that the electrical breakdown of the BLM does not always damage it permanently; as far as is known, the BLM can remain in an undamaged state physically. But in the breakdown region it is arguable that a metastable state can be formed, with anomalous electrical impedance.

A more complex behaviour is characteristic of molecular films of proteins produced by the Langmuir-Blodgett technique (Vincett and Roberts 1980). These "LB films" can be as thin as 4 nm for a monolayer, so that high fields can readily be produced. In the Langmuir-Blodgett technique, a monolayer is deposited on a solid surface by withdrawing it from water, on the surface of which an insoluble monolayer film has been floated. The floating film is prepared by dissolving the protein in a non-aqueous solvent and depositing a few microlitres of solution on the aqueous surface; the solvent evaporates whilst the molecules rapidly spread out on the aqueous surface.

The packing of molecules on the aqueous surface must be such that the LB film has repeatable properties, which can be shown to be those of a monolayer. The aqueous surface must be of particular cleanliness and freedom from vibration, and the dependence of the monolayer surface pressure p upon its area A must be of particular form.

The surface pressure p is very weak when the area A is large, and at first fails to increase as the area is reduced by movement of a barrier; the molecules are very far from being close-packed, and the many holes in the film are eliminated. Continuous reduction of the film area ultimately causes the surface pressure to rise with decreasing area, approximately as

$$(p - p_0)(A - A_0) = RT \,, \tag{8}$$

where R is the gas constant, T the temperature, and p_0 and A_0 constants. This is a regime in which the molecules are close-packed, and are steadily compressed as the area is decreased by movement of the barrier. Further decrease of area causes a collapse of the film, with little change of pressure, a regime corresponding to the moving of molecules over the monolayer until a second layer is built up. Following this, continued area reduction can cause a second and even a third layer to be built up. Nevertheless, it is the monolayer regime which is usually studied. It is found that the molecular film thickness increases somewhat as the area is reduced, corresponding to a variation in the packing and arrangement of the individual molecules. It is therefore necessary to deposit LB films under standard thermodynamic conditions of the floating film, with surface pressure and temperature controlled.

When a glass surface is slowly dipped under or withdrawn from the water, with the film floating on its surface, a monomolecular layer is formed on the glass, and is then allowed to lose excess free water by drainage. This process can be repeated many times. Not all surfaces will receive the films; for example, it is not always possible to mount films on gold. Metals such as aluminium and zinc, however, will readily receive films, and such metals can be evaporated onto the glass, to serve as electrodes, before dipping. Furthermore, metals can be evaporated onto the deposited films so that electric fields can be applied across the molecular layers.

The capacitance C of N layers of real permittivity ϵ', area A and thickness d is given by

$$C = A\epsilon'\epsilon_0/dN \,, \tag{9}$$

where ϵ_0 is the vacuum permittivity; therefore the thicknesses of LB films having N deposited monomolecular layers can be studied by means of the dependence of their capacitances on N^{-1}, if ϵ' is known. Since most electrodes are covered with a thin layer of oxide, which is of known permittivity, there is a well-understood correction to the $C(N^{-1})$ slope analysis. This type of experiment is a mandatory test of the performance of Langmuir-Blodgett technology.

Dielectric relaxation spectroscopy applied to correctly prepared LB films shows that structural changes are brought about by high electric fields (Hasted et al. 1981, 1985). Structural changes are, of course, well known when energy is supplied to proteins, denaturing by heat being the most obvious example; but after the changes brought about by electric fields the molecules slowly revert to normal.

The question must be posed: is this an example of the predicted metastable state?

Although it is not yet possible to give an unequivocal answer, it is certainly possible to show what experiments must be carried out in order to be certain that this is the phenomenon predicted.

Low frequency relaxation spectra have been obtained not only by conventional bridge techniques, but by frequency response analysis, a technique in which electrical energy is supplied in the form of a potential step function, and the response to different frequencies is measured within a range of narrow bands. It is possible to vary the intensity of the energy supplied; electric fields within the range $1-200 \times 10^5$ V/m have been maintained, and a wider range could if desired be achieved.

Relaxation spectra of humid haemoglobin LB films exhibit a low frequency relaxation process (a peak in the dielectric loss function with a corresponding fall in the permittivity), together with a higher frequency process which is incomplete and has not shown complete consistency. As the protein is dried, the low frequency process becomes weaker and takes longer. The low frequency process is not displayed at accessible frequencies by the completely dry protein, although there is still electrical conduction. At low fields, the activation energy for the processes in the wet haemoglobin is very large (~ 1.8 eV), which shows that the dipolar reorientation, taking place with the assistance of the bound water, involves the "breakage" of quite a number of hydrogen bonds. This is in line with measurements on protein cast film (Eden et al. 1980).

It is the field dependence of the relaxation process which is of particular significance. As the alternating field is increased, the relaxation peak moves to higher frequencies, and at the same time it is broadened, as though a number of different characteristic relaxation times were possible for the molecule in its environment. The following table shows the dependence.

The films exposed to the highest fields (1.7×10^7 V/m) retain their 500 Hz relaxation frequency even when this is re-measured with low fields (3.4×10^5 V/m). The lengths of time over which this anomalous relaxation frequency is retained are as long as 40 days at laboratory temperature; but by gentle heating without denaturation, these times can be reduced considerably, and the original state is restored.

Table 1. Variation of relaxation frequency with applied field

Field (10^5 V/m)	Frequency (Hz)
3.4	0.05
17	0.05
34	0.5
68	2
170	500

Films already affected by exposure to X-rays have been found to exhibit relaxation changes after exposure to high electric fields. After many hours the normal relaxations were again exhibited, as though the affected films had been restored by the electric fields.

This behaviour may be interpreted as the production of a metastable state; the applied electric fields are, of course, oscillatory, albeit at low frequencies. A different method of monitoring the vibration modes of the molecule might well show sensitivity of the polymer to the application of low frequency electric fields of $\sim 10^7$ V/m intensity. Such a method might be found in far infra-red absorption, where there are a variety of characteristic vibration modes, as will be discussed below.

In addition to the dielectric relaxations there are electrical conduction processes in the protein LB films. These are investigated by the technique of applying a slowly increasing potential across the film (a "ramp voltage") and monitoring the current. For sufficiently fast ramp speeds ($>$ 170 mV/s), displacement current dominates, and the current-voltage characteristic is linear. For slower speeds, however, there is time for a space-charge of carriers to build up within the solid; its movement produces a hysteresis effect, the space charge field affecting the release and capture of carriers at the metal electrode surfaces. This hysteresis might also indicate the formation of a metastable state.

For sufficiently large fields the relationship between current I and field E is approximately exponential; this is characteristic of ionic conduction (Mott and Gurney 1940), with proportionality

$$I \propto \sinh eEa/2\,kT \tag{10}$$

for carrier charge e, and the Boltzmann constant k. The ions are supposed to hop between fixed centres separated by distance a. For $eEa \gg 2\,kT$, the approximate exponential behaviour allows a calculation of hopping distance to be made, provided that the temperature dependence is correct, which it is within 2%, at least down to 233 K. The result, a = 2.5 nm, for haemoglobin, is significant because it is approximately the distance between the haem groups in the molecule. This represents a confirmation that the pre-breakdown conduction, as opposed to the dielectric relaxation, takes place by hopping between centres within the haemoglobin molecule.

The trap depths can be determined by thermally stimulated conduction measurements (Hasted et al. 1985). An LB film is polarized by the application of a high field at laboratory temperature, and cooled in the field to liquid nitrogen temperature before disconnection of the potential; the temperature is then allowed to rise at a known rate and the charge released and measured. According to an equation similar to (7), the exponential rise rate is a measure of the trap depths, which for haemoglobin (Hasted et al. 1985) are found to be 0.38 eV and 0.5 eV.

The trapping of charge is attributed to its retention on a "bound water" molecule. The relaxation processes are sensitive not only to structural change induced by electric fields, but to the hydration of the protein. The strongly bound water, occurring at low hydrations, cannot readily be observed with convential dielectric spectroscopy, since the relaxation frequencies are sub-Hertzian. Thermally stimulated depolarization studies show that the activation energy at low hydration is as high as 0.55 ± 0.04 eV for crystallized lysozyme (Celaschi and Mascarenhas 1977); but subsequent water layers are bound with an energy of only 0.34 ± 0.02 eV.

4 Biological Electrets and Piezoelectrics

The lowest energy state for an assembly of electric charges is that in which there is the best spatial electric balance, with the positive and negative charges mutually closest, so that the electrical macroscopic homogeneity extends down to the smallest dimensions. This is true of ionic crystals as well as of electrolytes and of fully ionized plasma. When there is complete freedom of movement of an assembly of positive and negative charges, there are at any moment more of the opposite sign than of similar sign in the neighbourhood of any given charge; this is the "ionic atmosphere" which retards the transport of ions in electric fields, according to the Debye Hückel theory of strong electrolytes (Falkenhagen 1938) and also the theory of fully ionized plasma. If Boltzmann statistics are applied to this charge distribution, it is found that the electric field around any given charge, e, falls off exponentially, according to the relations

$$E = E_0 \exp(-r/r_D) \tag{11}$$

$$r_D = (kT/8\pi ne^2)^{1/2} , \tag{12}$$

where n is the charge number density. Only within the "Debye radius", r_D, is there appreciable electric field in an electrolyte or a gaseous plasma; macroscopically, no field can penetrate, since the plasma charges would rearrange in order to accommodate it. The lowest energy state is macroscopically electrically neutral, and has negligible polarization.

It follows that if there are materials with permanent or semi-permanent polarization, they are in a higher energy state, which may be either metastable, or, to all intents and purposes, stable. Such stable materials are termed electrets (by analogy with magnets), and have been known for many years. For example, a specimen of carnauba wax showing electret properties has been on display in a museum in Tokyo since 1912; it has been regularly monitored and had, by 1986, lost more than half of its "power". The fact that electrets can lose their polarization, with time, is an indication that they are metastable states. Electret polymers include many biopolymers and biological specimens, such as the dorsal structures of cockroaches.

Electrets are arrangements of polar molecules orientated in such a way that the dipoles reinforce each other, resulting in permanent polarization. The thermally stimulated depolarization studies of protein LB film discussed above show that at liquid nitrogen temperatures these materials could be regarded as electrets. Some protein films can retain residual polarization even at laboratory temperature, and even without previous "poling" in strong electric fields.

When a protein LB film is laid down on glass, with metal electrodes at each surface, the standing potential across it can be measured without taking appreciable current, using an electrometer. In order to obtain reproducible conditions, the film must previously have been short-circuited so that the potential across it is very close to zero. On removal of the short circuit, the potential rises whilst electrometer measurements are taken. It may take many hours for the equilibrium static potential to be reached; this equilibration time depends upon the ambient conditions, since if neutralizing charge is required by the distorted (short-circuited) specimen, then its rate of uptake could be affected by the concentration of atmospheric ions, which may initially be present to a small extent even within screened containers or vacuum systems. Under equilibrium conditions, in an electrically neutral atmosphere, the rates of arrival and loss of positive and negative ions must balance; it is possible under unscreened laboratory conditions that there may be appreciable fluctuations in the "static" electric potential of polar biopolymers.

The significance of the static potential of the globular protein molecule lies in its relation to the molecular dipole moment, which is known from measurements in solution. The static potentials would be expected to be approximately proportional to the dipole moments, implying that the protein molecules are laid down with their dipole moments making on the average the same angle with the surface. Molecules are presumably distributed about different angles, although just how wide this distribution is, and how it is affected by LB film thermodynamics and conditions of laydown is not yet known.

If a protein LB film could be regarded as an electret, attention must be paid to the conditions under which the metastable state may be destroyed and regenerated. There is scope for using "static" LB film potential measurements as a method of monitoring both structural and energy states of protein molecules. The technique is passive, in that nothing is applied to the molecule that might change its energy; it produces its own signal.

However, there are large interfacial potentials which complicate this simple picture; they arise from the reaction of water vapour with the metal or oxide electrode interface (Jones et al. 1980). Furthermore, biomolecules such as porphyrins which show electronic conduction exhibit potentials arising from Schottky band-bending at these interfaces (Jones et al. 1985).

In addition to their properties as electrets, some solid biomolecular specimens can exhibit piezoelectricity (Cochran et al. 1968). Potentials generated by mechanical stress have been studied not only in bone but in proteins such as keratin (Fukada et al. 1975). Thermally induced structural transitions such as denaturation can be studied simultaneously by TSD and by thermal variation of the piezoelectric constants. For example, thermally stimulated currents in the axial direction of porcupine quills show maxima at 210 °C and 235 °C, where the two helicoil transformations in the molecular conformation of keratin take place.

5 Vibration Frequencies of Biopolymers

We have seen that biomolecules absorb and interact with electromagnetic energy at radiofrequencies, by non-resonance processes. It is only as Gigahertz frequencies, i.e. millimetre and sub-millimetre wavelengths, are reached, that resonance processes begin to dominate electromagnetic wave absorption. It is important to search for all such characteristic vibrations, in order that when energy is supplied to the molecules, the appropriate exit channels can be monitored, either in absorption or possibly even in emission.

In the absence of readily available tunable submillimetre wave sources, use must be made of Fourier Transform Spectroscopy, with a wide band source of continuous radiation from a mercury lamp source. Although reflection has been used for dispersive measurements, absorption is the appropriate mode for investigating resonance processes.

Small biomolecules, such as amino acids and oligosaccharides (Husain et al. 1984a,b) exhibit a wealth of sharp resonances, some of which are lattice vibrations, and some of which are torsional and similar modes, for example of the CO_2^- group.

Sharp lattice vibrations would not normally be found in biopolymers, but other bands, such as protein backbone modes, are expected.

Soon after the development of the coherent oscillation theory, Carreri and his colleagues started a search for protein sub-millimetre absorption spectra, and found a broad band around 200 cm^{-1} in lysozyme and haemoglobin (Buontempo 1970). Subsequent experiments (Hasted et al. 1983; Genzel et al. 1983) have shown that there is structure on this broad band, which is found in all the best-known protein biopolymers.

The involvement of these absorption modes in coherent oscillation theory has not been demonstrated.

However, some resonant frequencies in the centimetre wave lengths have been reported for nucleic acids (van Zandt 1986). Confirmation is necessary before detailed interpretation can be made.

6 Conclusions

The most encouraging evidence for metastable states of biopolymers comes from the low frequency dielectric studies of molecular films of protein molecules and of biomolecular electrets.

It is allowable that some states, particularly as seen in the context of thermally stimulated depolarization or conduction experiments, may indicate specific structural changes rather than coherent oscillation theory phenomena. In thermal stimulation experiments it is a steady rather than an oscillatory electric field which "poles" the specimen and produces the temporary state which is maintained at low temperatures but is destroyed by raising the temperature, that is, by the application of energy. This is not necessarily the metastable state of coherent oscillation theory.

However, the non-linear behaviour of LB protein films under the action of low frequency electric fields is probably a different phenomenon. It occurs at laboratory temperature and with the protein in its natural moist state. It requires the absorption of electromagnetic energy above a certain threshold. The dielectric properties of the molecules are then altered significantly. They remain in the altered state whilst subthreshold energy is supplied, for the purpose of measurement. Eventually this altered state reverts to the original state, and various factors affect its lifetime. But very little is known about the altered state other than its dielectric absorption; visual inspection shows that its optical reflection coefficient is drastically changed; but detailed measurements have not been reported. And although hydrated proteins are very sensitive to thermal energy absorption, denaturing of a dry protein is not usually considered to be a readily reversible process; electromagnetic effects are demonstrably different from denaturation.

Acknowledgement. Dr. Dennis Rosen has helped with the preparation of this chapter, and is gratefully acknowledged.

References

Buontempo V, Carrerl G, Fasella P (1970) Phys Lett 31A:543

Celaschi S, Mascarenhas S (1977) Biophys J 20:273

Cochran GVB, Pawluk RJ, Bagset CAL (1968) Clin Ortho No 58, J.P. Lippincott

Cole KS, Cole RH (1941) J Chem Phys 9:341

Eden J, Gascoyne PRC, Pethig R (1980) J Chem Soc, Faraday Trans 2, 76:426

Edwards GS, Davis CC, Saffer JD, Swicord ML (1984) Phys Rev Lett 53:1284

Fröhlich H (1958) Theory of dielectrics. Clarendon, Oxford

Fukada E, Zimmerman RL, Mascarenhas S (1975) Biochem Biophys Res Commun 62:415

Genzel L, Kremer F, Poglitsch A, Bechtold G (1983) In: Fröhlich H, Kremer F (eds) Coherent excitations in biological systems. Springer, Berlin Heidelberg New York, p 58

Hasted JB (1973) Aqueous dielectrics. Chapman and Hall, London

Hasted JB, Millanny HM, Rosen D (1981) J Chem Soc Faraday Trans 2, 77:2289

Hasted JB, Husain SK, Ko AY, Rosen D, Nicol E, Birch JR (1983) In: Fröhlich H, Kremer F (eds) Coherent excitations in biological systems. Springer, Berlin Heidelberg New York, p 71

Hasted JB, Ko AY, Al-Baker Y, Kadifachi S, Rosen D (1985) J Chem Soc Faraday Trans 2, 81:463

Husain SK, Hasted JB, Rosen D, Nicol E, Birch JR (1984a) Infrared Phys 24:201

Husain SK, Hasted JB, Rosen D, Nicol E, Birch JR (1984b) Infrared Phys 24:209

Jones R, Tredgold RH, O'Mullane JE (1980) Photochem Photobiol 32:223

Jones R, Tredgold RH, Hoorfar A (1985) Thin solid films 123:307

Klug DD, Kranbuehl DE, Vaughan WE (1969) J Chem Phys 50:3904

Kremer F, Poglitsch A, Genzel L (1984) In: Ross Adey W, Lawrence AF (eds) Non-linear electrodynamics in biological systems. Plenum, New York

Mott NF, Gurney RW (1940) Electronic processes in ionic crystals. Oxford University Press

Pethig R (1979) Dielectric and electronic properties of biological materials. Wiley, New York

Rosen D, Sutton AM (1968) Biochim Biophys Acta 163:226

Van Zandt LL (1986) Phys Rev Lett 57:2085

Vincett PS, Roberts GG (1980) Thin solid films 68:135

Physical Aspects of Plant Photosynthesis

F. DRISSLER[1]

1 Introduction

1.1 Equation and Bases of Gross Reaction

Chemical Overall Reaction

The production of high-energy organic matter and oxygen from low-energy inorganic substances in the presence of light in plant cells, i.e. photosynthesis, satisfies the reaction equation:

$$n \cdot CO_2 + n \cdot H_2O + \text{light} \rightarrow (CH_2O)_n + n \cdot O_2 + 112\ \text{kcal/Mol}\,. \quad (1)$$

Investigations of the chemical gross conversion have revealed that the overall reaction [Eq. (1)] can be separated in fast photoprocesses and complicated sequences of comparatively slow, chemical dark reactions (Govindjee 1975; Drissler and Hägele 1974; Hägele and Drissler 1975; Richter 1969; Emerson and Arnold 1932; Hoffmann 1975; Bassham 1962; Renger 1977).

Primary Processes

The conversion described by Eq. (1) takes place if the energy required for all chemical dark processes is passed to the biological system via the light reactions. As primary processes of photosynthesis, the absorption of light, the transport of electronic excitation energy and charge separations caused by these purely physical processes are preconditions for the chemical fixing of light energy.

Biological Bases

Photosynthetic activity of plant cells (diameter approximately 50 μ) is located in the so-called chloroplasts (diameter approximately 5–10 μ). These contain lamella structures which result from joined spherical cavities, the thylakoids (Bishop 1974; Lumry and Spikes 1960). The walls of these cavities are biological membranes (Richter 1969), in which photoactive pigment systems are stored together with numerous other functional elements (Table 1).

The so-called pigment systems consist of 300–500 molecules such as chlorophyll a (Chl a), chlorophyll b (Chl b) and carotenoids (Car), which after absorption of light can transfer their excitation energy to a special chlorophyll a arrangement with trap

1 Gaußstraße 40, D-7000 Stuttgart 1, FRG

Table 1. Photosynthetic pigments of higher plants and green algae (Hoffmann 1975)

Photoactive pigment	Part (%)	Main group
α-Carotene β-Carotene	9	Carotenes
Lutheine Violaxanthine Neoxanthine Kryptoxanthine Zeaxanthine	25	Xanthophylls
Chlorophyll b Chlorophyll a	16 50	Chlorophylls

character called reaction centre (Govindjee 1975; Witt and Moraw 1959; Witt and Müller 1959; Raoinowitch 1959; Diges 1956; Sauer and Brotz 1969). The excitation of a reaction centres is followed by primary charge separations, to which secondary chemical processes are added in the case of active photosynthesis (Drissler and Hägele 1974). All primary processes, as well as the related competitive processes, are summarized in Table 2.

Table 2. Primary processes of photosynthesis and related competitive processes (indices 0 and 1 represent singlet states S_0 and S_1)

Primary processes of photosynthesis	Description
Absorption of light by:	
Carotenoids	$Car_0 + h\nu \rightarrow Car_1$
Chlorophyll b	$(Chl\ b)_0 + h\nu \rightarrow (Chl\ b)_1$
Chlorophyll a	$(Chl\ a)_0 + h\nu \rightarrow (Chl\ a)_1$
Transport of absorbed light energy:	
From carotenoids to chlorophyll a	$Car_1 + (Chl\ a)_0 \rightarrow (Chl\ a)_1 + Car_0$
From chlorophyll b to chlorophyll a	$(Chl\ b)_1 + (Chl\ a)_0 \rightarrow (Chl\ b)_0 + (Chl\ a)_1$
From chlorophyll a to chlorophyll a	$(Chl\ a)_1 + (Chl\ a)_0 \rightarrow (Chl\ a)_0 + (Chl\ a)_1$
From chlorophyll a to reaction centre R	$(Chl\ a)_1 + R_0 \rightarrow (Chl\ a)_0 + R_1$
Charge separation at the reaction centres:	
Reduction of primary acceptor A	$(D\ R_1\ A) \rightarrow (D\ R^+\ A^-)$
Oxydation of primary donator D	$(D\ R^+\ A^-) \rightarrow (D^+\ R\ A^-)$
Emission of light by:	
Chlorophyll a	$(Chl\ a)_1 \rightarrow (Chl\ a)_0 + h\nu$
Reaction centre	$R_1 \rightarrow R_0 + h\nu$
Radiationless decay of:	
Chlorophyll a	$(Chl\ a)_1 \dashrightarrow (Chl\ a)_0 + vib.$
Chlorophyll b	$(Chl\ b)_1 \dashrightarrow (Chl\ b)_0 + vib.$
Carotenoids	$Car_1 \dashrightarrow Car_0 + vib.$
Charge recombinations:	
Of primary separated charges	$(D^+\ R\ A^-) \rightarrow (D\ R\ A) + h\nu$

1.2 Energy Scheme

General

The endergonic water splitting reaction [Eq. (2)],

$$H_2O \rightarrow H\cdot + \cdot OH\,, \tag{2}$$

which can be considered as part of the gross conversion [Eq. (1)], cannot occur with the energy of a single absorbed light quantum ($14{,}000/cm^{-1}$–$25{,}000/cm^{-1}$) only, since its activation energy amounts to approximately $26{,}000/cm^{-1}$ (Fong 1975). The combination of two light reactions is therefore necessary to create the basic energetic preconditions.

Questions related to the special kind of energetic coupling have been examined and discussed in detail. Based on these considerations, a so-called Z-scheme is generally assumed at present, in which two separate photosystems (PS I and PS II) are linked together by a chemical chain in a way that two light reactions can occur behind one another (Fig. 1).

Light Reaction One

The reaction centre of PS I is called P 700 since it has a long-wavelength absorption maximum at 700 nm (Drissler and Hägele 1974; Kok 1961). An excitation of the P 700 is followed by the release of an electron to the primary acceptor Z, which is the starting point of a chemical electron transport chain.

A series of coupled redox reactions finally leads to the formation of reduced nicotinamide-adenine-dinucleotide-phosphate ($NADPH + H^+$), which enters the complex cycle of chemical secondary processes (Drissler and Hägele 1974), as reduction equivalent. P 700 remains in an oxydized state, but can be reduced again by the primary donor D_I of PS I (Govindjee and Govindjee 1965; Witt 1971; Rumberg and Witt 1964; Rumberg 1964; Siedow et al. 1973; Ke 1973; Warden and Bolton 1974a,b;

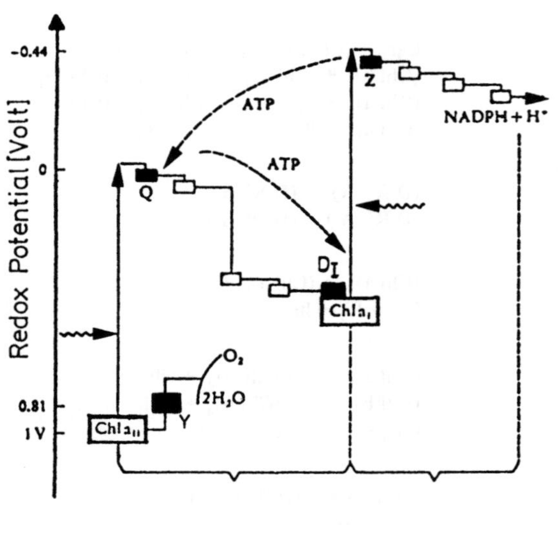

Fig. 1. Energy scheme of two coupled light reactions (energies are given in terms of normal potentials of redox pairs involved).
Chl a_I Reaction centre of Photosystem I (P 700). *Chl a_{II}* Reaction centre of Photosystem II (P 690). *Q, Z* Primary electron acceptors. *D_I, Y* Primary electron donors

Evans and Sihra 1976; Rurainski et al. 1971; Rumberg et al. 1965; Butler et al. 1973; Breton and Geakintov 1970):

$$D_I\ (P\ 700)\ Z + h\nu \rightarrow D_I\ (P\ 700)^+\ Z^- \rightarrow D_I^+\ (P\ 700)\ Z^- . \qquad (3)$$

Light Reaction Two

The release of molecular oxygen is considered as the consequence of the second light reaction, which occurs in PS II. The mechanisms here correspond to those known from PS I (Boardman 1972; Vermealic and Mathis 1973a,b; Renger and Wolff 1975; Bonaventura and Myers 1969; van Gorkom and Donze 1973). However, the reaction centre has its absorption maximum at approximately 690 nm and is accordingly termed P 690. Its excitation is followed by the transfer of an electron to the primary acceptor Q. This causes the primary donor Y to pass an electron from the water splitting reaction to P 690 (Govindjee 1975):

$$Y\ (P\ 690)\ Q + h\nu \rightarrow Y\ (P\ 690)^+\ Q^- \rightarrow Y^+\ (P\ 690)\ Q^- . \qquad (4)$$

Electrons flow from the acceptor Q via a further redox chain to all the P 700 particles which have contributed towards a reduction of NADP in the course of the first light reaction.

Photophosphorylization

Processes according to Eq. (3) and Eq. (4) also open up two possibilities for a production of the biochemical energy carrier ATP (adenosine triphosphate) which is required during the cycle of chemical dark reactions (Drissler and Hägele 1974; Bassham 1962).

At first a development of ATP is observed in the course of the first light reaction (Richter 1969; Hoffmann 1975). In addition to electron transport chain leading to the reduction of NADP, a second possible path is open to the electrons taken over from Z which finally passes them back to P 700. This type of ATP formation is called cyclic phosphorylization.

A further kind of ATP production is possible in a non-cyclic manner. In this case the electrons pass through the whole Z-scheme: the potential difference between Q and P 700 is sufficient for the occurrence of phosphorylization.

1.3 Photoactive Pigments

According to Table 1 the photosystems of higher plants and of green algae contain two main groups of active pigments: chlorophylls and carotenoids.

Chlorophylls

Based on their molecular structure (Richter 1969) the chlorophylls belong to the tetrapyrrole compounds (Fig. 2): four pyrrole rings (I-IV), which are nitrogen-containing fiver-rings, are linked together via methine bridges (–CH=), thus forming the porphin ring. The high number of conjugated double bonds which are part of the ring structure are responsible for a strong light absortion of the chlorophylls.

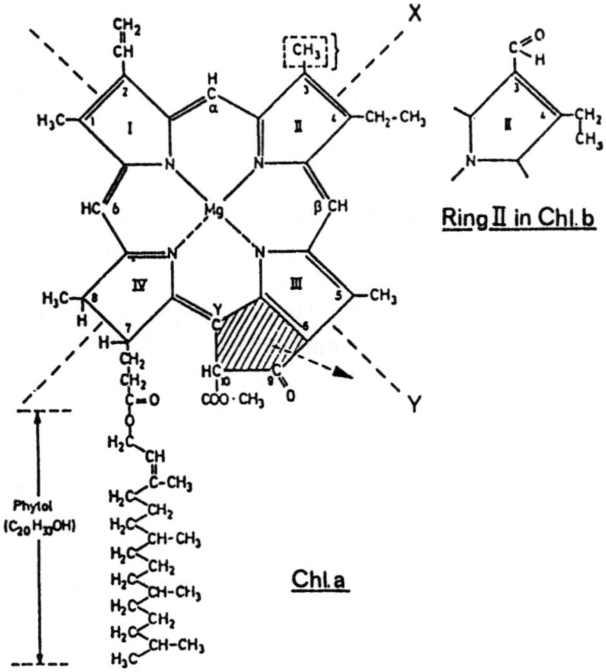

Fig. 2. Molecular structures of chlorophyll a and chlorophyll b. *x* and *y* represent the polarizations of the two optical transitions Qx and Qy (cf. Figs. 3 and 4)

Even a slight alteration of the ring structure can therefore cause drastic changes in absorption by varying the number of conjugated double bonds, as e.g. in the cases of Chl a and Chl b (Figs. 3, 4). Moreover, the optical properties of the chlorophylls are strongly determined by the magnesium atom bound in the centre of the porphin ring.

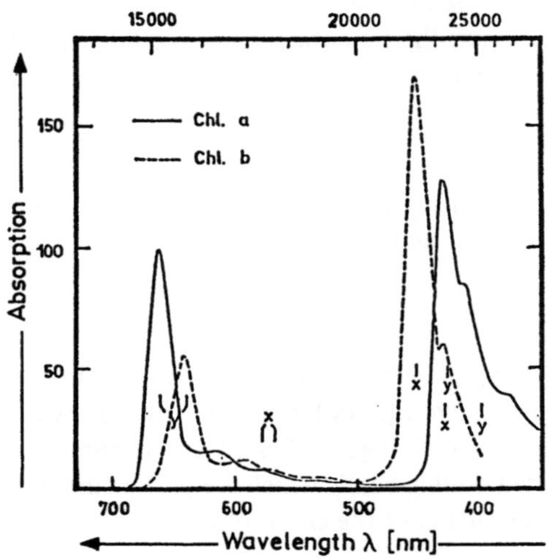

Fig. 3. Absorption spectra of chlorophyll a and b dissolved in ether (see Fig. 2 for parameters x and y) (Goedheer 1964)

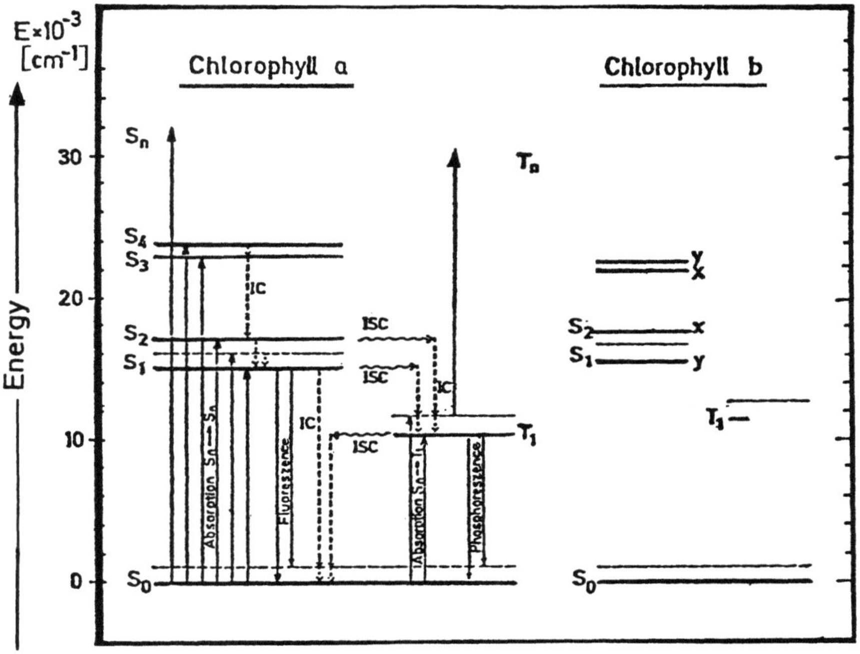

Fig. 4. Simplified energy scheme for the monomeric form of chlorophyll a and b (Goedheer 1964). S_n Singlet levels. T_n Triplet levels. *IC* Internal conversion processes. *ISC* Intersystem crossing processes

A strong influence of the molecular environment is due to the polar properties of the whole chlorophyll molecule: the non-polar phytol reside is bound as a water-insoluble hydrocarbon to the polar porphyrin ring (porphin ring and magnesium centre atom), thus forming the whole chlorophyll molecule of which the physical properties are directly influenced by the neighbouring particles.

If chlorophyll, for example, is in a polar environment, the magnesium atom preferably leads to complex formation. Located in a non-polar neighbourhood, a direct interaction of chlorophyll-porphyrin rings leads to the creation of aggregates in which the magnesium central atom of one chlorophyll molecule is bound to the ketocarbonyl group of another. In this way dimers and – where many particles are integrated – oligomers can develop (Vernon and Seely 1966).

Altogether, the variable molecular properties of the chlorophylls allow their varied functions in the photosynthetic process: as antenna particles Chl a and Chl b absorb light (govindjee 1975; Drissler and Hägele 1974), and transfer the excitation energy with extremely high efficiency to the reaction centre in which Chl a appears as electron donor, acceptor and possibly also sensitizer for electron exchange processes.

Investigations of absorption spectra from chlorophyll a and b in various polar, organic solvents and similar in vivo studies have provided differing results. That is why it has to be supposed that either chlorophyll molecules are exposed to very unusual environs in the plant membrane, or that specific interactions can lead here to spectral changes (Cotton 1976).

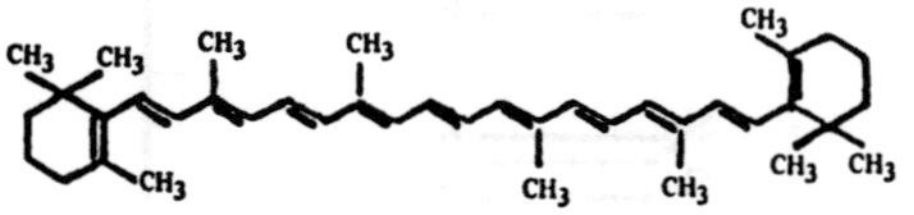

Fig. 5

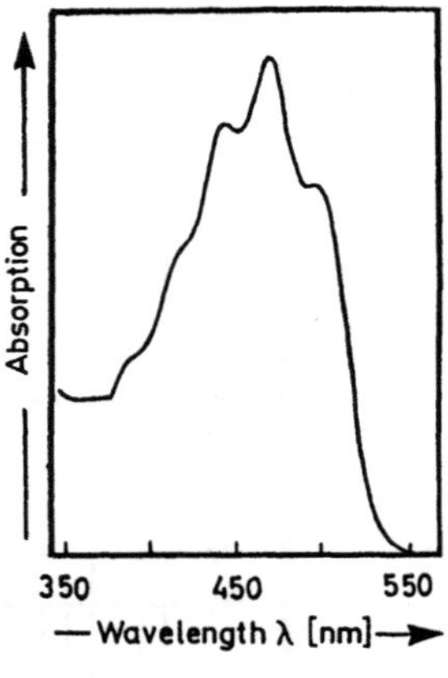

Fig. 6.

Fig. 5. Molecular structure of β-carotene (Chapman et al. 1967)

Fig. 6. Optical absorption of β-carotene dissolved in benzene (Hermann and Ricard 1973)

Carotenoids

Like the molecular structures of the chlorophylls, those of the carotenoids (Fig. 5) also consist of a system of conjugated double bonds (Chapman et al. 1967; Hermann and Ricard 1973), which enables them to absorb light strongly (Fig. 6). However, in contrast to the ring structure in the porphyrins, this consists mainly of linear chains in which only carbon and hydrogen atoms are present.

The carotenoids can be subdivided into the two main groups of carotenes and xanthophylls: while the carotenes comprise all isomer compounds of the kind $C_{40}H_{56}$, the xanthophylls represent carotene derivates with an oxygen content.

Little is known about the exact role of carotenoids within the context of photosynthesis. However, it is certain that they transfer 50-70% of their excitation energy on Chl a particles (Govindjee 1975). A protective function for chlorophylls against irreversible photooxydation has also been observed (Hoffmann 1975).

1.4 Questions Related to Physics

General Questions

By virtue of their nature, unresolved problems in photosynthesis which should be accessible to methods of experimental physics are to be found mainly in the area of the physical primary processes. For the understanding of some functional mechanisms and structural conditions, questions such as summarized below have to be clarified:

- Analysis of the highly effective energy transport mechanisms which transfer excitation energy from the absorbing pigments to the reaction centre particles.
- Analysis of structural environments for pigments in the photoactive systems (PS I and PS II) as well as of the molecular arrangement in reaction centre particles.
- Analysis of the role of singlet, triplet, acceptor, donor and vibrational states.
- Understanding of the general role of carotenoid molecules.

Contrary to usual physical samples, biological organisms, as for example the cells of monocellular green algae, are mixed systems of very complex chemical and structural composition which, nevertheless, show a stable functional order. The understanding of this general order, of its regulatory mechanisms, and of its molecular basis, is in some respects one of the most challenging problems today.

An approach to these questions by methods of physics can only provide clear results if some aspects are included which appear to be typical for living structures in general (Fröhlich 1977; Szent-Györgyi 1978; Witt 1979). Possible changes of the sample due to experimental conditions, the occurrence of collective phenomena which are absent after extraction of subsystems as well as the presence of extremely high electric membrane fields during biological activity ($\sim 10^6$ V/cm) are of importance for correct interpretations.

Theoretical models, which have been developed for an explanation of general regulatory mechanisms in biological systems are still without experimental confirmation. In this context one of the approaches of Fröhlich (1968a,b,c, 1973, 1975a,b, 1977) should be mentioned: according to his considerations coherent polar vibrations are strongly, overthermally excited during biological activity. They cause particular long-range interactions which lead to collective properties of the whole biological multicomponent system, thus creating functional order.

The presence of pigment molecules in the living structure of monocellular green algae offers from an experimental point of view the important possibility of applying well-known optical methods for obtaining information about the questions and problems mentioned above.

Optical Experiments

At room temperature, living cells of *Chlorella pyrenoidosa* show prompt (Govindjee 1975) and delayed (Strehler and Arnold 1951) emission of light in the spectral range from 640 to 800 nm if irradiated with visible light. Whereas the lifetime of the former is of the order of nanoseconds (Govindjee 1975; Mar et al. 1972), the latter shows a non-exponential decay in the time range from microseconds to hours (Govindjee 1975; Arnold and Azzi 1971). The total intensity of the prompt emission is about 10^3 to 10^5 times higher than that of delayed emission (Strehler 1957). Both kinds of radiation are attributed primarily to singlet transitions ($S_1 \rightarrow S_0$) of chlorophyll a. In order to explain the delayed fluorescence, various models have been suggested, some of them involving triplet states and some based on charge-recombination mechanisms (Stacy et al. 1971; Bertsch and Lurie 1971; Rabinewitch and Govindjee 1969).

Low temperature measurements of prompt as well as of delayed fluorescence down to 4.2 K have been reported in early works (Cho et al. 1966; Cho and Govindjee 1970a,b; Warden 1974), but were not comparable.

Furthermore, Raman investigations of monocellular green algae at low temperature have shown that Stokes spectra of photosynthetic samples can provide important information on structural arrangements of photoactive pigments as well as on their chemical composition (Lutz 1979).

In the following, results of optical experiments with samples of whole *Chlorella pyrenoidosa* cells are summarized which include emission of light as well as Raman experiments. The measurements were carried out in a way which can provide comparative experimental information about:

- Influence of temperature on experimental data. Temperature is of great with biological significance.
- Enhancements of vibrational population as predicted by the theoretical approaches mentioned above.
- Possible correlations between results and biological activity (which in the case of photosynthesis can be triggered by light).

Since the living state itself is, of course, correlated with temperatures above the melting point of water, attention was concentrated, in particular, on warming-up experiments.

2 Materials and Methods (Drissler 1977)

2.1 Growth of Algae

The starting culture used to produce the *Chlorella* cells was obtained from the algae collection at the University of Göttingen. This material was held in a nutrient solution as given by Lorenzen (Lorenzen 1968; Kuhl and Lorenzen 1964), exposed to a carbon dioxide-air mixture (1.5% CO) and illuminated with fluorescent lights.

2.2 Sample Preparation

For room temperature measurements on living cells, a flow arrangement, as shown in Fig. 7, was used in order to keep the photosynthetic activity of the algae efficient. It reduces the time of exhibition of cells to strong excitation light to values below 2 ms, which proved to be negligible in terms of damage under the applied experimental conditions. Furthermore, the flow set-up allows a control of biological conditions such as temperature and supply with CO_2.

For low temperature investigations, the cells were separated from the nutrient solution by repeated washing and centrifuging. They could then be attached as a paste to an optically inactive sample holder and frozen in the dark. This preparation technique was very important in order to avoid the use of cuvettes. All available cuvette materials showed a comparatively strong interfering emission upon excitation in the blue region of the visible spectrum, thus falsifying the optical investigations. As a consequence of using such highly concentrated specimens, the shapes of the observed

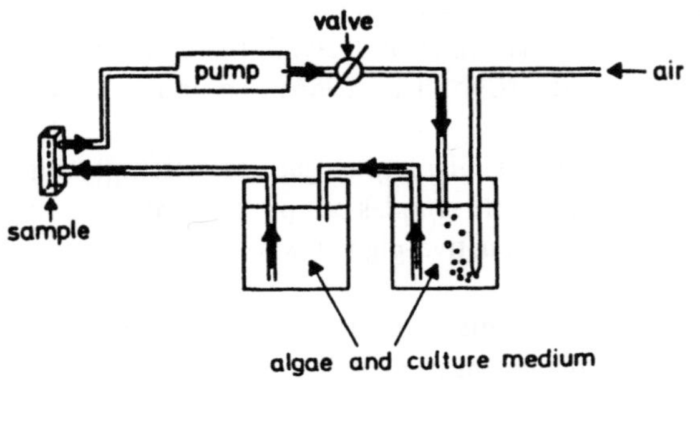

Fig. 7. Flow set-up for measurements with suspension of living cells at room temperature

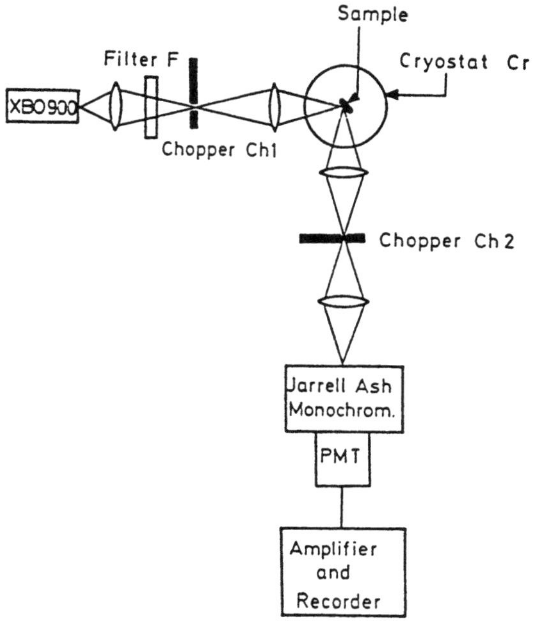

Fig. 8. Experimental set-up for investigations of prompt and delayed emission

spectra are distorted by reabsorption processes. This must be taken into account in a conclusive discussion of the experimental results.

2.3 Experimental Arrangements

Prompt Emission

Figure 8 shows schematically the experimental set-up as used for investigations of prompt and delayed emission processes. The cryostat allowed investigations between 1.7 and 310 K. Temperatures below 77 K were measured using a carbon resistor; an iron-constantan thermo-couple was used between 77 and 310 K. In this way the sample temperature could be determined with an accuracy of +/− 2 K. The experiments at intermediate temperatures were performed while the sample was warming up. To ensure a constant temperature over the entire specimen, the warm-up rate was very slow (0.3 K/min). The optical windows of the cryostat exhibited a slight and temperature-dependent emission which was corrected for in experimental results.

A Xenon high-pressure lamp XBO 900 W was used for excitation in combination with a filter-set F. Unless otherwise specified, F consisted of a heat reflector, a Schott GG 420 cut-off filter and a $CuSO_4$-liquid filter, yielding an excitation bandwidth from 420 to 560 nm.

Prompt and delayed emission studies were performed with a 0.25 m monochromator (Jarrell-Ash) and monitored using a cooled EMI 9658 photomultipler tube. The spectra, as reported in the following, were corrected for the spectral sensitivity of this detection scheme.

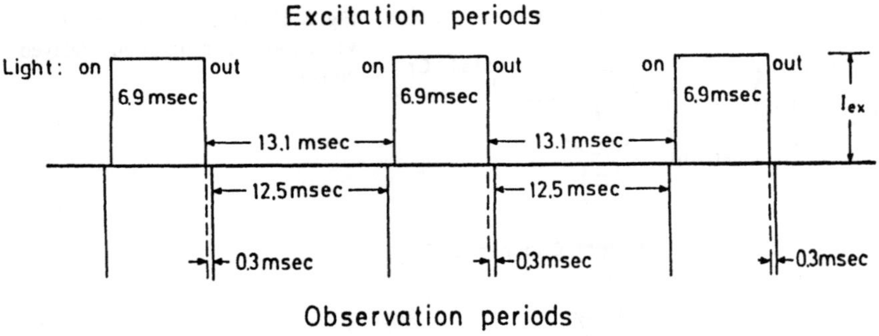

Fig. 9. Sequence of the excitation and observation periods used for measurements of delayed emission (I_{ex} intensity of excitation). One cycle includes: excitation (6.9 ms) – dark period (0.3 ms) – observation (12.5 ms) – dark period (0.3 ms)

The set-up described above permitted investigations of prompt fluorescence spectra. Delayed emission, which is always superimposed, is at room temperature roughly 10^3–10^5 times weaker and is therefore negligible. At low temperatures this factor is even greater.

Delayed Emission

In order to investigate the spectra of delayed emission separately from prompt fluorescence, two choppers, Ch 1 and Ch 2 (Fig. 8), operated synchronously at 3000 rpm in such a way that either the excitation or the observation path was clear (Fig. 9). Using this sequence, only decays lasting longer than 0.3 ms could be observed in the spectra of delayed emission. Thus, this arrangement eliminates not only the prompt fluorescence but also the faster components of delayed emission.

Raman Scattering

Raman investigations were carried out with a set-up as sketched in Fig. 10 using the same sample preparation techniques and conditions as described before. Spectra of scattered light were recorded at ~5 cm^{-1} resolution using Ar-laser excitation, a 1-m double monochromator and photon-counting detection (Drissler and Mcfarlane 1978).

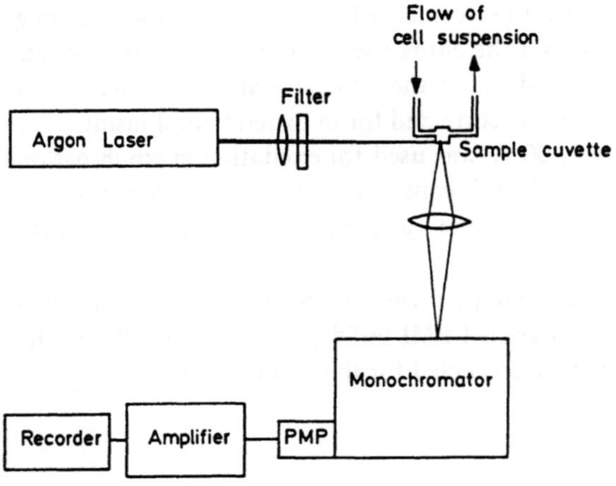

Fig. 10. Experimental set-up for studies of Raman scattered light

3 Experimental Results

3.1 Prompt and Delayed Emission (Drissler et al. 1977)

Spectra Between 4.2 K and 300 K

Some emission spectra are shown in Figs. 11-14. Using a suspension of living, photosynthesizing cells at room temperature an emission as sketched in Fig. 11 can be observed (cell concentration: 10^7 cells/ml). Figures 12-14 show the prompt and delayed emission spectra, respectively, for a paste of algae cooled to 4.2 K. Some spectra observed during warming up are shown in Fig. 15 for 4.2 K, 70 K, 173 K, and 279 K.

The spectra of prompt and delayed emissions of living cells suspended in nutrient solution at room temperature (Fig. 11) are identical, within the limits of error, and exhibit a pronounced maximum at 687 +/- 2 nm. These results are in agreement with previous reports (Govindjee 1975; Strehler and Arnold 1951; Azzi 1966).

If a paste of algae is cooled to 4.2 K the shapes of the prompt and delayed emission spectra are distinctly changed (Figs. 12 and 13). The short-wavelength parts between 670 and 700 nm are rather weak, while intense peaks occur between 700 and 750 nm. Long-wavelength tails, that can be observed up to 820 nm, are attached to these peaks. This spectral structure is at least partially due to reabsorption processes which occur in highly concentrated samples as used here (Szalay et al. 1967).

The spectrum of prompt fluorescence (Fig. 12), which is more structured than that at room temperature (Fig. 11), indicates the superposition of a large number of emission bands. This is evident, especially in Fig. 14, which shows the prompt fluorescence between 670 and 700 nm on an expanded scale. The band positions coincide within the limits of error with those from *Chlorella pyrenoidosa* in dilute suspensions (Govindjee 1975; Cho et al. 1966; Cho and Govindjee 1970a). This correspondence shows that conclusions which are based only on the spectral position of individual bands are not significantly influenced by additional reabsorption effects in the paste.

The structure of the delayed emission spectrum at 4.2 K (Fig. 13) is less clearly resolved than that of the prompt fluorescence. Furthermore, the short-wavelength wing of the main bands is more intense for delayed emission so that its spectrum appears broadened (Fig. 15).

To obtain more information on the sub-band structure, both types of emission spectra have been measured at various temperatures between 4.2 K and room temperature (Fig. 15). Such investigations indicate clearly a strong change of their shape with temperature. The comprehensive analysis of the complex band structure requires further experimental investigations. Nevertheless, it can be stated that the positions of the main peaks in the delayed emission spectra are the same as those in the prompt fluorescence.

Intensities

Dependence on Excitation. The intensity dependence of the delayed emission on the exciting intensity at low temperatures (4.2 K and 77 K) is shown by Fig. 16 in a double logarithmic scale. A straight line of slope 1 fits the measured data points over the entire range of exciting intensities.

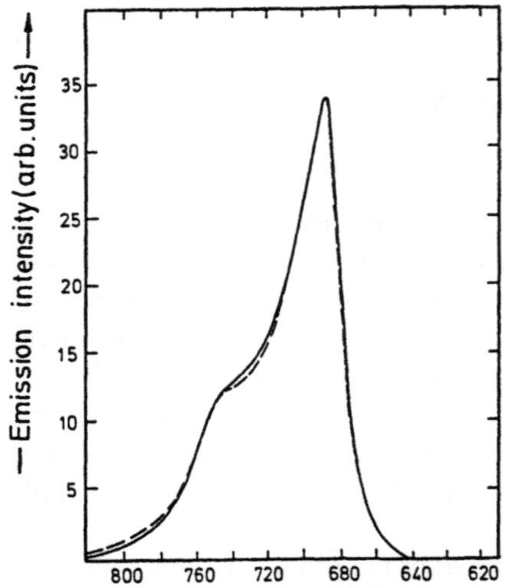

Fig. 11

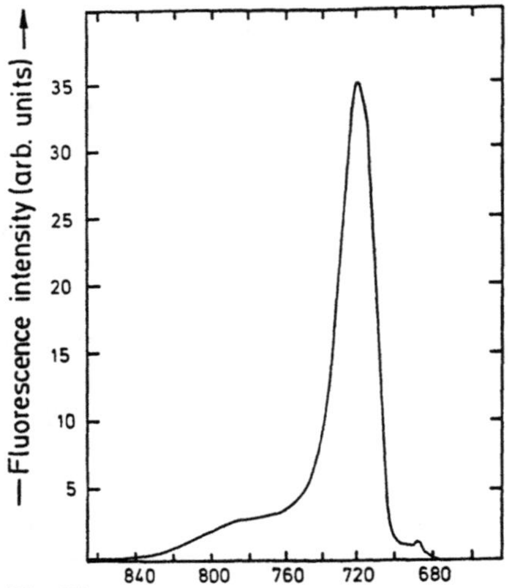

Fig. 12

Fig. 11. Prompt (———) and delayed (– – – – –) emission spectra from a suspension of photoactive cells at 298 K under saturating conditions for photosynthesis (concentration of algae: 10^7 cells/ml; excitation wavelengths: 420–560 nm). The dominant peaks of both spectra are located at 687 +/– 2 nm

Fig. 12. Prompt fluorescence spectrum of a paste of cells at 4.2 K (excitation wavelengths: 420 to 560 nm)

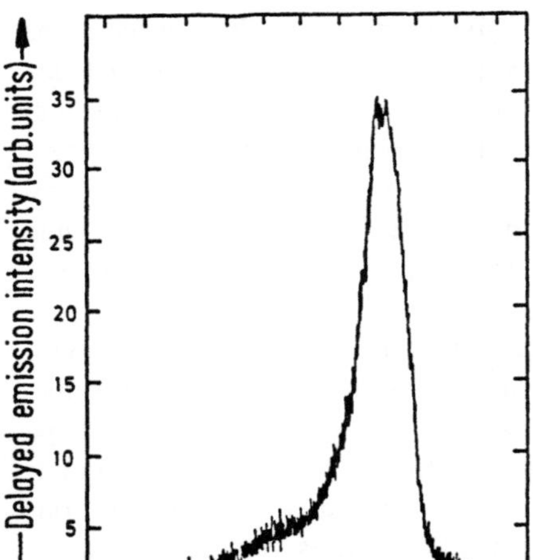

Fig. 13

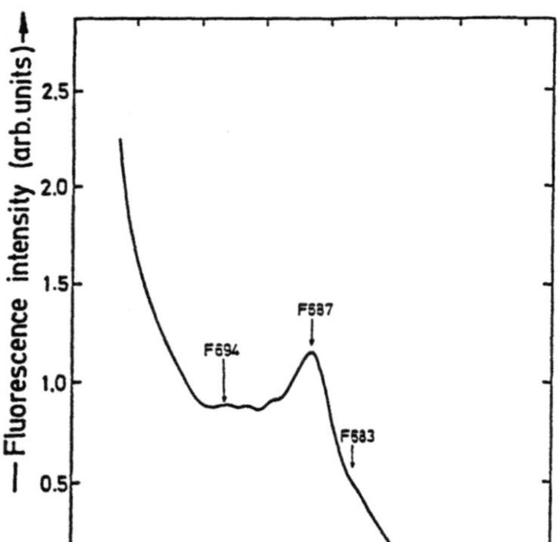

Fig. 14

Fig. 13. Delayed emission spectrum of a paste of cells at 4.2 K (lifetimes: < 0.3 msec; excitation as in Fig. 12)

Fig. 14. Short-wavelength part of the prompt fluorescence spectrum as sketched in Fig. 12 shown on an expanded scale. The positions of some emission bands are marked

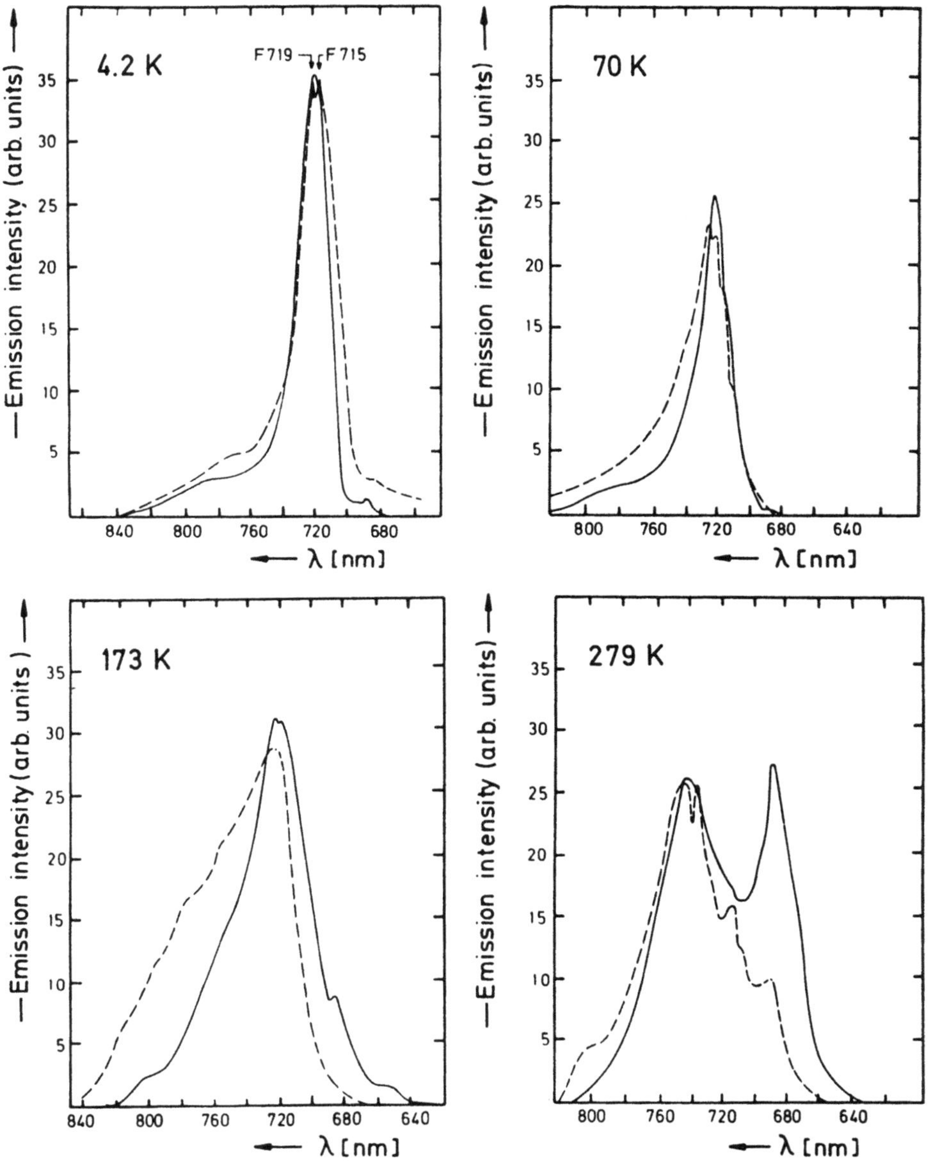

Fig. 15. Comparative sketch of prompt (———) and delayed (– – – – –) emission spectra at 4.2 K, 70 K, 173 K and 279 K measured during warming-up of a paste of cells from 4.2 K to room temperature. The spectra are only representative as for their shape but not for the total intensity

Temperature Dependences. The total intensity of both prompt and delayed emission shows a complicated temperature dependence between 1.7 and 310 K respectively (Figs. 17 and 18). Prompt emission can be observed over the entire range of temperature; delayed emission, on the other hand, is not detectable between 248 K and 270 K.

Below 200 K the temperature dependencies of both types of emission intensities are similar. In contrast to this, they are basically different above 200 K.

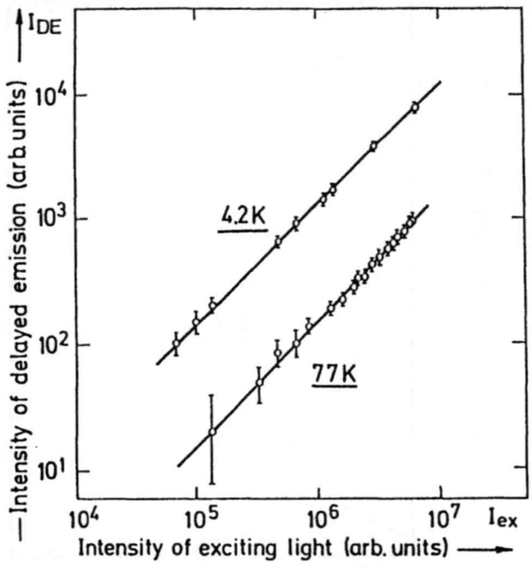

Fig. 16. Dependence of the total intensity of delayed emission at 4.2 K and 77 K on the intensity of excitation

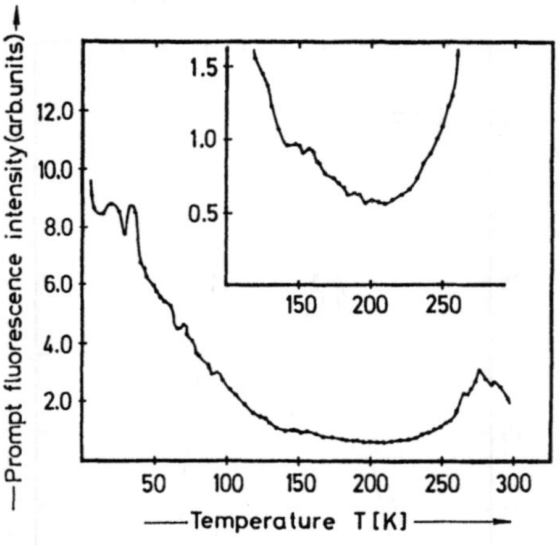

Fig. 17

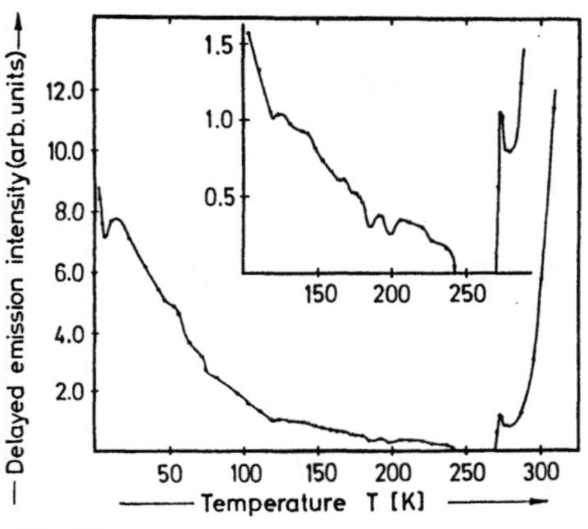

Fig. 18

Fig. 17. Total emission intensity of prompt fluorescence measured with a paste of cells at various temperatures during warming up from 4.2 K to room temperature

Fig. 18. Total emission intensity of delayed emission measured as in Fig. 17

3.2 Raman Scattering (Drissler 1981, 1982)

Stokes Spectra Between 4.2 K and 300 K

The Stokes Raman spectrum of living *Chlorella pyrenoidosa* cells suspended in nutrient at room temperature is shown in Fig. 19a. Its main features are four groups of strong bands (width ~10 cm^{-1}) which are superimposed on a rather structureless background. All lines can be assigned to carotenoid vibrations (Fawcett and Long 1976).

As the laser wavelength was changed between 514.5 nm and 454.5 nm, substantial changes in the scattered Stokes intensities occurred which overwhelmed the non-re-

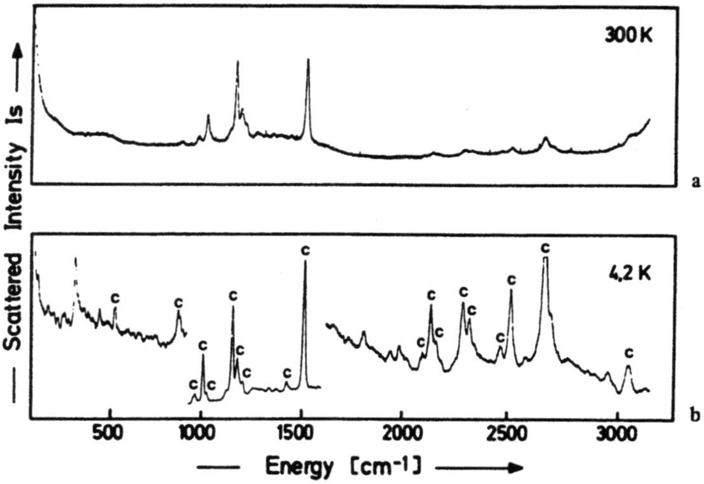

Fig. 19. a Stokes Raman spectrum from a suspension of photoactive *Chlorella pyrenoidosa* cells measured at room temperature (laser excitation: 488.0 nm; power: 500 mW). **b** Stokes Raman spectrum measured with a paste of algae at 4.2 K (laser excitation and power as in **a**. C: carotenoid vibrations)

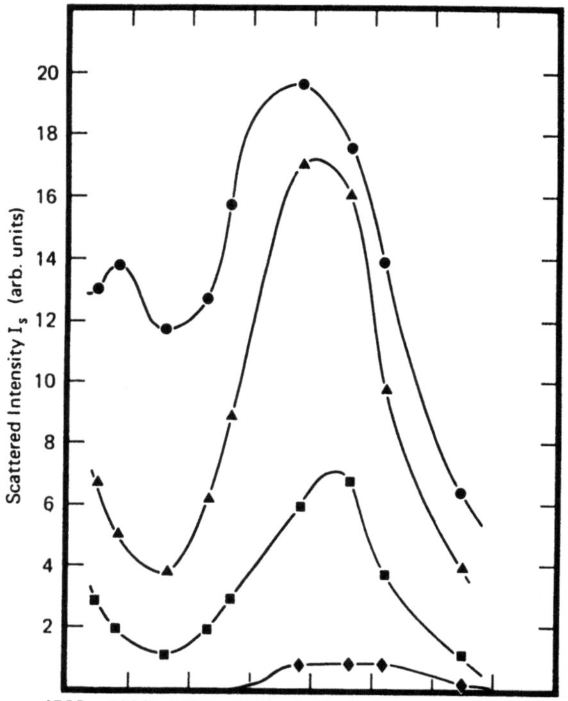

Fig. 20. Excitation profiles of the Stokes shifted bands at 868 cm^{-1} (♦), 1006 cm^{-1} (■), 1157 cm^{-1} (▲) and 1527 cm^{-1} (●)

sonant ω4-behaviour (Fig. 20), and are attributed to resonances with the absorption band of carotenoid pigments.

The complete absence of pronounced chlorophyll bands at room temperature conditions (Fig. 19a) is in striking contrast with the low temperature result (Fig. 19b), which also shows carotenoid bands (labelled C), but in addition exhibits narrow lines which are attributed mainly to chlorophyll b vibrations (Lutz 1979).

This spectral difference between 4.2 K and room temperature can occur due to a thermal broadening of lines during warming up, but can be of biological relevance as well. Stokes spectra were therefore investigated at various intermediate temperatures in order to obtain further information on the temperature behaviour of Raman bands.

Figure 21 indicates that the spectral shape of the scattered light is roughly similar below 230 K. Due to a disappearance of the whole spectral structure between 230 K and 235 K, neither carotenoid nor chlorophyll bands can be observed at further increasing temperatures. Simultaneously, a strong structureless background appears. Above 260 K the carotenoid bands which are known from room temperature studies (Drissler and Macfarlane 1978) appear and the background intensity decreases.

Figure 22 shows the temperature-dependent intensity of several carotenoid and chlorophyll bands.

Stokes/Anti-Stokes Ratios (Drissler 1980)

Since any population enhancement in vibrational states (see Sect. 1.4) should lead to enhanced Anti-Stokes Raman scattering, experiments which measure the ratios R of corresponding Stokes- and Anti-Stokes intensities have been performed with suspensions of *Chlorella pyrenoidosa* cells in nutrient during photosynthetic activity (Figs. 19a, 23).

The intensity ratios R of the main Anti-Stokes bands which can be attributed to vibrations of carotenoid molecules deviate considerably from the thermal population factor (see Table 3). An investigation of the dependence of Anti-Stokes intensities on laser power showed that local heating is not responsible for this enhancement (Fig. 24).

Table 3. Evaluation of Anti-Stokes/Stokes intensity ratios measured for the vibrations at 868 cm^{-1}, 1006 cm^{-1}, 1157 cm^{-1} and 1527 cm^{-1} with living *Chlorella pyrenoidosa* cells. The population enhancement is given by P. Resonances with the carotenoid absorption band have been taken into account according to Fig. 20

Vibrational energy ν_V [cm^{-1}]	Observed ratio R 19436/19436 [cm^{-1}]	Anti-Stokes enhancement R/n/(n + 1)	Res. factor 19436 + Ω/19436 From Fig. (3)	$\frac{P}{n/(n+1)}$
868	0.1	10.3	4.0	2.6 ± 0.5
1006	0.060	7.2	5.6	1.3 ± 0.1
1157	0.020	5.3	4.1	1.3 ± 0.1
1527	0.0024	3.5	2.7	1.3 ± 0.2

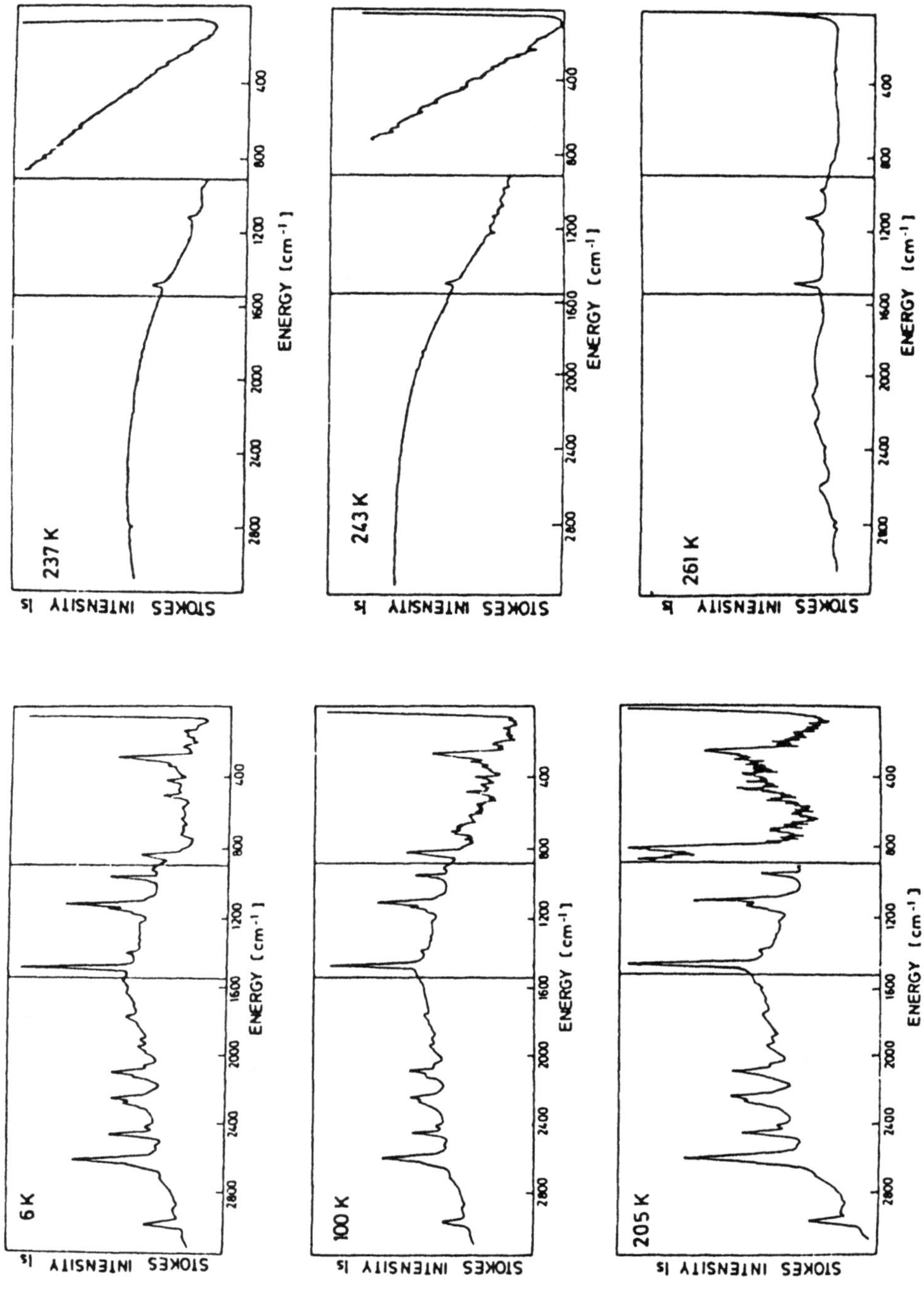

Fig. 21. Stokes Raman spectra observed with a paste of *Chlorella* cells at 6 K, 100 K, 205 K, 237 K, 243 K and 261 K

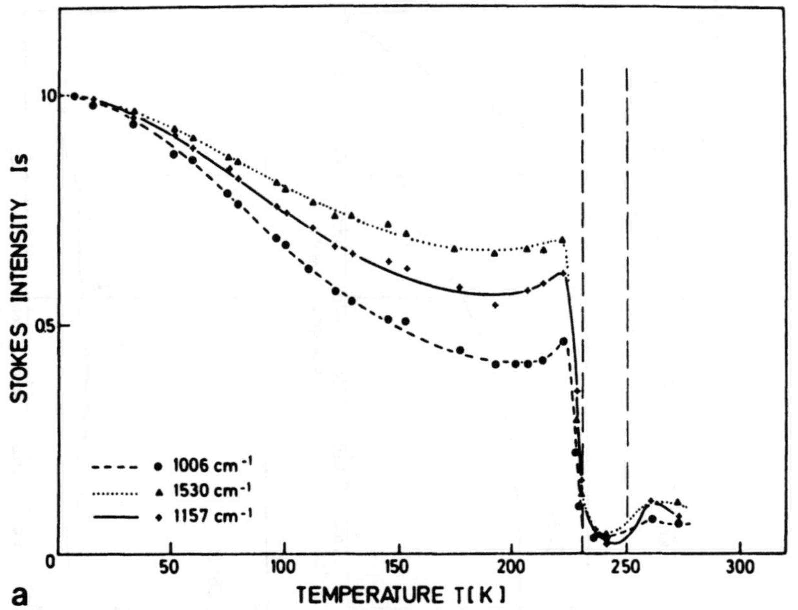

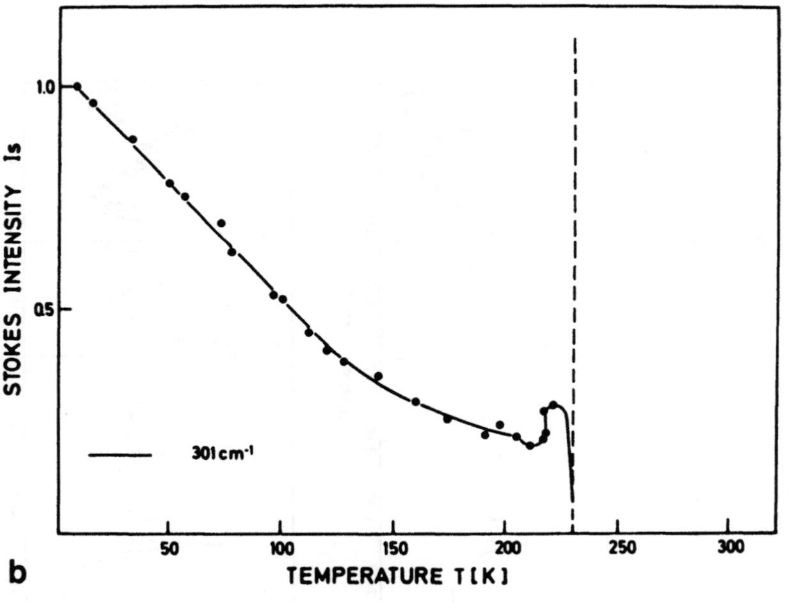

Fig. 22. Temperature dependent Stokes intensities of carotenoid (**a**) and chlorophyll (**b**) bands at 1006 cm^{-1}, 1157 cm^{-1}, 1530 cm^{-1} and 301 cm^{-1}. Striking features are the simultaneous decrease in intensity at 230 K for all bands as well as the reappearance of only the carotenoid lines at 261 K

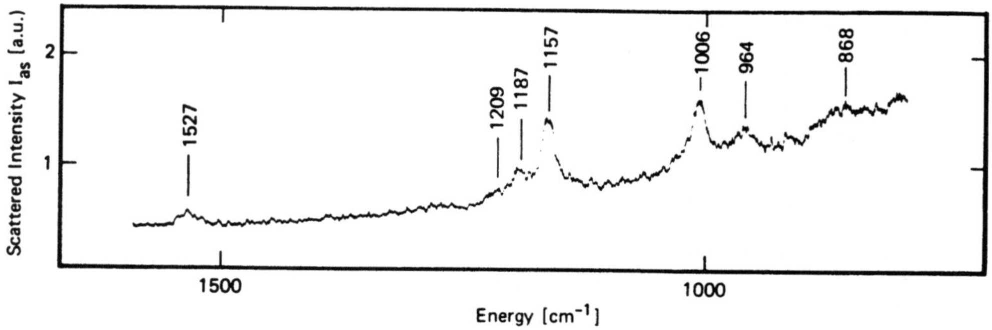

Fig. 23. Anti-Stokes Raman scattered light from living *Chlorella pyrenoidosa* cells at room temperature (laser excitation at: 514.5 nm; power: 1.0 W)

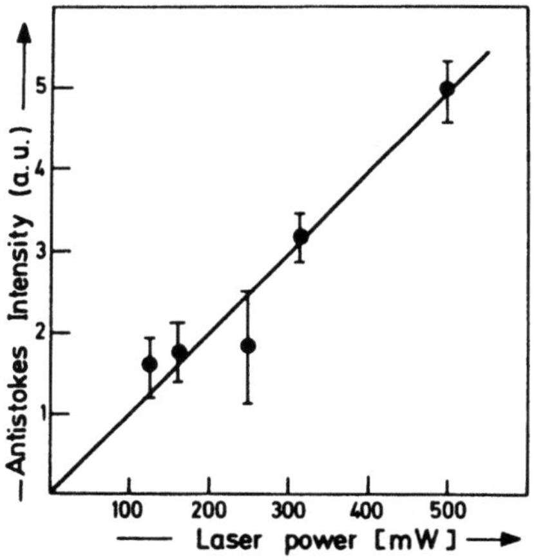

Fig. 24. Dependence of the Anti-Stokes intensity on laser power (excitation at 514.5 nm)

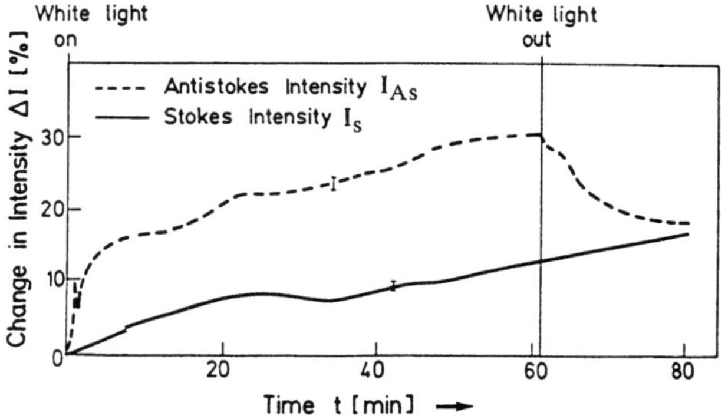

Fig. 25. Time-dependent changes of scattered Raman intensities I_S and I_{AS} at 1006 cm^{-1} during additional irradiation of the reservoirs (see Fig. 7) with white background light.
The CO_2-concentration of the cell suspension was kept constant by a supply with a CO_2-air mixture

Effect of Photobiological Activity

In an attempt to investigate correlations between photosynthetic activity and Raman intensities, white light was used for illumination of the reservoirs which are part of the flow system (see Sect. 2, Fig. 7). It should keep the whole photoprocess (biological activity) active when the cells are not exposed to the laser light.

The experimental results are summarized below in detail:

- During additional illumination with white background light, both Stokes- and Anti-Stokes intensity increased slowly. This increase amounts after 60 min to about 13% of the Stokes intensity and 31% for the Anti-Stokes intensity.
- The kinetics of the background light effect is different for Stokes and Anti-Stokes scattering, which indicates a time-dependent ratio R.
- The increase in Anti-Stokes intensity is partially reversible if background illumination is switched off. A reversibility of the enhanced Stokes intensity was not observed. This may be due a synthesis (increase) of scattering centres during illumination, which is a well-known effect.

It is important to note that an influence of additional white light was observed only if cells were simultaneously supplied with CO_2. This clearly shows a dependence on the photobiological activity (Fig. 25).

4 Discussion

4.1 Prompt and Delayed Emission (Drissler et al. 1977)

The identical spectral shapes of prompt and delayed emission which are observed with living, photosynthesizing cells at room temperature (Fig. 11) confirm that delayed emission originates from the same electronic levels as the prompt. These states are short-lived. The storage of energy, therefore, occurs in other, metastable states, for which a possible way of deactivation is the one via the above-mentioned short lived emitting levels. The same statement is true at low temperatures (Figs. 12–15), since the main peaks of delayed emission appear with identical spectral position in the prompt fluorescence spectra. This requires that the luminescence originates in the first excited singlet level of the radiating particles, thus excluding phosphorescence (triplet emission) to acount for the long-lived emission.

The temperature response of the total emission intensities (Figs. 17 and 18) is similar for the short- and long-lived fluorescence below 200 K. This suggests that only mechanisms acting in the short-lived emission process are influenced by the temperature since they are involved in both types of emission. No evidence is given for a significant dependency on temperature of the processes populating the emitting singlet levels for delayed emission of light. This result favours the assumption of a tunnelling process being involved in the population mechanism of delayed fluorescence. Since such processes do not require a thermal activation, back reactions within the primary photochemistry, the detrapping of electrons or of excitons clearly account for the occurrence of any delayed emission at low temperatures (Drissler 1977; Drissler et al. 1977). This is confirmed by the observed dependence of delayed fluorescence at these temperatures on excitation intensity (Fig. 16) which favours one-quantum processes in general.

At temperatures above 200 K, the total emission intensities of prompt and delayed emission are basically different. This can be accounted for by a distinct temperature dependence of the population mechanism, since both types of fluorescence are fed by different processes while the emitting events are identical. This conclusion indicates that processes which are not dependent on temperature (such as tunnelling processes) are no longer dominant in the creation of radiative states for delayed fluorescence. The experimental results (e.g. the disappearance of delayed emission between 248 and 270 K) can be explained by assuming that the lifetime of the metastable states is shortened by thermally activated processes, thus reducing the lifetime of the delayed fluorescence below the limits of the experimental set-up (0.3 ms). This is in agreement with other observations of a glow peak at 265 K (Desai et al. 1975), which could be explained assuming a thermally activated reoxydation of primary acceptors Q (PS II) by oxydized chlorophylls.

The distinct increase of delayed fluorescence intensity above 270 K indicates that additional processes become dominant in this temperature range. It seems that the electronic pathways of photosynthesis open at the melting point of water, at least partially. Additional states which are included in the transport chain are available to support delayed fluorescence.

4.2 Temperature Effect on Raman Scattering

Investigations of the temperature-dependent absorption of *Chlorella pyrenoidosa* cells (Beck 1978) clearly indicate that the number of scattering centres, as well as the resonance conditions for Raman scattering, are constant between 200 K and 270 K.

The usual equations for Raman intensities (Lutz 1974; Szent-Györgyi 1978) show that the observed behaviour of Stokes Raman intensities at 230 K and 261 K in this case has to be due to changes of essential transition matrix elements.

Furthermore, the simultaneous disappearance of both carotenoid and chlorophyll Raman bands at 230 K suggests that the experimental observations cannot be explained by a particular molecular property. On the contrary, it is strongly indicated that the symmetry of the whole particle system dominates its scattering properties and therefore shows a collective behaviour.

The symmetry of the scattering pigment system thus changes at 230 K in a way which cancels its Raman activity. This structural change represents a kind of phase transition (Fig. 22) which needs to be further investigated.

At 261 K another transition leads to the spectrum which is known from investigations with living cells at room temperature (Fig. 19a). This further rearrangement may be related with the approach of the melting point of water and/or other compounds in the photosynthetic membrane.

The increase in background intensity between 230 K and 261 K is in agreement with a well-known thermoluminescence peak which has been observed with various photosynthetic samples (Sane et al. 1974).

It should be noted in the context of Stokes-Raman warming-up experiments that the room temperature shape of the spectra was observed only at temperatures between

261 K and 323 K, which clearly is in the range where biological activity can occur. Further Raman investigations have to show whether a correlation between the photosynthetic function and the observed spectral shape can be confirmed.

4.3 Vibrational Population Enhancement

According to the theoretical considerations mentioned in Sect. 1.4, a high, non-thermal population of certain polar vibrations can occur in living structures as a collective behaviour of the multi-component system if it is sufficiently supplied with energy. These vibrations are coherent and may cause selective long-range interactions.

Since any population enhancement should lead to enhanced Anti-Stokes Raman scattering, experiments which measure the ratios R of corresponding Stokes- and Anti-Stokes intensities have been strongly indicated (see Sect. 3.2).

The intensity ratios R of the main Anti-Stokes bands (Fig. 23) with respect to their Stokes counterparts (Fig. 19a) deviate considerably from the thermal population factor (Table 3). As the investigation of the dependence of Anti-Stokes intensities on laser power has shown (Fig. 24), local heating is not responsible for this enhancement. The bands which are observed can be attributed to vibrations of carotenoid molecules which are part of the photoactive pigment system in algae (Lutz 1974).

The analysis of the experimentally measured intensity ratios R from Raman equations which include resonance effects (Fig. 20) of scattered photons is given by Table 3. The major part of Anti-Stokes enhancement appears to be due to resonances with the carotenoid absorption band at $\approx$ 480 nm, which are difference for Stokes and Anti-Stokes photons. The population appears to be enhanced by 30 to 160% if compared with the normal thermal equilibrium.

Although the major part of the experimentally measured Anti-Stokes enhancement can be explained as due to a resonance with the carotenoid absorption band at $\approx$ 480 nm, a real enhancement in population could not be ruled out. This result was discussed in the literature (Kinoshita et al. 1982) but can be reproduced only if the biological conditions support the photobiological activity. This was clearly confirmed by the effect of white background light on the Anti-Stokes intensity and is also in agreement with requirements by theoretical considerations (Fröhlich 1968a,b,c, 1973, 1975a,b, 1977).

5 Conclusions

It was the goal of this chapter to use well-known methods of optical spectroscopy for the analysis of excited states which are involved in the photosynthetic pathway.

Raman spectra have been measured in an attempt to clarify questions related to the role of coherent vibrational states in the biological membrane as predicted by theoretical considerations: since an enhanced population of vibrational states could not be ruled out and was even increased during activation of the photosynthetic process, further more detailed Raman investigations are clearly indicated.

The results of prompt and delayed fluorescence studies have shown that their detailed analysis can provide useful information on the properties of excited electronic states, as well as on the states of primary charge separation.

Particular attention was given to warming-up experiments, since the transition from very low temperatures with frozen chemistry into the temperature range where biological processes can basically occur could give new insight into parameters depending on the living process.

As a result of the investigations at various temperatures, the remarkable similarity between the behaviour of Stokes Raman scattering and delayed emission must be pointed out.

References

Arnold W, Azzi J (1971) Photochem Photobiol 14:233
Azzi JR (1966) Oak Rdige National Laboratory Technical Memo, no 1534
Bassham JA (1962) Sci Am 206:88
Beck J (1978) Diplomarbeit am Physikalischen Institut, Part 3. University of Stuttgart
Bertsch W, Lurie S (1971) Photochem Photobiol 14:251
Bishop DG (1974) Photochem Photobiol 20:281
Boardman NB (1972) Biochim Biophys Acta 283:469
Bonaventura C, Myers J (1969) Biochim Biophys Acta 189:366
Breton J, Geacintov NE (1970) Service de biophysique, Department de Biologie, Centre d'Etudes Nucleaires
Butler WL, Visser JWM, Simons HL (1973) Biochim Biophys Acta 325:539
Chapman D, Cherry RJ, Morrison A (1967) Proc Soc A 301:173
Cho F, Govindjee (1970a) Biochim Biophys Acta 205:371
Cho F, Govindjee (1970b) Biochim Biophys Acta 216:139
Cho F, Spencer J, Govindjee (1966) Biochim Biophys Acta 126:174
Cotton ThM (1976) Spectroscopic investigations of chlorophyll a as donor and acceptor: A basis for chlorophyll a interactions in vivo. Northwestern University, Evanston, USA
Desai TS, Sane PV, Tatake VG (1975) Photochem Photobiol 21:345
Drissler F (1977) Dissertation at the University of Stuttgart, Institute of Physics (Part 3)
Drissler F (1980) In: Pratesi R, Sacchi CA (eds) Lasers in photomedicine and photobiology (Springer series in optical sciences). Springer, Berlin Heidelberg New York
Drissler F (1981) J of Collect Phenom 3:147
Drissler F (1982) Proc of laser 81. In: Waidelich W (ed) Optoelectronics in medicine. Springer, Berlin Heidelberg New York
Drissler F, Hägele W (1974) Photosynthese (1. Teil). In: Physik in unserer Zeit 6:164
Drissler F, Macfarlane RM (1978) Phys Lett 69A:65
Drissler F, Hägele W, Schmid D, Wolf HC (1977) Z Naturforsch 32a:88
Duysens LNM (1956) Annu Rev Plant Physiol 7:25
Emerson R, Arnold W (1932) J Gen Physiol 15:391
Evans MCW, Sihra CK (1976) Biochem J 158:71
Fawcett V, Long DA (1976) In: Molecular spectroscopy, vol 4. A specialist periodical report. The Chemical Society, London
Fong FK (1975) Appl Phys 6:151
Fröhlich H (1968a) Int J Quant Chem 2:641
Fröhlich H (1968b) Phys Lett 26a:402
Fröhlich H (1968c) Nature 219:743
Fröhlich H (1973) Collect Phenom 1:101
Fröhlich H (1975a) Proc Natl Acad Sci USA 72, 11:4211
Fröhlich H (1975b) Phys Lett 51a:21
Fröhlich H (1977) Nuovo Cimento 7:399
Goedheer JC (1964) Biochim Biophys Acta 88:304
Govindjee (1975) Bioenergetics in photosynthesis. Academic Press, New York

Govindjee, Govindjee R (1965) Photochem Photobiol 4:793
Hägele W, Drissler F (1975) Photosynthese (2. Teil). In: Physik in unserer Zeit 1, 2
Hermann JP, Ricard D (1973) Appl Phys Lett 23:178
Hoffmann P (1975) Photosynthese. Akademie-Verlag, Berlin
Katz JJ, Daugherty RC, Boucker LJ (1966) In: Vernon LP, Seely GR (eds) The chlorophylls. Academic Press, New York, p 185
Ke B (1973) Biochim Biophys Acta 301:1
Kinoshita S, Hirata K, Kushida T (1982) Osaka University, Department of Physics, Toyonaka, Osaka 560, Japan
Kok B (1961) Biochim Biophys Acta 48:527
Kuhl A, Lorenzen H (1964) In: Methods in cell physiology. Academic Press
Lorenzen H (1968) Phykos 7:50
Lumry R, Spikes JD (1960) Photobiol 20:281
Lutz M (1974) Raman Spectroscopy 2:497
Mar T, Govindjee, Singhal GS, Merkelo H (1972) Biophys J 12:797
Rabinowitch E (1959) Dep Bot. University of Illinois, Urbana, Illinois
Rabinowitch E, Govindjee (1969) Photosynthesis. John Wiley, New York
Renger G (1977) In: Biophysik. Springer, Berlin Heidelberg New York, p 415
Renger G, Wolff Ch (1975) Z Naturforsch 30c:161
Richter G (1969) Stoffwechselphysiologie der Pflanzen. Thieme, Stuttgart
Rumberg B (1964) Z Naturforsch 19b:707
Rumberg B, Witt HT (1964) Z Naturforsch 19b:693
Rumberg B, Schmidt-Mende P, Skerra B, Vater J, Weikard J, Witt HT (1965) Z Naturforsch 20b: 1086
Rurainski HJ, Randles J, Hoch GE (1971) FEBS Lett 13:98
Sane PV, Tatake VG, Desai TS (1974) FEBS Lett 45:290
Sauer K, Dratz EA (1969) Prog Phot Res II:837
See e.g., Lutz M (1979) In: Chlorophyll organization and energy transfer in photosynthesis. CIBA Found Symp, p 61
Siedow JN, Curtis VA, Pietro AS (1973) Arch Biochem Biophys 158:889
Stacy WT, Mar T, Swenberg CE, Govindjee (1971) Photochem Photobiol 14:197
Strehler BL (1957) In: Gaffron H et al. (eds) Research in photosynthesis. Interscience, New York, p 118
Strehler BL, Arnold W (1951) J Gen Physiol 34:809
Szalay L, Török M, Govindjee (1967) Acta Biochim Biophys Acad Sci Hung 2(4):425
Szent-Györgyi A (1978) The living state and cancer. Marcel Dekker, New York
Van Gorkom HJ, Donze M (1973) Photochem Photobiol 17:333
Vermeglio A, Mathis P (1973a) Biochim Biophys Acta 314:57
Vermeglio A, Mathis P (1973b) Bioelectrochem 1:364
Warden JT (1974) FEBS Lett 42:61
Warden JT, Bolton JR (1974a) Photochem Photobiol 20:251
Warden JT, Bolton JR (1974b) Photochem Photobiol 20:263
Witt HT (1971) Rev Biophys 4, 4:365
Witt HT (1979) Biochim Biophys Acta 505:355
Witt HT, Moraw R (1959) Physik Chem 20:283
Witt HT, Müller A (1959) Z Physik Chem 21:1

Emission of Radiation by Active Cells

J. K. POLLOCK and D. G. POHL [1]

1 Introduction

The possibility that living cells are capable of high frequency electrical oscillations was first suggested by Fröhlich in a series of papers (Fröhlich 1968, 1980, 1982). His theoretical development noted that an extremely high electric field exists across all biological membranes (10^7 V/m) and that the highly polarized molecules within these membranes could interact cooperatively under certain conditions. Fröhlich calculated that if the thermal vibrational modes that existed in the membrane received energy above a threshold energy, then the modes would condense into one coherent mode having a frequency in the range 10^{11}–10^{12} Hz. Further work (Fröhlich 1984) revealed that polar modes in biological molecules, and larger systems, exist with both higher and lower frequencies. Stimulated by this proposal, a number of different lines of investigation have given evidence that these oscillations do, indeed, exist.

Rowlands and his co-workers (Rowlands et al. 1981; Rowlands et al. 1982; Rowlands and Sewchand 1982; Sewchand et al. 1982) reported that human erythrocytes formed rouleaux at a rate that is not consistent with random diffusive collisions. The observed accretion rate of these rouleaux, the stacks of these disk-shaped cells that resemble rolls of coins, required that an attraction between the cells with a range between 5 and 10 μm be assumed in the kinetics equations. This long-range interaction seems to be mediated by extended macromolecules in the suspension, it does not occur in solutions lacking extended macromolecules. Nor does it occur in solutions containing globular molecules, albumin for example. However, if the albumin has been heated, causing it to extend, the long-range interaction is observed.

The interaction does not occur in cells that have been metabolically depleted or in cells with no membrane potential, indicating that it is tied to cell metabolism. A Fröhlich-type interaction of the AC fields produced by the cells has been proposed as a mechanism for this cell stacking (Sewchand et al. 1982; Paul et al. 1983).

Some investigators have reported success in detecting ac oscillations from cells or collections of cells. The power spectra of the mormyrid fish *Pollimyrus isidori* ranges from 1 to about 50 kHz and is involved with the electric organs (Westby and Kirschbaum 1978, 1982). Other high frequency fields have been reported from single cells by Jafary-Asl and Smith (1983), and Pohl (1985), Jafary-Asl and Smith reported detecting oscillations from the yeast *Saccharomyces cerevisiae* at around 8 MHz. Pohl, studying the large alga *Netrium digitus* found weak signals at around 7 and 33 kHz.

1 Department of Chemistry and Physics, Georgia College, Milledgeville, GA 31061, USA

The phenomenon of dielectrophoresis was used by Pohl and his co-workers (Pohl 1980, 1981, 1982, 1984; Pohl and Lamprecht 1985; Pohl et al. 1981; Rivera et al. 1985; Phillips et al. 1987) to detect the presence of fields about cells. Dielectrophoresis is the motion of neutral matter in non-uniform electric fields (Pohl 1951). Pohl used the cells as sources of non-uniform fields and measured their ability to attract or repel small dielectric particles with two techniques. The hanging drop technique (Rivera et al. 1985; Pohl and Lamprecht 1985; Phillips et al. 1987) produces patterns of particles about the cells reminiscent of the patterns of iron filings about the poles of a magnet. The technique of micro-dielectrophoresis (μ-DEP) uses the same principle but gives a more quantitative measure of the particle gathering ability of a cell.

2 Micro-Dielectrophoresis

In a non-uniform electric field, a particle with an effective dielectric constant higher than the suspending medium will be attracted to the region where the electric field is stronger, while a particle with an effective dielectric constant lower than the suspending medium will be repelled from that same region (Pohl 1978). This effect was used by Pohl (Pohl 1980; Pohl 1981; Pohl et al. 1981) to detect fields generated by cells.

In the technique of μ-DEP, the cell is the source of the non-uniform electric field, and the small detecting particles (dia $< 2\ \mu m$) are either attracted to, or repelled from the cell surface, depending on their dielectric constant. The procedure consists of placing the cells to be studied in a de-ionized suspension of particles of either high or low relative dielectric constant (K′) and allowing them to settle to the bottom of a thin, flat capillary tube. A visual count is then made of the number n, the number of particles in contact with the cell surface, and of the number p, the number of particles in a fixed area surrounding the cell. The ratio of $\langle n/p \rangle$ thus gives a measure of the relative particle-collecting ability of the cell.

The particles used in these studies were chosen to be insoluble in aqueous solution and biologically inert. These particles are listed in Table 1. Note that the relative dielectric constants (K′) of the first group of particles are considerably higher than the suspending water K′ = 78) and should be strongly attracted. The second group of particles have relative dielectric constants much less than water and should be repelled from the cells.

Table 1. Dielectric constant of particles used in μ-DEP

Particle	Relative dielectric Constant, K′, at 1 kHz
$BaTiO_3$	2000
$SrTiO_3$ (Sr reduced)	5000
$NaNbO_3$	650
$BaSO_4$	11.5
SiO_2	3.8
Al_2O_3	7.5

Table 2. Particle-gathering ability of cells; or micro-dielectrophoresis

Cell	Ratio of high K′ to low K′ particles
Bacteria: *Bacillus cereus* (synchronized)	4.0
Fungi: *Saccharomyces cerevisiae* (yeast)	2.0
Nematode: *Monochoides changii* (worm)	2.0
Aves: Chicken red blood cells	2.0
Mammalian:	
Mouse "L" fibroblasts	2.0
Mouse ascites (tumor) fibroblasts	2.2
Mouse fetal fibroblasts	2.0

The results from these μ-DEP experiments (Pohl 1980, 1981, 1982, 1984; Pohl et al. 1981), are summarized in Table 2. They indicate that cells are producing electric fields and, thus, attracting particles of high dielectric constant in greater numbers than particles of low dielectric constant. The numbers given in the table are the ratio of $\langle n/p \rangle$ for high K' particles to $\langle n/p \rangle$ for low K' particles for each given cell type. Note that particles with a high K' are collected in at least twice the numbers as particles with a low K'. This effect exists for all cell types examined to date. Such a universal phenomenon would indicate that these fields are involved in a basic cellular process either as an indicator that the process is going on, or as an integral, necessary part of that process.

The dependence of the DEP force on the solution conductivity can be used to estimate the frequency of the cellular fields being detected in the following manner. The DEP force has been given by Jones (1979) as:

$$F_{DEP} = 2\,\pi a^3\, K_e \nabla\, |E_0^2|\,, \tag{1}$$

where K_e is the total effective dielectric permittivity, given as:

$$K_e = \mathrm{Re}\left\{\frac{\epsilon_0 K_1'\,(K_2 - K_1)}{(K_2 + 2\,K_1)}\right\}, \tag{2}$$

and the individual complex permittivities of the cell, K_2, and of the aqueous medium, K_1, can be written as a combination of the relative dielectric constant, K', the conductivity, σ', the angular frequency of the applied electric field $\omega = 2\,\pi f$, and ϵ_0, the absolute permittivity of free space; where $j = \sqrt{-1}$

$$K = K' - j\sigma'/(\omega\epsilon_0)\,. \tag{3}$$

As can be seen from the above equation, increasing the conductivity of the suspending medium increases the effective dielectric constant of the medium. The effective dielectric constant can be easily calculated using the relation:

$$K_{effective} = \{(K')^2 + [\sigma'/(\omega\epsilon_0)]^2\}^{1/2}\,. \tag{4}$$

So, as the conductivity of the water increases, its effective dielectric constant eventually becomes greater than the dielectric constants of the particles and they are no longer attracted to the cell.

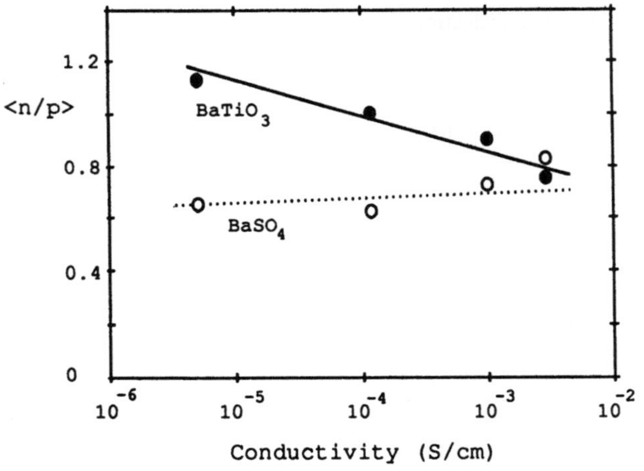

Fig. 1. The effect of increasing solution conductivity on the μ-DEP (⟨n/p⟩) of mouse ascities cells: $BaTiO_3$ and $BaSO_4$ particles. (Pohl et al. 1981)

Figures 1 and 2 show the ⟨n/p⟩ ratio for mouse ascites cells as a function of increasing solution conductivity. Figure 1 compares the ⟨n/p⟩ ratio of $BaTiO_3$, a high dielectric constant particle, and $BaSO_4$, a low dielectric constant particle. Figure 2 compares the ⟨n/p⟩ ratio of $NaNbO_3$, a high dielectric constant particle, and SiO_2, a low dielectric constant particle for the same cells. Note that as the conductivity increases in both figures, the difference between high K' and low K' particles disappears. Using Eq. (4) and the conductivity at which the ⟨n/p⟩ ratios are the same, an upper limit for the frequency of the field generated by the cell can be calculated.

From Fig. 1, the conductivity at which the ⟨n/p⟩ ratios for $BaTiO_3$ and $BaSO_4$ coincide is 10^{-2} S/cm. Using the value for $BaTiO_3$ given in Table 1, Eq. (4) gives 9 MHz as an upper limit of the cell's frequency. From Fig. 2 the conductivity is 1.5×10^{-3} S/cm and, using the value of K' from Table 1, Eq. (4) gives a comparable vaue for the upper limit of the frequency, 4.2 MHz.

In additional experiments (Pohl 1980; Pohl et al. 1981), the particle-gathering ability of cells seemed to be at a maximum in cells that were rapidly dividing. In studying yeast cells *(S. cerevisiae)* at different stages in their life cycle, the ratio of ⟨n/p⟩

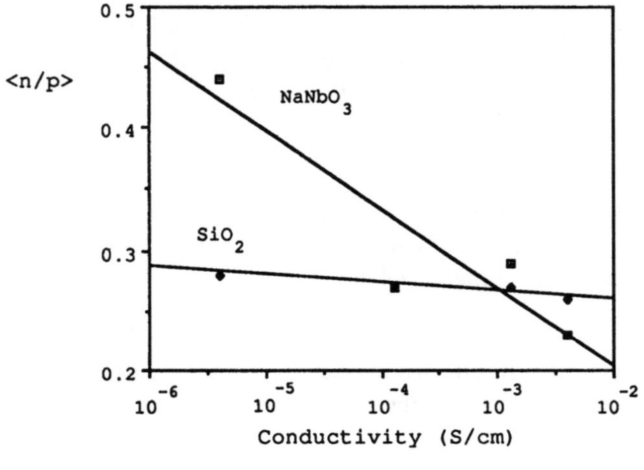

Fig. 2. The effect of increasing solution conductivity on the μ-DEP (⟨n/p⟩) of mouse ascities cells: $NaNbO_3$ and SiO_2 particles. (Roy et al. 1981)

numbers was a maximum at or near mitosis. Figure 3 shows the ⟨n/p⟩ ratio for both high K′ and low K′ particles as a function of life cycle. The maximum collection of the particles occurs at the point in the life cycle that corresponds to the onset of mitosis. This effect is present even when the curves are corrected for cell size.

The association of these fields with reproduction is apparent for other cells, too. As shown in Table 2, cells of the bacteria *B. cereus*, taken from a synchronized culture so that all the cells were in the growth phase, showed an unusually high ⟨n/p⟩ ratio. Studies on mouse "L" fibroblasts showed that cells from a rapidly dividing culture had twice the particle gathering ability on high K′ particles as cells from a confluent culture that were no longer dividing (Pohl 1980). The results from these experiments give further indication that the fields being produced are associated with cell reproduction.

3 Patterns

In a further series of experiments, the μ-DEP technique was modified to produce visible patterns of particles about the cells. Rivera et al. (1985) used a hanging drop technique and a much higher concentration of particles to produce patterns analogous to those observed with iron fillings about a magnet. They detected the presence of fields about several species of large algae and about the fertilized eggs of the African clawed frog *(Zenopus laevis)*. Their work was confirmed in Germany by additional studies on algae (Pohl and Lamprecht 1985). Phillips et al. (1987) later extended the hanging drop technique to the smaller mammalian cells.

This study (Phillips et al. 1987) used a wide variety of mammalian cells: VERO and BSC-1, two African green monkey kidney lines; AK-D, feline lung line; BFK, primary culture of bovine fetal kidney; and WI-38 and VA-13A, two human lung cell lines. All cell types showed the ability to collect or repulse particles similar to the algae previously studied.

The hanging drop technique has an advantage over μ-DEP in that repulsion can be observed in the hanging drop. It can be seen from Eq. (2) that K_e can be either positive or negative, and thus the DEP force can be either positive or negative. From the values in Table 1, it should be expected that the low dielectric constant particles would be repelled from the cells, not merely be unaffected. μ-DEP is unable to show this repulsive force, since it only measures particles collected on the cells. In the hanging drop, however, the cells show clear zones around themselves where they have repelled the low dielectric constant particles. This pattern is distinctly different from the case of dead cells where the particles are unaffected by the cell and settle about it indiscriminately. Attraction patterns are clearly observed with high dielectric constant particles, with certain cell types showing dipole or quadrapole field patterns.

Figures 3, 4, and 5 show examples of the types of patterns formed about mammalian cells. Figure 3 shows a view containing a number of cells. Each cell has collected high dielectric constant particles about itself, indicating the presence of a cellular electric field. Figure 4 shows a view of the same types of cells (VERO) under the same conditions, only the particles here have a low dielectric constant. In this case, the particles are repelled from all the cells as expected from dielectrophoretic theory.

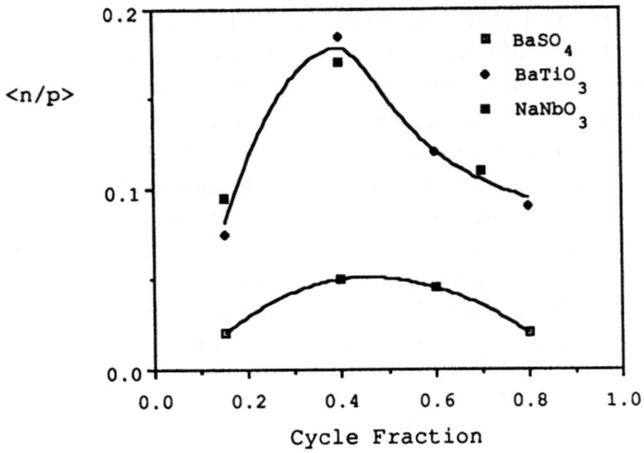

Fig. 3. Variation of μ-DEP (⟨n/p⟩) through the life cycle of *Saccharomyces cerevisiae*: $BaTiO_3$, $NaNbO_3$, and $BaSO_4$ particles. (Pohl et al. 1981)

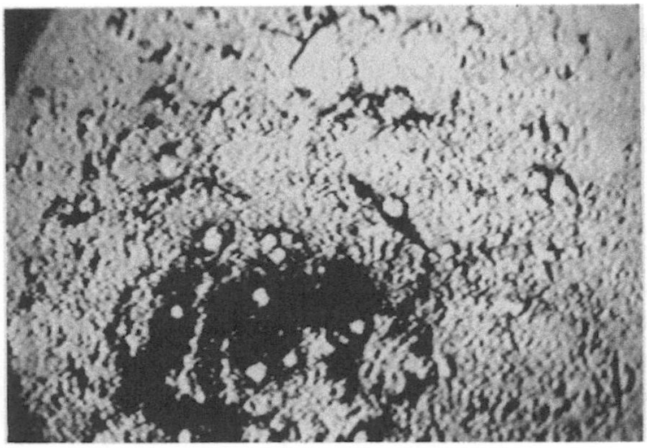

Fig. 4. Attraction pattern: VERO cells (African green monkey kidney cell line) and $BaTiO_3$ particles. Medium: Isoosmolar dextrose, conductivity, $s = 10^{-3}$ S/cm

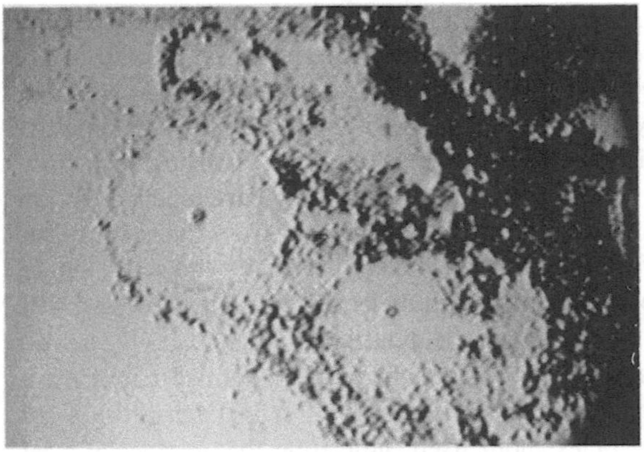

Fig. 5. Repulsion pattern: VERO cells (African green monkey kidney cell line) and $BaSO_4$ particles. Medium: Isoosmolar dextrose, conductivity, $s = 10^{-3}$ S/cm

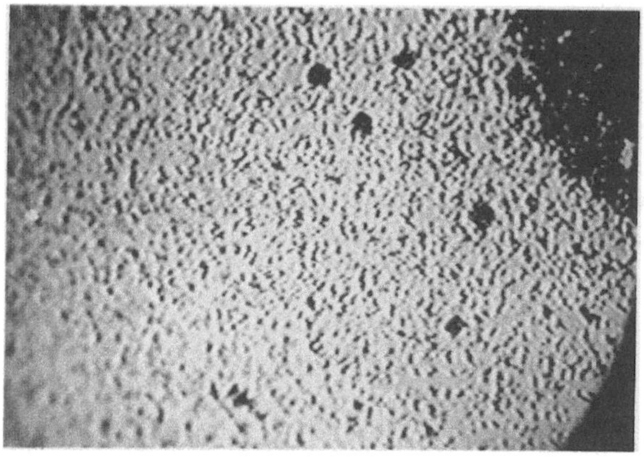

Fig. 6. Neutral pattern: Heat-killed BSC-1 cells (African green monkey kidney cell line) and $SrTiO_3$ particles. Medium: Iso-osmolar dextrose, conductivity, $s = 10^{-3}$ S/cm

Figure 6 shows how heat-killed cells have no effect at all on the distribution of the particles in the hanging drop.

As expected, the patterns being formed in these experiments were sensitive to changes in the conductivity of the solution just as was observed in the μ-DEP experiments. The attraction patterns not only diminished and eventually disappeared with increasing solution conductivity, but at a sufficiently high conductivity a repulsive pattern developed with even high dielectric constant particles.

4 Summary

The evidence from the μ-DEP experiments and from the closely related pattern experiments consistently indicate that cells are producing radio frequency electric fields. These experiments have shown that:

1. High dielectric constant particles of all kinds ($BaTiO_3$, $SrTiO_3$, $NaNbO_3$) are attracted to cells much more strongly than particles of low dielectric constant ($BaSO_4$, SiO_2, Al_2O_3). In fact, the pattern experiments show that low dielectric constant particles are repelled from the cells rather than being attracted.
2. The attraction disappears as the solution conductivity increases as expected from dielectric theory.
3. The effect is observed for a wide range of organisms: bacterial, fungal, algal, and mammalian cells.
4. The effect is shown by living cells but not by dead cells.
5. The effect is maximal at or near mitosis.
6. Current dielectric theory can be used to predict an upper limit for the frequency of the oscillation. For the case of mouse ascites cells the theory predicts a frequency between 4 and 9 MHz.

The evidence for oscillations in the Mega Hertz range is at the lower end of the range predicted by Fröhlich, but agrees with Smith (Jafary-Asl and Smith 1983), who found oscillations from yeast cells between 7 and 8 MHz. A mechanism has been proposed by Pohl (1985) to explain these relatively low frequency oscillations.

The model (Pohl 1985) assumes the existence of oscillating chemical reactions of a type that is well known (Noyes and Field 1974, Noyes and Field 1977; Epstein et al. 1983). Part of the cyclic reaction would involve an ionic phase, so that charge waves would be possible. Pohl suggests that the longitudinal charge waves of the reaction could be collimated by regular structures, such as the endoplasmic reticulum, within the cell. The collimated charge waves would then interact electrically to form a set of coherent charge waves involving large areas of the cell and creating an oscillating dipole or multipole.

References

Epstein IR, Kustin K, de Kepper, Orban M (1983) Oscillating chemical reactions. Scientific American 248(3):112–123

Fröhlich H (1968) Long range coherence and energy storage in biological systems. Int J Quantum Chem 2:641–649

Fröhlich H (1980) The biological effects of microwaves and related questions. Adv Electronics Electron Phys 53:85–152

Fröhlich H (1982) What are non-thermal electric biological effects? Bioelectromagnetics 3:45–46

Fröhlich H (1984) General theory of coherent excitations on biological systems. In: Adey WR, Lawrence AF (eds) Nonlinear electrodynamics in biological systems. Plenum, New York, p 491–496

Jafary-Asl AH, Smith CW (1983) Biological dielectric in electric and magnetic fields. IEEE Annu Rep Conf Elec Insulation Dielectric Phenomena, p 350

Jones TB (1979) Dielectrophoretic force calculation. J Electrostatics 6:69–82

Noyes RM, Field RJ (1974) Annu Rev Phys Chem 25:95

Noyes RM, Field RJ (1977) Oscillations in chemical systems. Acc Chem Res 10(8):273–280

Paul R, Chatterjee R, Tuszynski JA, Fritz OG (1983) Theory of long-range coherence in biological systems. I. The anomalous behavior of human erythrocytes. J Theor Biol 104:169–185

Phillips W, Pohl HA, Pollock JK (1987) Evidence for AC fields produced by mammalian cells. Bioelectrochem Bioenerg 17:287–296

Pohl HA (1951) The motion and precipitation of suspensoids in divergent electric fields. J Appl Phys 22:869–871

Pohl HA (1978) Dielectrophoresis. The behavior of neutral matter in nonuniform electric fields. Cambridge University Press, Cambridge

Pohl HA (1980) Micro-dielectrophoresis of dividing cells. In: Keyzer H, Gutmann F (eds) Bioelectrochemistry. Plenum, New York, p 273–295

Pohl HA (1981) Natural electrical RF oscillation from cells. J Bioenerg Biomembr 13:149–169

Pohl HA (1982) Natural cellular electrical resonances. Int J Quantum Chem: Quantum Biol Symp 9:399–409

Pohl HA (1983) Natural AC fields in and about cells. In: Adey WR, Lawrence AF (eds) Nonlinear electrodynamics in biological systems. Plenum, New York, p 87

Pohl HA (1985) AC field effects of and by living cells. In: Chiabrera A, Niccolini C, Schwan HP (eds) Interactions between electromagnetic fields and cells. NATO ASI Series A, Life Sciences. Plenum, New York, p 435–457

Pohl HA, Lamprecht I (1985) Wechselfelder umgeben wachsende Zellen. Die Umschau 6:366–367

Pohl HA, Braden T, Robinson S, Piclardi J, Pohl DG (1981) Life cycle alterations of the micro-dielectrophoretic effect of cells. J Biol Phys 9:133–154

Rivera H, Pollock JK, Pohl HA (1985) The ac field patterns about living cells. Cell Biophys 7: 43–55

Rowlands S, Sewchand LS (1982) A quantum mechanical interaction of human erythrocytes. Can J Physiol Pharmacol 60:52–59

Rowlands S, Sewchand LS, Lovlin RE, Beck JS, Ennis EG (1981) A Fröhlich interaction of human erythrocytes. Phys Lett 82A:436

Rowlands S, Sewchand LS, Ennis EG (1982a) Further evidence for a Fröhlich interaction of erythrocytes. Phys Lett 87A:256

Roy SC, Braden T, Pohl HA (1981) Possibility of pseudoferoelectric state in cells: some experimental evidence. Phys Lett 83A:142–144

Sewchand LS, Roberts D, Rowlands S (1982) Transmission of the quantum interaction of erythrocytes. Cell Biophys 4:253–259

Westby GWM, Kirschbaum F (1978) Emergence and development of the electric organ discharge in the mormyrid fish, *Pollimyrus isidori* II: Replacement of the larval by the adult discharge. J Comp Physiol 127(1):45–60

Westby GWM, Kirschbaum F (1982) Sex differences in the waveform of the pulse type electric fish *Pollimyrus isidori* (Mormyridae). J Comp Physiol 145(3):399–404

Physiological Signalling Across Cell Membranes and Cooperative Influences of Extremely Low Frequency Electromagnetic Fields

W. R. ADEY[1]

1 Introduction

From the first applications of the light microscope in biology more than 300 years ago, there has been a progressive development of the concept of a limiting membrane that defines the physical boundaries of a living organism, whether this be a bacterium, or a single-celled protozoan leading an independent existence, or a cell in the tissues of higher organisms. From initial concepts of this membrane as merely an enclosing device, emphasis has shifted to its role as a window through which the cell as a unitary biological element can sense its chemical and electrical environment.

Knowledge of the structural and functional organization of cell membranes has evolved by the use of increasingly sophisticated research tools. At the limits of resolution of the light microscope, cell membranes appear only as an envelope of fat molecules (the plasma membrane). With the electron microscope, the plasma membrane is redefined as a lipid bilayer about 40 Å in width, with polar head groups on external and internal surfaces of the double layer.

More importantly, these studies of membrane ultrastructure have revealed numerous strands of protein (intramembranous particles, IMPs) inserted into the plasma membrane. These IMPs span the membrane from outside to inside. They have external protrusions into the fluid surrounding the cell with terminal glycoprotein strands that sense electric fields and form receptor sites for chemical stimuli, including hormones, antibodies and neurotransmitters. Inside the cell, they make functional contact with key elements in the cell machinery, including enzymes and the numerous fine tubes and filaments of the cytoskeleton. They "float" in the sea of lipid molecules of the plasma membrane, leading to the generally accepted *fluid mosaic* model of the cell membrane (Singer and Nicolson 1972). Thus, these intramembranous protein strands form signalling pathways by which external stimuli are sensed and conveyed to the cell interior.

Long-accepted functional models of the plasma membrane have emphasized the existence in most cells of a major electrical gradient (the membrane potential) across the 40 Å width of the lipid bilayer. In these models, this gradient of 10^5 V/cm has been considered an effective electrical barrier against cell stimulation by weak electromagnetic (EM) fields in surrounding fluid. However, much recent research has shown that imposed weak low frequency EM fields many orders of magnitude weaker than

1 Pettis Memorial Veterans Hospital and Loma Linda University School of Medicine, Loma Linda, California 92357, USA

the membrane potential gradient in the pericellular fluid can modulate actions of hormone, antibody and neurotransmitter molecules at their cell surface receptor sites. These modulating actions of EM fields suggest highly cooperative processes in the underlying physical mechanisms. The observed sensitivities are as low as 10^{-7} V/cm in the ELF spectrum. Moreover, many of these interactions are "windowed" with respect to field frequency and amplitude, and to duration of field exposure. Use of imposed EM fields in these studies has allowed identification of some of the sequences and energetics of steps that couple signals from the outside to the inside of the cell, pointing the way to further studies designed to elucidate their cooperative nature. The windowed character of many of these responses points to their nonlinear and non-equilibrium character, and focuses current research on physical substrates for these interactions (Adey 1975, 1977, 1981a,b, 1983, 1984, 1986; Lawrence and Adey 1982; Maddox 1986). Thus, a carefully designed spectrum of ELF fields offers unique tools in evaluating the physical ordering of cell membrane signal coupling.

This paper first reviews the use of EM fields width ELF components to reveal co-operative aspects of transductive coupling of surface molecular stimuli to the cell interior. It also addresses physical and physiological models developed to describe these processes.

2 Experimental Approaches to the Sequence and Energetics of Cell Membrane Transductive Coupling

We shall note that cell membranes function as powerful amplifiers of their first weak interactions with both EM fields and humoral stimuli; and that, as revealed by field effects, these interactions are consistent with quantum processes involving long-range interactions between electric charges on cell surface macromolecules. Moreover, since our studies have shown similar sensitivities in a wide range of tissues and cell types, we conclude that these electrochemical sensitivities may be a general biological property of all cells.

Based on the Singer-Nicolson fluid mosaic cell membrane model discussed above, there is a minimum sequence of three steps in transductive coupling (Adey 1984), and each is calcium-dependent: (a) cell surface glycoproteins that are stranded protrusions from intramembranous helical proteins (IMPs) sense the first weak electrochemical events associated with binding of neurohumoral molecules, hormones and antibodies at their receptor sites; (b) transmembrane portions of IMPs signal these events to the cell interior; (c) internally, there is coupling of the signal to the cytoskeleton (and thus to the nucleus and to other organelles).

The numerous fixed charges on the terminals of surface glycoprotein strands cause the cell surface to behave as a polyanionic sheet (Elul 1966, 1967). It moves visibly in response to focal current pulses that simulate synaptic action, so that the cell surface may be likened to a field of waving corn, responding to an infinite variety of faint electrochemical breezes that blow along the membrane surface.

The polyanionic surface sheet of stranded glycoproteins attracts a cationic counterion layer, principally of calcium and hydrogen ions, which compete for these anionic binding sites. This ionic bilayer is now recognized as a major element in a membrane surface compartment (Blank 1986). It contributes a negative surface potential of many millivolts that can modulate the action of drugs, such as acetyl choline and curare (Van der Kloot and Cohen 1979). Bass and Moore (1968) predicted a "local alkalosis" in the course of excitation, due to the depletion of the hydrogen ion atmosphere as these ions displace calcium from binding sites on surface proteins. Amongst these proteins on the surface of nerve cells are calcium-dependent protein kinase enzymes (ecto-kinases) that are selectively activated during cell membrane depolarization. They phosphorylate other surface proteins, including those necessary for cell adhesion in neuronal development and maturation (Ehrlich et al. 1986).

We shall examine the role of calcium ions in the first steps of transductive coupling at the cell membrane surface and in the ensuing steps of calcium-dependent transmembrane signalling to calcium-dependent intracellular enzyme systems. There are inward and outward signal streams at cell membranes and recent studies have addressed the pathophysiology of transmembrane signalling in cancer promotion (Adey 1986; Byus et al. 1987a,b).

The inward signal stream is the result of a complex sequence of events. Binding of humoral stimulating molecules at their specific binding sites elicits a ripple effect extending along the membrane surface. It is manifested in altered calcium binding to the surface glycoproteins (glycocalyx) (Bawin and Adey 1976; Bawin et al. 1975; Bawin et al. 1978a; Lin-Liu and Adey 1982), and in a concurrent ripple in membrane-related proteins that serve both receptor and enzyme functions (Nihsizuka 1983, 1984). Following this first sequence of events at cell surface receptor sites initiated by weak electrical and chemical stimuli, signals are transmitted to the cell interior along coupling proteins (Luben et al. 1982; Luben and Cain 1984). This may involve nonlinear vibrational modes in the amide spines of helical proteins that span the membrane from the surface to the cell interior (Adey and Lawrence 1984; Lawrence and Adey 1982).

There is also an outward signal stream at cell membranes. It mediates organization of the cell surface in allogenic cytotoxicity in lymphocytes targeted against tumor cells (Lyle et al. 1983, 1987), in secretion of neurotransmitters (Kaczmarek and Adey 1973, 1974), hormones (Jolley et al. 1983), connective tissue elements (Luben et al. 1982), and bone (Fitzsimmons et al. 1986). All these processes are calcium-dependent and all are sensitive to one or more types of imposed EM field, including the spatial arrangement of the surface receptor mosaic (Lin-Liu et al. 1984).

From these studies, there is the prospect of identifying major physical and chemical elements in the pattern of inward signals, and of distinguishing between normal and abnormal signal streams. The cell membrane has been identified as a prime site of many EM field interactions (Adey 1983). We will examine the role of calcium ions in the first responses to weak EM fields and to chemical stimuli at cell surfaces. We will then consider the coupling of transmembrane signals to intracellular enzyme systems.

2.1 Cooperative Modification of Calcium Binding with Amplification of Initial Signals

If this model of transmembrane signalling is correct, it should meet certain experimental criteria, particularly with respect to the strong emphasis that it places on the role of calcium in the initial events of stimulus recognition at the cell membrane surface. These criteria are posed in the following questions. Is the binding of calcium in cerebral tissue influenced by pericellular fields at the same tissue electric gradients as the EEG (20–100 mV/cm)? Does altered calcium binding also occur with far weaker fields shown to influence circadian rhythms in man and birds (Wever 1975), time estimation in monkeys (Gavalas-Medici and Day-Magdaleno 1976), and navigation and predation in sharks and rays (Kalmijn 1971) at typical tissue gradients of 10^{-7} to 10^{-8} V/cm? What is the evidence that sensitivity of calcium binding to imposed fields is consistent with nonequilibrium processes, involving long-range interactions between charge sites on surface macromolecules? Would this nonequilibrium behavior manifest itself in some form of frequency sensitivity to the "biological spectrum" below 100 Hz? Would there also be "windows" of sensitivity with respect to the intensity of the imposed field?

Initial stimuli associated with weak pericellular EM oscillations and with binding of humoral molecules at their receptor sites elicit a highly cooperative modification of calcium binding to glycoproteins along the membrane surface. A longitudinal spread is consistent with the direction of extracellular current flow associated with physiological activity and with imposed EM fields. It would also be consistent with spreading calcium-dependent enzymatic activation in cell membranes proposed by Nishizuka (1983) and recently observed on the surface of cultured neurons (Ehrlich et al. 1986).

This is an "amplifying" stage, with evidence from concurrent manipulations of these initial events by imposed fields that there is a far greater increase or decrease in calcium efflux than is accounted for in the energy of the imposed field or in the events of the receptor-ligand binding (Bawin and Adey 1976; Bawin et al. 1975; Lin-Liu and Adey 1982). There is further striking evidence for the nonlinearity of this modification in calcium binding in its occurrence in quite narrow frequency and amplitude "windows" (Adey 1981a,b; Bawin et al. 1978b; Blackman et al. 1979, 1985a,b).

2.1.1 Effects of ELF and ELF-Modulated Radiofrequency Fields on Tissue Calcium Binding

Field effects on tissue calcium binding have been studied with RF fields amplitude-modulated at ELF frequencies, with ELF electric fields, with ELF electromagnetic fields, and with combined ELF electromagnetic and static magnetic fields. Most of these studies have used cerebral tissue. Calcium levels are typically high in fluid around cells (2.0 mM) and low in the cell interior (10^{-7} M). This high calcium gradient across the cell membrane is maintained by high affinity transport systems. Since calcium levels in the cytoplasm are so low, entry of micromolar amounts of calcium into the cell are powerful stimuli to intracellular systems, including activation of major enzyme systems.

a) Effects of RF Fields with ELF Amplitude-Modulation on Tissue Calcium Binding. For ELF-modulated RF fields that induce tissue gradients of approximately the same amplitude as the electroencephalogram (EEG) in the fluid around brain cells (10 to 100 mV/cm), studies in different laboratories are consistent in reporting an increased calcium efflux from neural tissue that depends on the ELF modulation frequency and field intensity. The pioneering observations of Bawin and her colleagues (Bawin

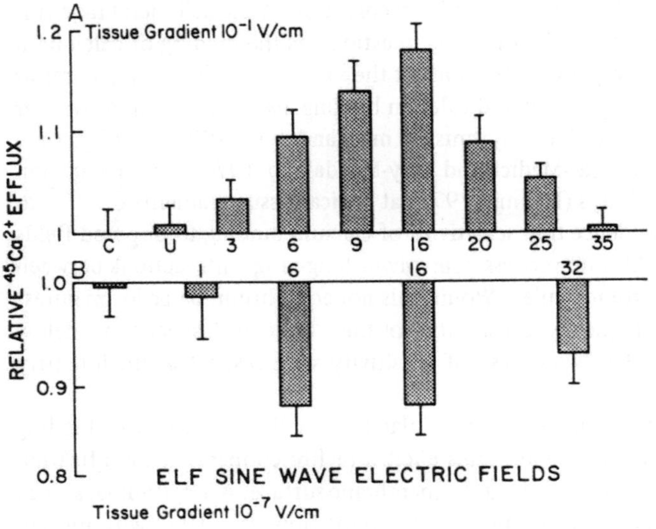

Fig. 1. *Upper curve* $^{45}Ca^{2+}$ efflux from isolated chick cerebral hemispheres exposed to a weak RF feld (147 MHz, 0.8 mW/cm²), amplitude-modulated at low frequencies (*abscissa*). (From Bawin et al. 1975). *Lower curve* effects of far weaker electric fields (56 V/m in air) in the same frequency spectrum from 1–32 Hz (Bawin and Adey 1976). Tissue gradients differ by about 10^6 between upper and lower curves

et al. 1975; Bawin et al. 1978b) were the first to show a "tuning curve" of increased calcium efflux as a function of modulation frequency, with a maximum effect around 16 Hz and less at higher and lower frequencies (Fig. 1). Field carrier waves at 147 MHz and 450 MHz produced similar results and unmodulated carriers had no effect. These studies were performed in awake, intact cat cerebral cortex (Adey et al. 1982), in isolated chick cerebral cortex (Bawin et al. 1975, 1978b; Blackman et al. 1979), and in cerebral synaptosome fractions (Lin-Liu and Adey 1982). In cultured nerve cells in a 915 MHz field, increased calcium efflux occurred at specific energy absorption rates (SARs) of 0.05 and 1.0 mW/g, but not at higher, lower or intermediate levels (Dutta et al. 1984). The response at 0.05 mW/g was dependent on 16 Hz modulation, but at the higher level was not. Bawin's finding of a low frequency modulation window were confirmed by Blackman et al. (1979), who also noted an intensity window, confirmed by Bawin et al. (1978b) for incident fields in the range 0.1–1.0 mW/cm² that would produce EEG-level gradients in cerebral tissue.

It is important to note that these sensitivities were observed over an enormous range of physical dimensions, from intact cerebral cortex to cultured neurons and finally in isolated terminals of cerebral nerve fibers (synaptosomes) with mean diameters around 0.7 μm. Clearly, neither size nor geometry are primary determinants of these interactions. Also, these fields induce gradients at physiological levels. For example, they are similar to those of the EEG in fluid around brain cells. They are 4 to 6 orders of magnitude less than the electric barrier of the membrane potential, emphasizing the role of an amplifying function in their ultimate effects on intracellular mechanisms discussed below.

b) Modulation of Tissue Calcium Binding by ELF Fields. For ELF fields that induce far weaker gradients around 10^{-7} V/cm, findings have been less consistent. ELF fields at frequencies below 100 Hz are coupled capacitively to test objects of the dimensions of animal bodies or tissue cultures, rather than by Maxwellian radiation mechanisms characteristic of interactions with RF fields, where antenna effects greatly increase induced field levels. Thus, a 60 Hz field with an electric gradient of 100 V/m couples to the human body to produce tissue gradients in the range 10^{-6}–10^{-7} V/cm. On the other hand, an RF field at 100 MHz with an incident energy of 1.0 mW/cm^2 (61 V/m) has an expected tissue level of the order of 10^{-1} V/cm (similar to the EEG in fluid surrounding brain cells).

Bawin and Adey (1976) reported reduced calcium efflux from isolated cerebral cortex that followed a reciprocal frequency tuning curve to that seen in response to stronger tissue fields induced by modulated RF fields at 1.0 mW/cm^2. Blackman et al. (1985a) failed to detect responses at these low levels. Fitzsimmons et al. (1986) found an 83% increase in embryonic bone matrix after exposure to a sinusoidal 10-Hz electric field at 10^{-7} V/cm. However, sensitivities to ELF fields at these intensities may involve combined effects of static magnetic fields and oscillating EM fields, a combination not considered in the design of earlier studies.

c) Effects of Combined Static Magnetic and Oscillating ELF EM Fields. Semm (1983) has reported a sensitivity of the pineal gland to the orientation of the head with respect to the earth's magnetic field. In pigeons, guinea pigs, and rats, about 20% of pineal cells respond to changes in both direction and intensity of the earth's magnetic field. The peptide hormone melatonin secreted by the pineal powerfully influences the body's circadian rhythms. During the night, experimental inversion of the horizontal component of the earth's magnetic field significantly decreased synthesis and secretion of melatonin and activity of its synthesizing enzymes (Welker et al. 1983). This magnetic sensitivity has been traced to nerve fiber pathways from the retina and to the optic tectum of the midbrain (Semm and Demaine 1986). Units in the basal optic root stimulated while the eyes were illuminated by light of different wavelengths exhibited peaks of magnetic responsiveness at 503 and 582 nm. Cells selectively sensitive to magnetic field direction were found in the midbrain optic tectum. They also showed directional sensitivity to pulsed photic stimuli. Response peaks varied with orientation of the pigeon's head in the horizontal plane. The results suggest that a substrate for detection and elaboration of environmental magnetic field information exists in the pigeon's visual system.

Blackman et al. (1985b) have shown interactions between the earth's magnetic field and a weak imposed low frequency EM field (40 V peak-to-peak in air, estimated tissue components 10^{-7} V/cm) in determining calcium efflux from chick cerebral tissue. For example, halving the local geomagnetic field with a Helmholtz coil rendered a previously effective 15-Hz field ineffective; and doubling the geomagnetic field caused an ineffective 30-Hz signal to become effective.

2.1.2 *Signal Coupling Along Transmembrane Proteins Linking Cell External Surface and Interior*

a) Inward and Outward Signal Streams. Intramembranous protein particles (IMPs) that span the plasma membrane form a variety of pathways for signalling and energy transfer across cell membranes. These signal streams are directed in both inward and outward directions. By imposing 72 Hz and other low frequency pulsed magnetic fields used in bone fracture therapy (Bassett et al. 1982), Luben (Luben et al. 1982; Luben and Cain 1984) has identified a role for intramembranous proteins in conveying signals from hormone receptor sites on the membrane surface to the cell interior. These studies examined field effects on stimulation of adenylate cyclase within the cell by parathyroid hormone (PTH) and actions of the same fields on collagen secretion.

In bone cells, PTH binds to specific receptor sites; adenylate cyclase is located on the internal surface of the membrane. The receptor site outside and the catalytic subunit inside are coupled by the N protein. Stimulation of adenylate cyclase by PTH was inhibited by about 90% by induced pericellular electric gradients of only 1-3 mV/cm (Fig. 2), eight orders of magnitude less than the gradient of the membrane potential. However, the adenylate cyclase was not inactivated, remaining fully responsive to NaF, with and without field exposure. Nor did the field interfere with binding of PTH to receptor sites, since I-125-labelled PTH showed the same binding levels in control and field-exposed cultures. By exclusion, the evidence points to events involving a protein that couples between the receptor and the adenylate cyclase as the probable site of an important field action.

Further evidence for cell membranes as prime sites of interactions with low frequency EM fields comes from field effects on collagen synthesis in bone cells stimulated with either PTH or 1,25-dihydroxy-vitamin D_3. Both substances inhibit collagen synthesis. Unlike PTH with a receptor site on the membrane surface, vitamin D_3 has a primary site of action within the cell, possibly at the nucleus; and whereas field exposure blocked inhibition of collagen synthesis by PTH, it had no action on the inhibition by vitamin D_3 (Fig. 3). Thus, the evidence from both adenylate cyclase and collagen synthesis studies points to the cell membrane as a prime site of field action.

b) Human Epidermal Growth Factor (EGF) and Nerve Growth Factor (NGF) Receptor Proteins as Models of Membrane Coupling Proteins. Much attention now focuses on membrane receptor proteins for the human epidermal growth factor (EGF) and the nerve cell growth factor (NGF) as models of coupling proteins in studies of the nature of transmembrane signals.

The entire 1210 amino acid sequence of the EGF receptor protein has been deduced by Ullrich et al. (1985), with striking findings on the sequences that make up the extracellular, intramembranous, and cytoplasmic portions of the chain. The molecule appears to cross the membrane only once. The salient and surprising finding is the extremely short length of the intramembranous segment of 23 amino acids, predominantly hydrophobic, and with only one single amino acid with a side chain capable of hydrogen bonding.

Subsequent studies have shown that the NGF receptor protein also has a strikingly similar segment of 23 hydrophobic amino acids within the membrane (Radeke et al.

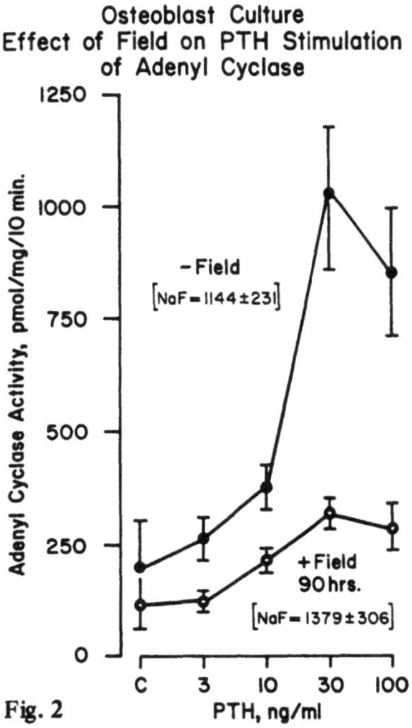

Fig. 2

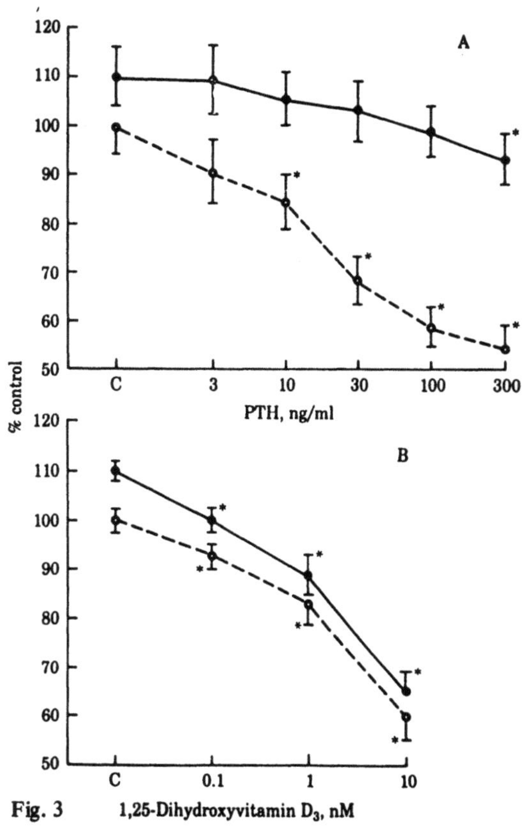

Fig. 3

Fig. 2. Adenylate cyclase activity in bone cells stimulated with parathyroid hormone (PTH), with and without an imposed 72 Hz pulsed magnetic field. (Luben et al. 1982)

Fig. 3. Effects of fields on collagen synthesis by bone cells. MMB-1 cells were grown in the presence of (*solid circles*) or absence (*open circles*) of a pulsed 72 Hz magnetic field for 12 h. Then various concentration of PTH (**A**) or 1,25-dihydroxyvitamin D3 (**B**) were added in fresh culture medium. Culture was continued for 48 h, with the final 4 h being in the presence of ^{3}H-proline at 1 μC/ml. Cells were disrupted and collagen synthesis measured by a collagenase technique. *C* control; * significantly different from control, $P < 0.05$. (Luben et al. 1982)

1987), suggesting that this configuration plays a fundamental role in processes of transmembrane signalling. This view is strengthened by studies with a chimeric protein constructed of the extracellular portion of the insulin receptor protein joined to the transmembrane and intracellular domains of the EGF receptor protein (Riedel et al. 1986). In this molecule, the EGF receptor kinase domain of the chimeric protein is activated by insulin binding. The authors conclude that insulin and EGF receptors employ closely related or identical mechanisms for signal transduction across the plasma membrane.

What can be inferred from these studies about the essential nature of the transmembrane signal? Ullrich et al. (1985) point out that this 23 amino acid sequence is probably too short to be involved in conformation changes; and that its hydrophobic character makes its participation in either ion or proton translocation unlikely. As an alternative, they suggest that an EGF-induced conformational change in the extracellular domain may be transmitted to the cytoplasmic domain by movement of this short intramembranous segment, or by receptor aggregation. As a further option, we have hypothesized that this transmembrane signalling may involve nonlinear vibration modes in helical proteins and generation of Davydov-Scott soliton waves (Lawrence and Adey 1982). Although evidence for soliton waves in DNA and helical proteins is not conclusive, we have strong supporting evidence for nonlinear, nonequilibrium processes at critical steps in transmembrane signal coupling, based on "windows" in EM field frequency, amplitude, and time of exposure that determine stimulus effectiveness.

Despite the hydrophobic character of this short transmembrane coupling segment, addition of EGF to human epidermal cell cultures causes a two-to-four-fold increase in cytoplasmic free calcium within 30–60 s (Moolenaar et al. 1986). This induced rise is completely dependent on extracellular calcium and is not accompanied by a change in membrane potential. This EGF-induced signal appears to result from calcium entry via a voltage-independent channel. This action is inhibited by cancer-promoting phorbol esters which have a specific membrane receptor (calcium-dependent protein kinase C). EM field interactions with these cancer promoters are discussed below.

2.1.3 Intracellular Enzymes as Molecular Markers of Transductive Coupling Through Cell Membranes

We have found three groups of intracellular enzymes that respond to signals initiated at cell membranes as a response to EM field exposure. These responses occur with or without concurrent interactions with humoral stimuli. These enzymes are: (1) membrane-bound adenylate cyclase involved in activation of protein kinases through conversion of ATP to cAMP, as discussed above the bone cells (Luben et al. 1982; Luben and Cain 1984); (2) cAMP-independent protein kinases that perform messenger functions; (3) ornithine decarboxylase (ODC), essential for growth in all cells by its participation in synthesis of polyamines essential for DNA formation.

a) Lymphocyte Protein Kinase Responses to RF Fields Amplitude-Modulated at Low Frequencies; Windows in Frequency and Time. Some intracellular protein kinases are activated by signals arising in cell membranes that do not involve the cAMP pathway. This group includes membrane protein kinases related to actions of cancer-promoting phorbol esters (see below). In human tonsil lymphocytes exposed to a low level RF field (450 MHz, 1.0 mW/cm^2), cAMP-independent protein kinases showed activity windowed with respect to exposure duration and modulation frequency (Byus et al. 1984). Reduced enzyme activity only occurred at modulation frequencies between 16 and 60 Hz, and only for the first 15–30 min of EM field exposure (Fig. 4). Unmodulated fields elicited no responses.

b) Lymphocyte Cytotoxicity Toward Tumor Cells and Windowed Effects of Modulation Frequency of RF Fields. Allogeneic T lymphocytes can be targeted against tumor cells, destroying the tumor cells by cell membrane rupture upon contact between the lymphocyte and the tumor cell (cytolysis). The same 450 MHz fields used in the lymphocyte protein kinase studies above reduced the cytolytic capacity of allogeneic

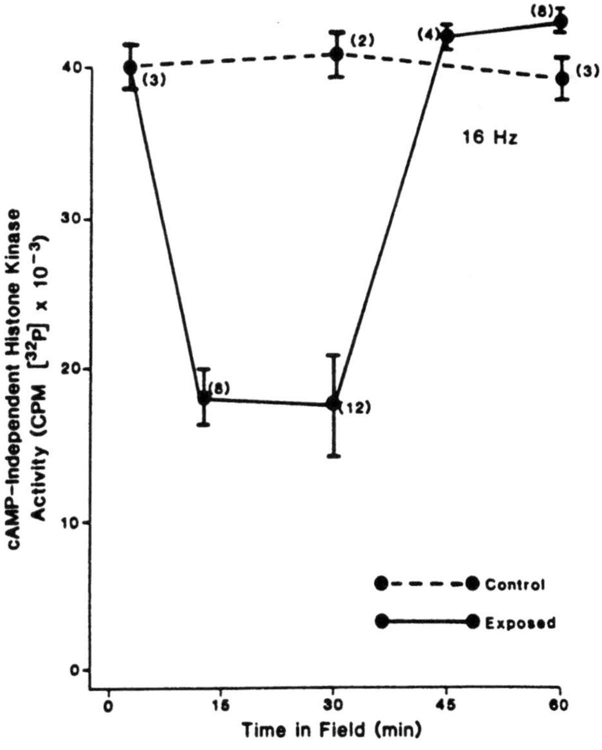

Fig. 4. cAMP-independent protein kinase activity in human lymphocytes is sharply reduced by a 450 MHz field (1.0 mW/cm², amplitude-modulated at 16 Hz) 15 to 30 min after onset of exposure, but returns to control levels at 45–60 min, despite continuing exposure – a window in time. (Byus et al. 1984)

T lymphocytes targeted against cultured lymphoma cells by 20% when modulated at 60 Hz (Fig. 5). Interactions were less at higher and lower frequencies (Lyle et al. 1983). Unmodulated fields were without effect. Similar effects occurred with 60 Hz EM fields in the range 0.1–10 mV/cm, with clear evidence of a threshold and field intensity effects (Lyle et al. 1987).

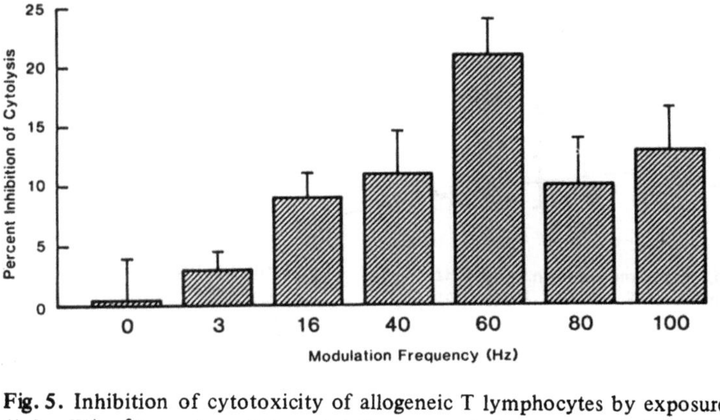

Fig. 5. Inhibition of cytotoxicity of allogeneic T lymphocytes by exposure to a 450 MHz field (1.5 mW/cm²), as sinusoidal modulation was varied between 0 and 100 Hz, showing a window of sensitivity to modulation frequency. (Lyle et al. 1983)

c) Ornithine Decarboxylase (ODC) Activity in Cultured Liver and Ovary Cells Exposed to EM Fields. ODC activity in synthesis of polyamines is an essential step in DNA synthesis in cultured Chinese hamster V79 cells is either enhanced or repressed by low frequency pulsed EM fields, depending on field parameters (Takahashi et al. 1986). Synthesis was enhanced in weak fields with specific combinations of pulse width (25 μs), frequency (10, 100 Hz) and magnetic intensity (2×10^{-5}, 8×10^{-5} T). Conversely, DNA synthesis by cells in stronger fields of 4×10^{-4} T was repressed to 80% of that in controls. RNA transcription was enhanced in Sciara salivary gland cells by sinusoidal EM fields as an inverse function of frequency in the range 72–4400 Hz (Goodman and Henderson 1986).

Clinically, ODC activity in cultures of suspected cancer cells (for example, human prostatic cancer), has proved a useful index malignancy. All agents that stimulate ODC are not cancer-promoting, but all cancer promoters stimulate ODC. Although its activation pathways are not well defined, binding of cancer-promoting phorbol esters at membrane receptors induces ODC. ODC activity in cultured liver and ovary cells was increased by 50% in a 3-h test period following a 1-h exposure to the same 450 MHz fields tested above, using sinusoidal amplitude-modulation at 16 Hz (Byus et al. 1987a) (Fig. 6). Similar sensitivities were noted with 60 Hz EM fields in the range 0.1 to 10.0 mV/cm in the culture medium (Byus et al. 1987b).

2.1.4 Enzymatic Markers of Joint Actions of EM Fields and Chemical Chancer Promoters at Cell Membranes: Multistep Models of Carcinogenesis

Based on animal tumor models, studies of multistep carcinogenesis have defined a series of chemicals or agents that are not mutagenic and thus are not cancer initiators by an action on DNA in the nucleus. However, they specifically promote growth of previously initiated cells or tumors, suggesting that cancer promotion may arise in events that are spatially and temporally separated from initiating events that damage nuclear DNA (Adey 1986, 1987). Nonionizing EM fields do not appear to function

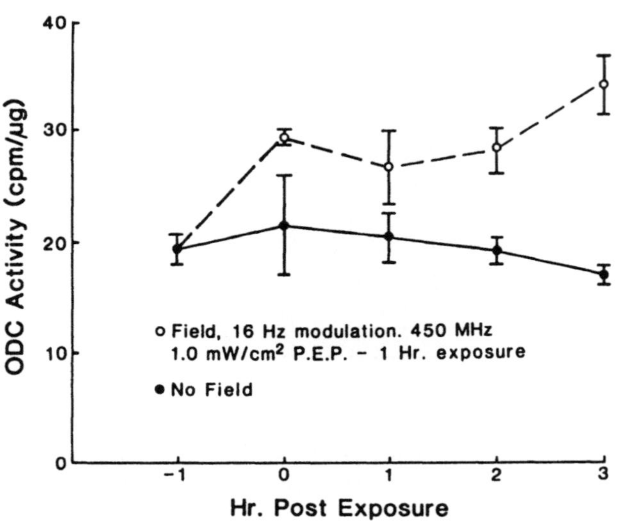

Fig. 6. Ornithine decarboxylase (ODC) activity in Reuber H35 hepatoma cells before and after exposure to a 450 MHz field (1.0 mW/cm² peak envelope power) amplitude-modulated at 16 Hz. (Byus et al. 1987a)

as classical initiators in the etiology of cancer by causing DNA damage and gene mutation. On the other hand, there are strong interactions at cell membranes between cancer-promoting substances, the phorbol esters, and EM fields. In consequence, we hypothesize that cancer promotion may involve a distorted inward signal stream from cell membranes directed to the nucleus and other organelles.

a) Action of Cancer-Promoting Phorbol Esters at Cell Surfaces. Phorbol esters have a specific receptor that occurs widely in cell membranes and functions both as a receptor and an enzyme; phosphatidylserine protein kinase, kinase C. It is calcium-dependent and normally activated by diacylglycerol formed from inositol phospholipids in response to cell surface stimuli. A single molecule of diacylglycerol can activate one molecule of protein kinase C. This appears to initiate a spreading domino effect which may activate all protein kinase C molecules around the whole membrane surface (Nishizuka 1983).

b) Protein Kinase C as a Receptor for Phorbol Esters; Synergic Actions of Phorbol Esters and EM Fields. Protein kinase C is activated irreversibly by phorbol esters and function as a specific receptor for them. Exhaustion of kinase C occurs with continued stimulation by phorbol esters and is associated with loss of cell division. This response is restored by intracellular injection of kinase C (Pasti et al. 1986), thus linking it to paths that signal from cell membranes to nuclear mechanisms mediating cell division. Protein kinase C belongs to a group of cAMP-independent protein kinases which we have identified above as sensitive to EM fields (Byus et al. 1984). We have also shown that induction of ODC follows stimulation of liver and ovary cells with a phorbol ester (Fig. 7), and that this response is sharply enhanced by 450 MHz fields (1.0 mW/cm^2), sinusoidally amplitude-modulated at 16 Hz (Byus et al. 1987a). Phorbol ester treatment of embryonic fibroblasts previously irradiated with X-rays and microwaves (2450 MHz, 34 mW/cm^2, 24 h) increased transformation frequencies above rates in cells irradiated only with X-rays (Balcer-Kubiczek and Harrison 1985). The findings are consistent with cell membranes as the site of persisting effects of prolonged microwave exposure. These effects may enhance promoter actions of the phorbol esters.

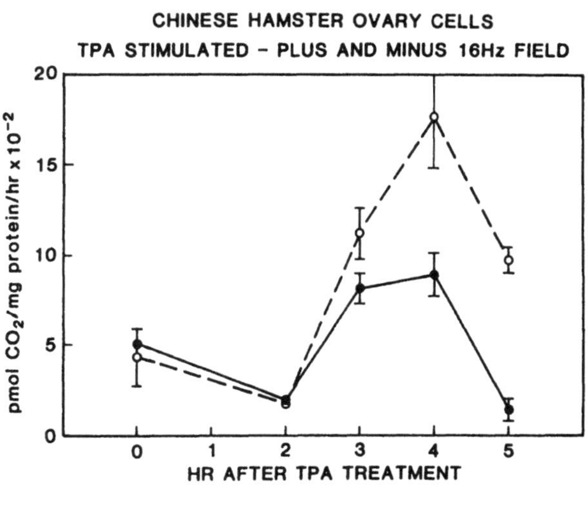

Fig. 7. Ornithine decarboxylase (ODC) activity in Reuber H35 hepatoma cells stimulated by a phorbol ester (tetradecanoyl phorbol acetate, TPA), with and without concurrent exposure to the same 450 MHz field as in Fig. 6. (Byus et al. 1987a)

3 Models of Cooperative Organization in Physiological Systems

It is implicit in the experiments cited above that interactions occurring at athermal levels between biological substrates and the low frequency components of EM fields must take place in biomolecular systems exhibiting dynamic patterns of organization. Turing's (1952) equations carry the prediction that a system of chemical reactions and diffusion may develop a dynamically maintained temporal and/or spatial pattern from an initially steady-state homogeneous distribution of matter. By choosing suitable pattern functions that are mathematically a complete set of eigenfunctions appropriate to the geometric configuration of the system of interest, Gmitro and Scriven (1966) extended Turing's analysis of the origins of patterns in a ring of cells to a whole gamut of line-like and surface-like configurations.

Othmer and Scriven (1971) have applied certain of these concepts to cellular networks, emphasizing how instability and pattern formation are influenced by preexisting pattern. In the context of developmental biology, they chose for a preexisting pattern the regular compartmentalizations of a system into discrete cells, each interchanging material with its nearest neighbor and with the surrounding bath of intercellular fluid. Here, in passing from unicellular to multicellular systems, one encounters a new level of organization and complexity, a level stemming from cell-to-cell interactions of various types. Through chemical contact, cells may interchange one or more of their constituents, thereby altering metabolic states (Fletcher et al. 1987). In this model, chemical transformations taking place within individual cells may be influenced by mass transport between cells, thereby establishing differences in cell composition. There are natural instabilities inherent in this type of dynamic interaction. They may lead to spatial organization and temporal oscillations within groups of cells in the genesis of form and rhythm. Othmer and Scriven focused on these possibilities, taking chemical interactions as their model, although there are other modes of intercellular communication, including electromagnetic interactions discussed above.

3.1 Dissipative Processes

There is much evidence that the molecular organization in biological systems needed to perform this sensing of weak stimuli, whether thermal, chemical or electrical, may reside in joint functions of molecular assemblies or subsets of these assemblies (Katchalsky et al. 1974). Katchalsky (1974) has reviewed historic developments from the time of Heraclitus (540–480 B.C.) of concepts of dynamic patterns that develop in a population of elements as a result of their complex flow patterns. These flow patterns can undergo sudden transitions to new self-maintaining arrangements that will be relatively stable over time.

Because these dynamic patterns are initiated and sustained by continuous inputs of energy, they are classed as "dissipative" processes. For this reason, they occur far from equilibrium with respect to at least one important parameter in the system (Katchalsky and Curran 1965). As nonequilibrium processes, they may be characterized by resonant or windowed phenomena, an important aspect of their occurrence in tissue interactions with weak EM fields, as discussed above. Also, two or more quite distinct mechanisms can give rise to the same dynamic pattern. A given pattern therefore need not relate to a unique mechanism; conversely, different mechanisms may generate a common pattern (Othmer and Scriven 1969, 1971, 1974). To biologists accustomed from their earliest training to consider cellular excitatory phenomena in terms of equilibrium processes, these concepts, although quite old in certain key areas of physics and chemistry (Rayleigh 1916), offer new insights on possible substrates for initial events in signal transductive coupling at cell membranes.

3.1.1 Concepts of Cooperative Processes in Biomolecular Systems

A strong theme characterizing many of these functional linkages between participating elements of a dynamic pattern is *cooperativity*, defined as ways in which components of a macromolecule, or a system of macromolecules, act together to switch from one stable state of a molecule to another. These joint actions frequently involve phase transitions, hysteresis, and avalanche effects in input-output relationships (Schmitt et al. 1975; Wyman 1948; Wyman and Allen 1951). Trigger signals to cooperative processes may be weak and the amplified response orders of magnitude larger, as in the sharply nonlinear release of $^{45}Ca^{2+}$ from binding sites in cerebral tissue by added Ca ions (Kaczmarek and Adey 1973) and by weak EM fields as discussed above (Bawin and Adey 1976; Bawin et al. 1975); and in a series of Ca-dependent processes at cell membranes that include the large generation of cAMP by glucagon binding to membrane receptors (Rodbell et al. 1974); the generation of cAMP by binding of parathyroid hormone to its membrane receptors and the modulation of this process by weak ELF fields (Luben et al. 1982); the amplification of immune responses in patching and capping at cell membranes (Edelman et al. 1973) and the modulation of cell-mediated cytotoxicity of lymphocytes by ELF components of weak EM fields (Lyle et al. 1983, 1987); and in swimming behavior of bacteria elicited by a small concentration gradient of an attractant (Aswad and Koshland 1974; Koshland 1975).

3.1.2 Amplification Effects in Cooperative Systems

Amplification effects in cooperative systems raise questions about thresholds and the minimum size of an effective triggering stimulus. In studies of cooperativity in biological systems, attention is usually focused on the effects of a change in an external parameter on the equilibrium constant for a given reaction (Blank 1976; Schwarz 1975). Although the sharp transition from one highly stable state to another such state that characterizes a cooperative process can also be achieved by noncooperative means, much larger transition energies would be required and the transition would occur more slowly. Sharp and fast transitions characteristic of many biological systems thus involve cooperative interactions, such as the individually weak forces in a series of hydrogen bonds or in hydrophobic reactions (Engel and Schwarz 1970; Schwarz 1975; Schwarz and Balthasar 1970; Schwarz et al. 1970).

Compared with tissue electric gradients induced by environmental EM fields, the requisite gradients for some known molecular transitions are very large. The helix-coil conformational change in poly(gamma-benzyl L-glutamate) can be induced by a gradient of 260 kV/cm (Schwarz and Seelig 1968). Long-lasting conformational changes occur in poly(A) · 2 poly(U) and in ribosomal RNA with pulsed electric fields of 20 kV/cm and with a decay time of 10 μs (Neumann and Katchalsky 1972). However, these sensitivities for nucleic acid chains in pure solutions contrast sharply with effects of ELF pulsed EM fields on DNA synthesis in cultured mammalian cells, where significant effects occur at field intensities in the range of 10^{-8}–10^{-4} T (Takahashi et al. 1986).

Therefore, observed EM field interactions of cells and tissues based on oscillating ELF tissue gradients between 10^{-7} and 10^{-1} V/cm noted above clearly would involve degrees of cooperativity many orders of magnitude greater than envisaged in the examples just cited. In part, this discrepancy appears to relate to far greater sensitivities of cellular systems to low frequency oscillating EM fields than to imposed step functions or DC gradients used in many electrochemical experiments and models to test levels of cooperativity in biological systems (Blank 1972).

A most important factor in determining this threshold for a low-level coherent oscillation to elicit a cooperative response is the thermal Boltzmann (kT) noise in the system. This is 0.02 eV at room temperature and is the basis of molecular collisional interactions. If modelled on this thermal threshold, the sensing of a gradient of 10^{-7} V/cm would require a cooperative molecular system extending over 300 m. The abundant evidence that extracellular gradients from 10^{-1} V/cm down to this level are biologically significant in systems of cellular dimensions is a salutary reminder of the importance of better understanding molecular and morphological substrates of this transductive coupling. Factors requiring consideration include temporal entrainment of activity in large systems of random generators by coherent oscillations far weaker than this random activity (Nicolis et al. 1973, 1974). A hypothesis as yet untested suggests that cell surfaces may act as extremely narrow-bandwidth low-pass filters in the transfer of thermal noise along the surface of micron-sized spheres and tubes (Bawin and Adey 1976), thus enhancing the signal-to-noise ratio of ELF oscillations in the pass band of the filtering system.

3.2 Chaotic Models for Nonlinear Oscillating Modes in Biological Systems

Self-sustained oscillations in biological systems may be modelled on the requirement for interaction of regular external perturbations with internal oscillations, resulting in synchronization of the system to the external drives (entrainment) (Kaiser 1984). A sharp frequency response results, exhibiting both frequency and intensity windows and rather irregular behaviour near the entrainment region.

A further increase in energy of external driving fields, both static and periodic, leads to sequences of period-doubling bifurcations, alternating with quasiperiodic and irregular regions (quasiperiodicity, chaos). As a consequence, a regularly driven self-oscillating system may exhibit intrinsic chaotic behaviour, even though the underlying dynamic is strongly deterministic. Finally, still higher levels of energy input destabilize the system (collapse), leading to the onset of propagating pulses (solitary waves or *solitons*). A nonlinear temporal structure is thus replaced by a nonlinear spatiotemporal structure.

Kaiser points out that systems having both periodic and chaotic states can exhibit completely different behaviours under periodic driving, and that therein lie additional possibilities for responses of large neuronal systems to imposed fields, as in the genesis of the electroencephalogram in cerebral cortex. Extreme sensitivity of chaotic systems to external driving may carry implications for brain functions in differences between regular and irregular EEG patterns and associated susceptibility to external fields. Circadian pacemaker functions of the suprachiasmatic nucleus have been modelled as a neural network exhibiting chaotic behaviour for certain functions (Carpenter and Grossberg 1983; Grossberg 1983).

Nicolis (1983) has examined the role of chaos in reliable information processing in simple nervous systems, as in the leech or cockroach, where complex behavioral repertoires exist in the absence of elaborate neuronal substrates. His chaotic model is based on the principle that information is produced not only by dissipating degrees of freedom in a system, but also by increasing resolution in systems with few degrees of freedom. For example, certain nonlinear, dissipative systems with just three degrees of freedom can exhibit random behaviour which is analogous to that produced by explicit stochastic equations. Instead of creating new degrees of freedom with increased bandwidth or dimensionality of state space, these systems generate iterative self-similar processes which decrease resolution or expand the dynamics of trajectories in a low-dimensional state space. Sensitivity to small differences in initial conditions gives rise to a probabilistic character in an otherwise simple deterministic system.

Space-time intermittency is a baffling phenomenon observed in spatially extended systems (Keeler and Farmer 1986). In general, space-time intermittency occurs where two qualitatively different types of behavior are intermittent in both space and time. A common manifestation occurs in fluid flow, where patches of turbulence are sometimes isolated in space. The interface between laminar and turbulent behaviour is dynamic and to the casual observer appears unpredictable; without changing parameters, turbulence can spread through the entire flow, or disappear, so that the entire flow becomes laminar. Modelled in a one-dimensional lattice of coupled quadratic maps, this system naturally forms spatial domains. Motion of the domain walls causes spatially localized changes from chaotic to almost periodic behaviour. The almost periodic phases have eigenvalues close to 1, resulting in long-lived laminar bursts with a 1/f low frequency spectrum.

3.3 Dispersive Models of EM Field Interactions at Cell Membranes with Initiation of Solitary Waves

As discussed above, we have hypothesized a three-stage model in transductive coupling from membrane surface to cell interior. The model proposes an initial interaction over domains of the membrane surface. This is dissipative in nature and highly cooperative in its modulation of Ca binding at sites on terminals of stranded glycoproteins. The ensuing step would be dispersive, and might involve solitary waves *(solitons)* similar to those proposed by Davydov (1979) moving in sequence down the length of glycoprotein and lipoprotein molecules (Adey and Lawrence 1984; Lawrence and Adey 1982).

These solitons may arise in interactions of phonons and excitons along linear molecules that result in nonlinear molecular vibrations. It is proposed that nonlinear interatomic forces (specifically in the hydrogen bond) can lead to robust solitary waves with greatly increased radiative lifetimes (Davydov 1979; Hyman et al. 1981). Davydov concludes that this would correspondingly increase the tendency for molecular vibrations to be the vehicle for energy transfer over long molecular chains, specifically over the amide "spines" in alpha-helix proteins and DNA with the bond sequence ---HNC=O---HNC=O–, etc. Davydov's nonlinear analysis shows that propagation of amide-I vibrations can couple to nonlinear sound waves in the alpha-helix, and the coupled excitation propagates as a localized and dynamically stable wave. The amide-I vibrations are the source of the longitudinal sound waves which stretch hydrogen bonds along the helix. Soliton formation exhibits a sharp threshold which is a function of energy coupling between the hydrogen bond stretching and the amide-I excitation (Scott 1981). With a further increase in coupling energy, there is a second threshold above which solitons will not form (Kaiser 1984). It remains to be determined whether the general properties of these nonlinear mechanisms can account for persisting molecular states in the needed time frame.

3.4 Aspects of Specific Cooperative Models of Biomolecular Systems

Imposed EM fields have proved unique tools in the developing awareness of the profound importance of the pericellular microenvironment. There is strong evidence that cell membranes are powerful amplifiers of weak electrochemical events in their immediate vicinity; and virtually all these sensitivities appear to involve maximum effects with natural or imposed fields at frequencies below 100 Hz, a spectral span that we have named the "biological spectrum".

Low frequency sensitivities have been modelled in terms of Lotka-Volterra (predator-prey) processes involving slow shifts in energy states of coherent populations of fixed charges on cell surface glycoproteins (Fröhlich 1975); in limit-cycle behaviour of calcium ions binding to cell surface macromolecules (Kaczmarek 1976); in chaotic behaviour of pseudorhythmic molecular oscillations (Kaiser 1984), discussed above; and in cyclotron oscillations of calcium ions exhibiting coherent states at the cell membrane surface (Liboff 1985; Polk 1984). Also, experimental data on microwave absorption by red cell membranes (Blinowska et al. 1985) and related models of enzyme-substrate activity based on quantum cooperative effects suggest cell membranes as a primary site for Fröhlich's coherent oscillations and a possible role for local superconductivity (Achimowicz et al. 1977).

3.4.1 Lotka-Volterra Models of Excitations in Cell Membranes

Based on the high sensitivity of biological systems to weak EM fields, Fröhlich has suggested that ELF electric oscillations of the EEG in brain tissue may relate to a storage mechanism through which a biomolecular system can store signal energy and, in so doing, overcome thermal noise (Fröhlich 1977). These electric vibrations cannot arise from a collective mode based on interactions of various molecular groups, since enormous tissue volumes would be involved to overcome thermal noise. Instead, Fröhlich has applied a general theory of coherent vibrations in biological systems (Fröhlich 1968, 1972, 1980, 1986a,b; Bhaumik et al. 1976) within the framework of the fluid mosaic membrane model.

Collective chemical oscillations that can be represented by the Lotka-Volterra (predator-prey) equations may occur between an entity formed of globular proteins and the ions and structured water surrounding them. These globular proteins would oscillate between a strongly electrically excited polar state and a weakly polar ground state. A slow chemical oscillation is thus connected with a corresponding electrical vibration. With strong electric interactions between the highly polar states and with considerable damping due to electric currents, limit-cycle conditions arise that make these oscillations highly sensitive to external electrical and chemical influences. In this model, expected coherent oscillations would occur at frequencies around 10^{11} Hz. Applying these models to the action of enzymes, there are possibilities for collective enzymatic reactions, based on long-range selective interactions between enzymes in excited polar states, others in the ground state, and substrate molecules. These collective coherent oscillations at 10^{11} Hz may initiate chemical oscillations at far lower frequencies in the populations of substrate and excited enzyme molecules, associated in turn with electrical oscillations at frequencies around 10 Hz.

3.4.2 Cyclotron Resonance Models of Calcium Ion Oscillations at Cell Membranes

Polk (1984) noted that free (unhydrated) calcium ions in the earth's geomagnetic field would exhibit cyclotron resonance frequencies around 10 Hz; and that these cyclotron currents would be as much as 5 orders of magnitude greater than Faraday currents if the calcium ions exhibited nearest-neighbor coherence. Experimental studies have shown that there are interactions between the earth's geomagnetic field and a weak imposed low frequency. EM field (40 V/m peak-to-peak in air, estimated tissue components 10^{-7} V/cm) in determining calcium efflux from chick cerebral tissue (Blackman et al. 1985b). For example, halving the local geomagnetic field with a Helmholtz coil rendered a previously effective 15-Hz field ineffective; and doubling the geomagnetic field caused an ineffective 30 Hz signal to become effective.

Liboff (1985) notes that for a mean value for the earth's geomagnetic field of 0.5 Gauss, most of the singly and doubly charged ions of biological interest have gyrofrequencies in the range 10–100 Hz. Liboff hypothesizes that imposed EM fields at frequencies close to a given resonance may couple to the corresponding ionic species in such a way as to selectively transfer energy to these ions. It is proposed that data from Blackman's experiments cited above may relate to cyclotron resonance in singly ionized potassium, with secondary effects on calcium efflux.

3.4.3 Cooperative Models of Enzyme Activation with Emphasis on Local Superconductivity

In Fröhlich's model of enzymatic activation discussed above, coherent pumping energy at certain power levels is required to initiate long range selective forces. Metabolic activity is proposed as a possible energy source. Achimowicz et al. (1977) have proposed an activating mechanism based on quantum cooperative effects that does not require this assumption. Their model proposes that macromolecular interactions destabilize the electron structure of the enzyme-substrate complex. This structure is determined by strong electron (exciton) and phonon-mediated electron-electron interactions. A structural transition occurs, analogous to a Peierls instability in crystal lattices, leading to a softening of the phonon spectrum and a modification of electron levels as electron and phonon systems are conjugated. Renormalization of electron and phonon spectra might lead to high temperature superconductivity.

Phenomenologically, this state would be equivalent to the existence of selective intermolecular forces.

4 Conclusions

One of the great but often unrecognized accomplishments in the quest for order in biological systems is the revelation of an exquisite succession of structural and functional hierarchies that interact within and between each other. It is at the atomic level that physical, rather than chemical events now appear to shape the flow of signals and the transmission of energy in biomolecular systems. These recent observations have opened doors to new concepts of communication between cells as they whisper together across barriers of cell membranes.

Use of weak EM fields to study the sequence and energetics of events that couple humoral stimuli from surface receptor sites to the cell interior has identified cell membranes as a primary site of interaction with these low frequency fields. Field modulation of cell surface chemical events indicates a major amplification of initial weak triggers associated with binding of hormones, antibodies, and neurotransmitters to their specific binding sites.

Calcium ions play a key role in this stimulus amplification, probably through highly cooperative alterations in binding to surface glycoproteins, with spreading waves of altered calcium binding across the membrane surface. Protein particles spanning the cell membrane form pathways for signaling and energy transfer. Fields millions of times weaker than the membrane potential gradient of 10^5 V/cm modulate cell responses to surface stimulating molecules. The evidence supports nonlinear, nonequilibrium processes at critical steps in transmembrane signal coupling.

Acknowledgments. Studies from this laboratory have been generously supported by the US Department of Energy, the US Environmental Protection Agency, the FDA Bureau of Radiological Health, the US Office of Naval Research, the Southern California Edison Company, and the US Veterans Administration.

References

Achimowicz J, Cader A, Pannert L, Wojcik E (1977) Quantum cooperative mechanism of enzymatic activity. Phys Lett 60A:383–384

Adey WR (1975) Evidence for cooperative mechanisms in the susceptibility of cerebral tissue to environmental and intrinsic electric fields. In: Schmitt FO, Schneider DM, Crothers DM (eds) Functional linkage biomolecular systems. Raven, New York, pp 325–342

Adey WR (1977) Models of membranes of cerebral cells as substrates for information storage. Bio Systems 8:163–178

Adey WR (1981a) Tissue interactions with nonionizing electromagnetic fields. Physiol Rev 61: 435–514

Adey WR (1981b) Ionic nonequilibrium phenomena in tissue interactions with electromagnetic fields. In: Illinger KH (ed) Biological effects of nonionizing radiation. Am Chem Soc., Washington, DC., pp 271–297

Adey WR (1983) Molecular aspects of cell membranes as substrates for interactions with electromagnetic fields. In: Basar E, Flohr H, Haken H, Mandell AJ (eds) Synergetics of the brain. Springer, Berlin Heidelberg New York, pp 201–211

Adey WR (1984) Nonlinear, nonequilibrium aspects of electromagnetic field interactions at cell membranes. In: Adey WR, Lawrence AF (eds) Nonlinear electrodynamics in biological systems. Plenum, New York, pp 3–22

Adey WR (1986) The sequence and energetics of cell membrane transductive coupling to intracellular enzyme systems. Bioelectrochem Bioenerg 15:447–456

Adey WR (1987) Cell membranes, the electromagnetic environment and cancer promotion. Neurochem Res (in press)

Adey WR, Lawrence AF (eds) (1984) Nonlinear electrodynamics in biological systems. Plenum, New York

Adey WR, Bawin SM, Lawrence AF (1982) Effects of weak amplitude-modulated microwave fields on calcium efflux from awake cat cerebral cortex. Bioelectromagnetics 3:295–307

Aswad D, Koshland DE (1974) Role of methionine in chemotaxis. J Bacteriol 118:640–645

Balcer-Kubiczek EK, Harrison GH (1985) Evidence for microwave carcinogenesis in vitro. Carcinogenesis 6:859–864

Bass L, Moore WJ (1968) A model of nervous excitation based on the Wien dissociation effect. In: Rich A, Davidson CM (eds) Structural chemistry and molecular biology. Freeman, San Francisco, pp 356–368

Bassett CAL, Mitchell N, Gaston SR (1982) Pulsing electromagnetic fields in ununited fractures and failed arthrodeses. J Am Med Assoc 247:623–627

Bawin SM, Adey WR (1976) Sensitivity of calcium binding in cerebral tissue to weak environmental electric fields oscillating at low frequency. Proc Natl Acad Sci USA 73:1999–2003

Bawin SM, Kaczmarek LK, Adey WR (1975) Effects of modulated VHF fields on the central nervous system. NY Acad Sci 247:74–91

Bawin SM, Adey WR, Sabbot IM (1978a) Ionic factors in release of $^{45}Ca^{2+}$ from chick cerebral tissue by electromagnetic fields. Proc Natl Acad Sci USA 75:6314–6318

Bawin SM, Sheppard AR, Adey WR (1978b) Possible mechanisms of weak electromagnetic field coupling in brain tissue. Bioelectrochem Bioenergetics 5:67–76

Bhaumik D, Bhaumik K, Dutta-Roy B (1976) On the possibility of Bose condensation in the excitation of coherent modes in biological systems. Phys Lett 56A:145–148

Blackman CF, Elder JA, Well CM, Benane SG, Eichinger DC, House DE (1979) Induction of calcium ion efflux from brain tissue by ratio frequency radiation. Radio Sci 14:93–98

Blackman CF, Benane SG, House DE, Joines WT (1985a) Effects of ELF (1–120 Hz) and modulated (50 Hz) RF fields on the efflux of calcium ions from brain tissue in vitro. Bioelectromagnetics 6:1–11

Blackman CF, Benane SG, Rabinowitz JR, House DE, Joines WT (1985b) A role for the magnetic field in the radiation-induced efflux of calcium ions from brain tissue in vitro. Bioelectromagnetics 6:327–338

Blank M (1972) Cooperative effects in membrane reactions. J Colloid Interface Sci 41:97–104

Blank M (1976) Hemoglobin reactions as interfacial phenomena. J Electrochem Soc 123:1653–1656

Blank M (1986) Electrical double layers in membrane transport and nerve excitation. Bioelectrochemical Soc, First Int School, Pleven, Bulgaria, Proceedings, p 26

Blinowska KJ, Lech W, Wittlin A (1985) Cell membrane as a possible site of Fröhlich's coherent oscillations. Phys Lett 109A:124–126

Byus CV, Lundak RL, Fletcher RM, Sadey WR (1984) Alterations in protein kinase activity following exposure of cultured lymphocytes to modulated microwave fields. Bioelectromagnetics 5:34–51

Byus CV, Kartun K, Pieper S, Adey WR (1987a) Microwaves act at cell membranes alone or in synergy with cancer-promoting phorbol esters to enhance ornithine decarboxylase activity. (Submitted Cancer Res)

Byus CV, Pieper S, Adey WR (1987b) The effect of environmentally significant low-energy 60 Hz electromagnetic fields upon the cancer-related enzyme ornithine decarboxylase. Carcigonesis 8(10)

Carpenter GA, Grossberg S (1983) Adaptation and transmitter gating in vertebrate photoreceptors. J Theor Neurobiol 1:1–42

Davydov AS (1979) Solitons in molecular systems. Physica Scripta 20:387–394

Dutta SK, Subramoniam A, Ghosh B, Parshad R (1984) Microwave radiation-induced calcium efflux from brain tissue, in vitro. Bioelectromagnetics 5:71–78

Edelman GM, Yahara I, Wang JL (1973) Receptor mobility and receptor-cytoplasmic interactions in lymphocytes. Proc Natl Acad Sci USA 70:1442–1446

Ehrlich YH, Davis TB, Bock DE, Kornecki E, Lenox RH (1986) Ecto-protein kinase activity on the external surface of neural cells. Nature 320:67–69

Elul R (1966) Applications of non-uniform electric fields. Part I. Electrophoretic evaluation of absorption. Faraday Soc. Trans, pp 3483–3492

Elul R (1967) Fixed charge in the cell membrane. J Physiol 189:351–365

Engel J, Schwarz G (1970) Cooperative conformational transitions of linear biopolymers. Angew Chem Int Ed 9:389–400

Fitzsimmons RJ, Farley J, Adey WR, Baylink DJ (1986) Embryonic bone matrix formation is increased after exposure to a low-amplitude capacitively coupled electric field, in vitro. Biochim Biophys Acta 882:51–56

Fletcher WH, Byus CV, Walsh DA (1987) Receptor mediated action without receptor occupancy: a function for cell-cell communication in ovarian follicles. In: Mahesh V (ed) Regulation of ovarian and testicular function. Plenum, New York

Fröhlich H (1968) Long-range coherence and energy storage in biological systems. Int J Quantum Chem 2:641–649

Fröhlich H (1972) Selective long range dispersion focres between large systems. Phys Lett 29A: 153–154

Fröhlich H (1975) The extraordinary dielectric properties of biological materials and the action of enzymes. Proc Natl Acad 72:4211–4215

Fröhlich H (1977) Possibilities of long- and short-range electric interactions of biological systems. Neurosci Res Program Bull 15:67–72

Fröhlich H (1980) The biological effects of microwaves and related questions. Adv Electronics Electron Phys 53:85–152

Fröhlich H (1986a) Coherent excitation in active biological systems. In: Gutmann F, Keyzer H (eds) Modern bioelectrochemistry. Plenum, New York, pp 241–261

Fröhlich H (1986b) Coherence and the action of enzymes. In: Welch GR (ed) The fluctuating enzyme. Wiley, New York, pp 421–449

Gavalas-Medici R, Day-Magdaleno SR (1976) Extremely low-frequency, weak electric fields affect schedule-controlled behavior of monkeys. Nature 261:256–258

Gmitro JL, Scriven LE (1966) A physicochemical basis for pattern and rhythm. In: Warren KB (ed) Intracellular transport. Academic Press, New York, pp 221–255

Goodman R, Henderson A (1986) Sine waves enhance cellular transcription. Bioelectromagnetics 7:23–29

Grossberg S (1983) Neural substrates of binocular form perception: filtering, matching, diffusion and resonance. In: Basar E, Flohr H, Haken H, Mandell AJ (eds) Synergetics of the brain. Springer, Berlin Heidelberg New York, pp 274–298

Hyman JM, McLaughlin DW, Scott AC (1981) On Davydov's alpha-helix soliton. Physica D30: 23–44

Jolley WB, Hinshaw DB, Knierim K, Hinshaw DB (1983) Electromagnetic field effects on calcium efflux and insulin secretion in isolated rabbit islets of Langerhans. Bioelectromagnetics 4: 103–107

Kaczmarek LK (1976) Frequency sensitive biochemical reactions. Biophys Chem 4:249–252

Kaczmarek LK, Adey WR (1973) The efflux of $^{45}Ca^{2+}$ and ^{3}H-gamma-aminobutyric acid from cat cerebral cortex. Brain Res 63:331–342

Kaczmarek LK, Adey WR (1974) Weak electric gradients change ionic and transmitter fluxes in cortex. Brain Res 66:537–540

Kaiser F (1984) Entrainment-quasiperiodicity-chaos-collapse: bifurcation routes of externally driven self-sustained oscillating systems. In: Adey WR, Lawrence AF (eds) Nonlinear electrodynamics in biological systems. Plenum, New York, pp 393–412

Kalmijn AJ (1971) The electric sense of sharks and rays. J Exp Biol 55:371–383

Katchalsky A (1974) Concepts of dynamic patterns. Early history and philosophy. Neurosci Res Program Bull 12:30–36

Katchalsky A, Curran PF (1965) Nonequilibrium thermodynamics in biophysics. Harvard University Press, Cambridge MA

Katchalsky A, Rowland V, Blumenthal R (eds) (1974) Dynamic patterns of brain cell assemblies. Neurosci Res Program Bull 12:1–195

Keeler JD, Farmer JD (1986) Robust space-time intermittency and l/f noise. Physica D23:413–446

Koshland DE (1975) Transductive coupling in chemotactic processes: chemoreceptor-flagellar coupling in bacteria. In: Schmitt FO, Schneider DM, Schneider DM (eds) Functional linkage in biomolecular systems. Raven, New York, pp 273–279

Lawrence AF, Adey WR (1982) Nonlinear wave mechanisms in interactions between excitable tissue and electromagnetic fields. Neurol Res 4:115–153

Liboff AR (1985) Cyclotron resonance in membrane transport. In: Chiabrera A, Nicolini C, Schwan HP (eds) Interactions between electromagnetic fields and cells. Plenum, New York, pp 281–296

Lin-Liu S, Adey WR (1982) Low frequency, amplitude-modulated microwave fields change calcium efflux rates from synaptosomes. Bioelectromagnetics 3:309–322

Lin-Liu S, Adey WR, Poo M-M (1984) Migration of cell surface concanavalin A receptors in pulsed electric fields. Biophys J 45:1211–1217

Luben RA, Cain CD (1984) Use of bone cell hormone responses to investigate bioelectromagnetic effects on membranes in vitro. In: Adey WR, Lawrence AF (eds) Nonlinear electrodynamics in biological systems. Plenum, New York, pp 23–33

Luben RA, Cain CD, Chen M-Y, Rosen DM, Adey WR (1982) Effects of electromagnetic stimuli on bone and bone cells in vitro: inhibition of responses to parathyroid hormone by low-energy, low-frequency fields. Proc Natl Acad Sci USA 79:4180–4183

Lyle DB, Schechter P, Adey WR, Lundak RL (1983) Suppression of T lymphocyte cytotoxicity following exposure to sinusoidally amplitude-modulated fields. Bioelectromagnetics 4:281–292

Lyle DB, Ayotte RD, Sheppard AR, Adey WR (1987) Suppression of T lymphocyte cytotoxicity following exposure to 60 Hz sinusoidal electric fields. (Submitted) Bioelectromagnetics

Maddox J (1986) Physicists about the hijack DNA? Nature 324:11

Moolenaar WH, Aerts RJ, Tertoolen LGJ, DeLast SW (1986) The epidermal growth factor-induced calcium signal in A431 cells. J Biol Chem 261:279–285

Neumann E, Katchalsky A (1972) Long-lived conformation changes induced by electric pulses in biopolymers. Proc Natl Acad Sci USA 69:993–997

Nicolis JS (1983) The role of chaos in reliable information processing. In: Basar E, Flohr H, Haken H, Mandell AJ (eds) Synergetics of the brain. Springer, Berlin Heidelberg New York, pp 330–344

Nicolis JS, Galanos G, Protonotarios EN (1973) A frequency entrainment model with relevance to systems displaying adaptive behaviour. Int J Control 18:1009–1027

Nicolis JS, Protonotarios E, Lianos E (1974) The role of noise in "self-organizing" systems. Univ of Patras, Greece, Dept Electrical Eng, Technical Report CSB-1, 55 pp

Nishizuka Y (1983) Calcium, phospholipid and transmembrane signalling. Philos Trans R Soc Lond B302:101–112

Nishizuka Y (1984) The role of protein kinase C in cell surface transduction and tumor promotion. Nature 308:693–696

Othmer HG, Scriven LE (1969) Interactions of reaction and diffusion in open systems. Ind Eng Chem 8:302–313

Othmer HG, Scriven LE (1971) Instability and dynamic pattern in cellular networks. J Theor Biol 32:507–537

Othmer HG, Scriven LE (1974) Nonlinear aspects of dynamic pattern in cellular networks. J Theor Biol 43:83–112

Pasti G, Lacal J-C, Warren BS, Aaronson SA, Blumberg PM (1986) Loss of mouse fibroblast response to phorbol esters restored by microinjected protein kinase C. Nature 324:375–377

Polk C (1984) Time-varying magnetic fields and DNA synthesis: magnitude of forces due to magnetic fields on surface-bound counterions. Proceedings Bioelectromagnetics Soc, 6th Annual Meeting, p 77 (abstract)

Radeke MJ, Misko TP, Hsu C, Herzenberg LA, Shooter M (1987) Gene transfer and molecular cloning of the rat nerve growth factor receptor. Nature 325:393–397

Rayleigh L (1916) On convective currents in a horizontal layer of fluid, when the higher temperature is on the under side. Philos Mag 32:529–546

Riedel H, Schlessinger J, Ullrich A (1986) A chimeric, ligand binding v-erbB/EGF receptor retains transforming potential. Science 236:197–200

Rodbell M, Lin MC, Salomon Y (1974) Evidence for interdependent action of glucagon and nucleotides on the hepatic adenylate cyclase system. J Biol Chem 249:59–65

Schmitt FO, Schneider DM, Crothers DM (eds) (1975) Functional linkage in biomolecular systems. Raven, New York

Schwarz G (1975) Sharpness and kinetics of cooperative transitions. In: Schmitt FO, Schneider DM, Crothers DM (eds) Functional linkage in biomolecular systems. Raven, New York, pp 32–35

Schwarz G, Balthasar W (1970) Cooperative binding of linear biopolymers. 3. Thermodynamic and kinetic analysis of the acridine orange-poly (L-glutamic acid) system. Eur J Biochem 12: 461–467

Schwarz G, Seelig J (1968) Kinetic properties and electric field effect of the helix-coil transition of poly (gamma-benzyl L-glutamate) determined from dielectric relaxation measurements. Biopolymers 6:1263–1277

Schwarz G, Klose S, Balthasar W (1970) Cooperative binding to linear biopolymers. 2. Thermodynamic analysis of the proflavine-poly (L-glutamic acid) system. Eur J Biochem 12:454–460

Scott AC (1981) The laser-Raman spectrum of a Davydov soliton. Phys Lett 62A:60–62

Semm P (1983) Neurobiological investigations on the magnetic sensitivity of the pineal gland in rodents and pigeons. Comp Biol Physiol 76A:683–692

Semm P, Demaine C (1986) Neurophysiological properties of magnetic cells in the pigeon's visual system. J Comp Physiol 159:619–625

Singer SJ, Nicolson GL (1972) The fluid mosaic model of the structure of cell membranes. Science 175:720–731

Takahashi K, Kaneko I, Date M, Fukada E (1986) Effect of pulsing electromagnetic fields on DNA synthesis in mammalian cells in culture. Experientia 42:185–186

Turing AM (1952) The chemical basis of morphogenesis. Philos Trans R Soc Lond B237:37–72

Ullrich A, Coussens L, Hayflick JS, Dull TJ, Gray A, Tam AW, Lee J, Yarden Y, Libermann TA, Schlessinger J, Downward J, Mayes ELV, Whittle N, Waterfield MD, Seeburg PH (1985) Human epidermal growth factor receptor cDNA sequence and aberrant expression of the amplified gene in A431 epidermoid carcinoma cells. Nature 309:428–421

Van der Kloot WG, Cohen I (1979) Membrane surface potential changes may alter drug interactions: an example, acetyl choline and curare. Science 203:1351–1352

Welker HA, Semm P, Willig RP, Wiltschko W, Vollrath L (1983) Effects on an artificial magnetic field on serotonin-N-acetyltransferase activity and melatonin content of the rat pineal gland. Exp Brain Res 50:426–431

Wever R (1975) The circadian multi-oscillatory system of man. Int J Chronobiol 3:19 55

Wyman J (1948) Heme proteins. Adv Protein Chem 4:407–531

Wyman J, Allen DW (1951) The problem of the heme interactions in hemoglobin and the basis of the Bohr effect. J Polymer Sci 7:491–518

The Interaction of Living Red Blood Cells

S. ROWLANDS [1]

1 Background

Shed human blood clots in a minute or so; that is the fluid turns into a jelly. If a clot is examined by microscope the red cells (erythrocytes), which outnumber the white cells a thousand times, are not oriented at random but display a pseudostructure. Red cells are biconcave discs 8 micrometres in diameter and the thickness of the rounded rim of the disc is about 2 micrometres. There is some spread in these dimensions in a single sample of blood. The pseudostructure comprises a stacking of the discs face to face, superimposed on a random packing of the stacks. This so-called rouleau-formation has been observed and studied for over a century (Fahraeus 1929; Rowlands and Skibo 1972).

Approximately half of blood is red cells, so they are close-packed even before rouleaux are formed. For the study of rouleaux it is convenient to remove most of the cells from the liquid part of blood. It is also convenient to stop the formation of the blood clot by anticoagulants (they are in common use in the practice of medicine). The behaviour of the cells can then be observed by microscope for prolonged periods as a single layer of objects which translate and gyrate in Brownian (thermal) motion; very slowly in respect of their mass (about 90 picograms).

When such a preparation is placed on the stage of a microscope, a few rouleaux consisting of two to four cells will already have formed. Under Brownian motion cells will collide with each other and form two-cell rouleaux; cells will collide with rouleaux, and rouleaux with rouleaux to form larger rouleaux (Fig. 1). The longest linear rouleau which I have seen in such experiments consisted of 28 cells. When two cells collide, or a cell or a rouleau touches the end of another rouleau, they adhere and then rather quickly (on the timescale of the Brownian movement) slide with respect to each other until the stack is aligned. The alignment is nearly, but not entirely, perfect. Sometimes a cell or the end of a rouleau will bump into the side of another rouleau in which case it stays there and will form a branch. In this way a network will eventually be the result. Few single cells or small rouleaux will remain after the lapse of some hours (Fig. 1).

Years ago in such experiments (Rowlands and Skibo 1972) I observed that when cells and rouleaux were a few microns apart they seemed to move towards each other quite quickly, although sometimes they failed to collide and would separate. I considered this to be an illusion, particularly as I knew of no mechanism for an attractive

1 Emeritus Professor, Faculty of Medicine, University of Calgary, Canada

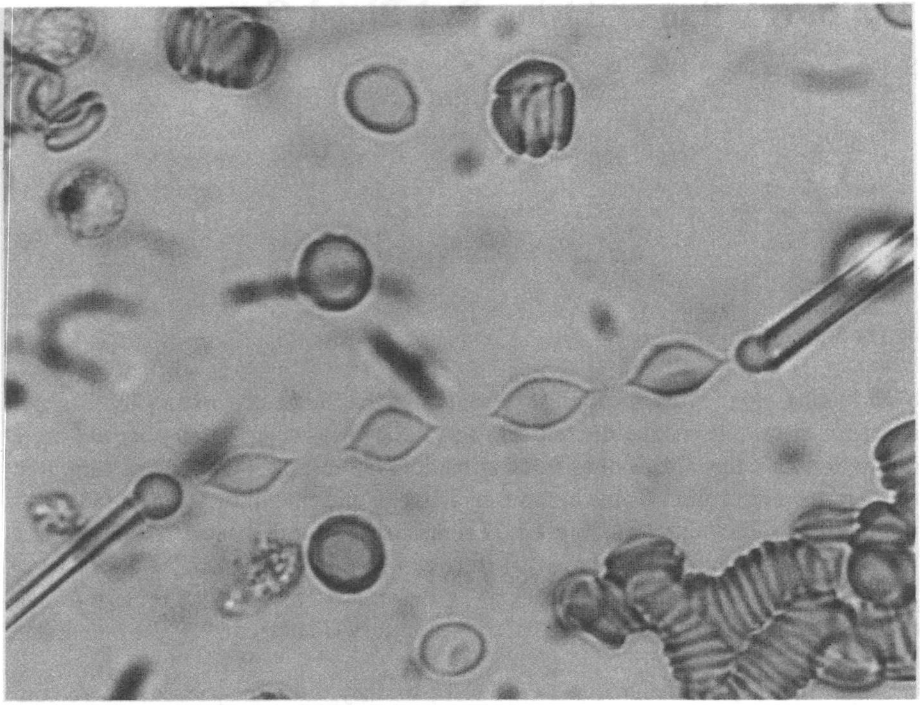

Fig. 1. A large branched rouleau is seen *lower right*. There is a small branched rouleau at *top right centre* and a two-cell rouleau, seen on end, is on its *left*. There are two three-cell rouleaux on end *below* this and a swollen single cell at *bottom centre*. The end cells of a linear six-cell rouleau have been drawn into the micropipettes seen *left* and *right*. The pipettes were then pulled apart. Contractils are not resolved by the light microscope. Their presence is inferred by the distortion of the shape of the cells in the chain. On release of the negative pressure in the pipettes a normal linear six-cell rouleau reforms

force. On the contrary, since it is well known that the cell surface carries a negative electric charge, there was reason to expect a repulsion. I made no attempt at measurement at that time and I did not publish the observation. Years later I came across Fröhlich's theory of coherent excitations in cells when I read his 1980 review (Fröhlich 1980) of the theory and of the supporting experiments. This theory postulated a mechanism for an attractive force between cells. Soon we had our first evidence for an attraction between human red cells and it depended on the presence of a supply of metabolic energy and of an electrical potential difference across the cell membrane (Rowlands et al. 1981) in accord with the requirements of the Fröhlich theory. (In addition to the surface charge with respect to the surrounding fluid, there is a potential difference of about 15 mV across the cell membrane).

Another unexplained observation resulted from the experimental manipulation of rouleaux (Rowlands and Skibo 1972). Small blood vessels such as those in the retina and in the conjunctiva of the eye can be seen quite easily with suitable magnification and illumination. No rouleaux appear in flowing blood in these vessels in healthy people. Quite a small shearing force will break rouleaux into single cells. In some diseases, however, particularly those involving the connective tissues, marked rouleau-formation is seen in the small veins. We thought that rouleaux might be more tightly bound in disease and to study this hypothesis we developed a method for measuring the force required to pull a rouleau apart. To our surprise a rouleau did not come apart. Rather, with increasing tension, it elongated with the cells taking on a lozenge shape and remaining attached to each other by invisible (to light microscopy) fibrils (Fig. 1). If the tension was removed the fibrils contracted and pulled the cells together, back into a rouleau indistinguishable from the original. Significantly, we also found that this phenomenon was sensitive to changes in the cellular membrane potential and to the supply of energy (Rowlands et al. 1983a).

2 The Ultra-Long-Range Attractive Force

We refer in this way to the forces which follow from the Fröhlich theory because their range (of the order of microns) is an order of magnitude greater than the range of known chemical forces. There follows a description of the experiments which established the existence of such a force between human (mammalian) erythrocytes. Further technical details are in the quoted references. The very first experiments (Rowlands et al. 1981) will not be described because the method of analysis used in them was soon superceeded.

2.1 The Brownian Movement of Erythrocytes in Anticoagulated Human Blood

Apart from the camera, the equipment (Rowlands et al. 1982a) was that used, before automation made it obsolete, for performing a "blood count" in routine medical haematological laboratories: a microscope and a haemacytometer chamber which fits the stage of the microscope. The chamber is a thick glass plate with an optically flat raised centre strip on which there is a precision-ruled grid. On either side of the centre strip there are two narrow strips on which is placed an optically-flat cover-slip. The two side strips are 0.1 mm higher than the centre strip so that, with the cover-slip in place, a chamber is formed, 0.1 mm deep, into which the blood sample (accurately and appropriately diluted) is placed. The grid consists of 50 μm squares and, depending on the magnification, 80–100 of the squares were in the field of view. The blood sample was depleted of red cells by centrifugation to give 300–600 red cells in the field of view. The occasional white cell was easily identified and excluded from the measurements. With this equipment the movement of a counted number of cells in an accurately known volume could be followed.

The camera was a time-lapse 16 mm movie camera set to take pictures at about 15-s intervals, but the precise timing was made by recorded frame counting against a stop watch. An experiment would last from 1/2 to 3 h giving 120–720 frames, many more than needed to keep error below the errors from other sources. Indeed a standard 36-frame 35-mm camera would have been quite adequate, except for the problem of the sticking of some cells to the floor of the chamber. Such cells are difficult to identify by examining still frames one by one. The 16-mm movie film with more than 100 frames could, however, be viewed at normal projection speeds and the stuck cells immediately identified from their lack of translation.

Sticking was a minor problem when the specimen was whole blood and was completely eliminated by siliconizing the floor of the chamber (Rowlands et al. 1982a). When the cells were suspended in artificial fluids sticking had to be carefully monitored (see below).

It was necessary to make sure that the haemacytometer chamber was horizontal, otherwise the cells would drift downhill. It was also necessary to seal the chamber against evaporation, which would produce a fluid current and a drift of cells. Both these errors were detected by viewing the film at normal projection speeds when a drift is obvious. Experiments with a drift were discarded.

2.2 Analysis of Results: Brownian Movement

We were recording a Brownian movement of cells in a plane on which was superimposed a small intercellular attraction. Einstein (1926) gives the mean square deviation of a sphere in Brownian motion in a plane as a function of time t:

$$\langle U^2 \rangle = 4\,kTt\mu^{-1}\ , \tag{1}$$

where k is Boltzmann's constant, T is the absolute temperature and μ is the Stokes' drag coefficient for rigid spheres:

$$\mu = 6\,\pi\eta r\ . \tag{2}$$

This coefficient relates the force necessary to give a sphere of radius *r* a small velocity v through a liquid of viscosity η.

We carried out a series of experiments each with about 30 cells in the field of view of the microscope and measured the Brownian movement of cells which remained isolated from other cells (Sewchand et al. 1982a). Averaged over many cells we found that $\langle U^2 \rangle$ increased linearly with time in accord with Einstein's Eq. (1). The drag coefficient μ calculated from (1) was, however, about double the value given by Eq. (2), despite corrections for the shape and deformability of the cells (Rowlands et al. 1982a). On the following grounds we concluded that there was an additional frictional force generated by contact of the cells with the floor of the chamber. When, instead of cells, we used microspheres which were nearly buoyant in the suspending fluid, the agreement with Eq. (2) was good. When the microspheres were more dense than the suspending fluid (as is the case of erythrocytes in blood) the drag coefficient was approximately double that of Eq. (2). This was the case also with cells in blood.

It will be seen in Sect. 2.3 that the drag coefficient for isolated cells (unperturbed Brownian motion) is the base from which the Fröhlich interaction is measured. Therefore in *every* experiment $\langle U^2 \rangle$ and *t* were observed, averaged over about 30 cells. In an experiment there are about 30 cells out of the several hundred in view which are more than 10 μm from any other cell. From these measurements the drag coefficient was calculated in *each* experiment. Correlation coefficients for the regression of $\langle U^2 \rangle$ on *t* averaged 0.94.

2.3 Analysis of Results: the Interaction

The cells are executing a slow Brownian movement until they touch, when they adhere to form a rouleau. The process is analogous to the coagulation of a colloid for which there is a well-authenticated theory by Smoluchowski. The theory is reviewed in shortened form by Chandrasekhar (1943). It invokes the possibility of an interaction at a distance between the colloid particles before contact. It assumes that the diffusion coefficient of the particles is given by

$$D = \frac{\mathrm{kT}}{6\,\pi\eta \mathrm{r}}\,, \tag{3}$$

where the denominator is the Stokes' drag coefficient for spheres [Eq. (2)]. The interaction at a distance is expressed as a zone of attraction $Z > r$ around each particle. If $Z = r$ there is no attraction. Under our experimental conditions it follows (at some length, Chandrasekhar 1943; Rowlands et al. 1982a) that if N is the total number of particles (single and multiple) in the field of view, the kinetics of rouleau formation is given by

$$\frac{1}{N_\mathrm{t}} - \frac{1}{N_0} = \frac{2\,\mathrm{kTZ}}{\eta \mathrm{r}} \circ t\,, \tag{4}$$

which predicts a linear relation between N_t^{-1} and time t. But it must be noted that the theoretical drag coefficient $6\,\pi\eta r$ must be replaced by the experimentally measured coefficient as described in the preceding section.

In practice the ciné-film was examined and, in the earliest frame at which the cells had settled after the chamber was filled and sealed, the total number of cells and rouleaux (N_0), defined by the grid, was counted. The count was repeated every fifth frame initially, and, later, at increasing intervals up to 20 frames. Correlation coefficients for the regression of N_t^{-1} on t averaged 0.98. This regression, the temperature, Boltzmann's constant and the *measured* drag coefficient were inserted into Eq. (4) to give an "interaction coefficient" which we define as

$$\Xi = \frac{Z}{r}\,.$$

With living blood Ξ is significantly greater than unity which implies an interaction at a distance.

2.4 Critique of the Method

The Smoluchowski method of analysis assumes that all the coagulating colloid particles are identical, which is only approximately true for red cells. The zone of attraction is also assumed to be the same for every particle, which is unlikely to be true for red cells; a blood sample consists of cells of all ages (life-time about 120 days) including cells about to die. When a particle reaches the periphery of the zone of attraction of another particle Smoluchowski assumes that the second particle instantaneously crosses the zone and that the two particles adhere. On the ciné-films the rate of approach is seen to be finite. There are other less important assumptions for which Chandrasekhar's paper (1943) should be consulted. A modified method of analysis (Enns et al. 1983), more in keeping with the experimental method, leads to the same conclusion. At all events the linearity of $\langle U^2 \rangle$ and of N^{-1} on t give substantial support to the validity of the original method, which is the one used in most of the reported experiments. But in view of the assumptions, the precise magnitude of the radius of the zone of attraction, as derived from the experimental results, must be treated with caution.

Perhaps the most important assumption of Smoluchowski is that if the cells touch, they adhere. The validity of this assumption can be checked in every experiment if the ciné-film is viewed with a film data analyzer (e.g. model 224A L-W International). This machine will stop at every frame for the making of measurements, but also the film can be run backwards and forwards at any rate up to 24 frames a second. So the time-sequence of all collisions can be monitored. In the experiments with blood every collision resulted in adhesion with the formation of a rouleau.

This was not always so in other experiments to be described. As an example, here is an extreme case. Red cells can be washed clean of all the liquid components of blood and then resuspended in a salt solution of appropriate osmotic pressure and pH [PBS-phosphate buffered saline (Rowlands et al. 1982a)]. If this suspension is experimented on, the cells never adhere and are mutually repelled, presumably by their surface charge. The Smoluchowski analysis cannot be invoked to calculate the magnitude of a "zone of repulsion" because the assumption of adhesion is not met with. We will see later that certain macromolecules, added to phosphate-buffered saline, will sustain rouleau-formation. If the concentration of macromolecules is low, some apparent collisions do not result in adhesion and values of $Z < r$ are the outcome.

2.5 The Interaction and the Fröhlich Theory

According to the theory (Fröhlich 1980) the ultra-long-range interaction requires (a) an intact cell membrane, (b) an electrical potential difference across the membrane to polarize the macromolecules which constitute the membrane and (c) a supply of metabolic energy to maintain coherent polar waves in the membrane.

2.5.1 The Cell Membrane

The structure of the membrane can be disorganized by heat, poisons etc. We used dilute glutaraldehyde because unpublished work had shown that cells so treated retain normal shape, area and volume. Moreover, although the cells become rigid, they adhere on collision and so satisfy that important criterion which was emphasized above. Whereas normal cells in blood resulted in an interaction coefficient $\Xi = 3$ (or a zone radius $Z = 3\ r = 12\ \mu\text{m}$), with glutaraldehyde-fixed cells (and with artificial microspheres) $\Xi = 1$ (Rowlands et al. 1982a,b). Over the series of experiments the difference between normal and fixed cells or microspheres was statistically valid ($p < 0.001$). In other words, both disorganized cells and microspheres coagulated at the rate predicted by Brownian theory, whereas living cells in blood coagulated three times as fast. This increased rate of coagulation is obvious to the eye when the films of these experiments are viewed at 24 frames per second.

2.5.2 The Membrane Potential Difference

The membrane potential of human red cells can be abolished by incorporating an "ionophore" (A23187) into the membrane. Associated with the membrane potential there is a substantial difference in the concentration of inorganic ions within and without the cell. The ionophore makes the membrane leaky to small ions, the difference in concentration vanishes and the membrane potential disappears. The treatment also causes a minority of the cells to change shape from discs to spheres but they still coagulate on contact. Experimentally, the coagulation rate is that predicted from Brownian motion; there is no ultra-long-range interaction when the membrane potential is abolished (Rowlands et al. 1982a,b).

A better way of reducing the membrane potential is by altering the pH of the blood by the addition of acid because the change can be reversed by alkali (Rowlands et al. 1982a,b). When the pH was reduced from its normal value of 7.6 down to 6.3 the membrane potential came close to zero. The interaction coefficient became unity, that is there was no interaction. But when the pH was returned to 7.6 in the *same* cells the former interaction was restored (Rowlands et al. 1982a,b). In addition to its reversibility, this method did not affect the shape of the cells. These experiments confirm that the ultra-long-range interaction requires a membrane potential difference.

2.5.3 The Energy Supply

There are many poisons which disturb the complex energy reactions in the cell, but most are irreversible. The simplest reversible method is to suspend the cells, after washing away all the surrounding nutrients, in phosphate-buffered saline and to stir them gently at 37 °C for about 24 h. In this time they use up their energy stores. These stores can be repleted in the same cells by the addition of adenosine followed by further gentle stirring for 3 h. In the depletion process some of the cells undergo a shape change, but they still adhere on contact, and the cells that are normal-looking form normal rouleaux. Depleted cells coagulate at the Brownian rate (no interaction) and the interaction returns when the cells' energy stores are restored (Rowlands et al. 1982a,b).

This series of experiments shows that the ultra-long-range interaction of human erythrocytes in blood is dependent on (a) an intact membrane, (b) a membrane potential difference and (c) a supply of metabolic energy in accord with the postulates of Fröhlich's theory of coherent excitations in biological systems.

3 The Interaction in Liquids Other Than Blood Plasma

Blood is a suspension of cells in a liquid named plasma. Plasma is a complex watery solution of many molecules and macromolecules. Albumin (MW ~60,000) is the predominant macromolecule with a concentration of about 60 g/l. At concentrations an order of magnitude less there is fibrinogen (which clots to the gel fibrin when blood is shed) and the immunoglobulins. Other molecules and macromolecules seem not to play a significant part in rouleau-formation.

It was noted above that if red cells are suspended in a solution consisting only of electrolytes they will survive but they do not form rouleaux or even adhere to each other. Some other factor or factors are needed.

3.1 Serum

If blood is allowed to clot in a test tube, the clot contracts (retracts) in about a quarter of an hour, leaving a pale yellow supernatant called serum. The plasma fibrinogen is now polymerized to fibrin in the clot but everything else remains in serum. When red cells are suspended in serum, in some experiments there is rouleau-formation and in others there is not (unpublished). With hindsight we attribute the presence or absence of rouleaux to the rather variable concentration of immunoproteins. Clearly, however, fibrinogen, which is absent from serum, is important to rouleau-formation. Albumin, present in serum, seems of lesser or no importance. These suggestions are elaborated in the following experiments.

3.2 Polyvinylpyrrolidone (MW 360,000) and Dextran 70 (MW 70,000)

It is known that rouleaux form when red cells are suspended in solutions of extended macromolecules in phosphate-buffered saline (Sewchand and Bruckschwaiger 1980). In such experiments with artificial solutions, a small concentration of albumin is necessary, for reasons unknown, otherwise the red cells lose their shape and become spheroidal or wrinkled (crenated). The necessary concentration is of the order of 1 g/l (one-sixtieth of the concentration in blood) but it is advisable to do an auxiliary experiment to determine the minimum concentration needed, as this may be different with different concentrations of the extended macromolecules (especially when dealing with red cells from other mammalian species).

There are several macromolecules which induce rouleaux, and many others that may, but the above-named are the only ones we have used apart from poly(ethylene oxide) in some unpublished work. At concentrations less than 0.5 g/l of polyvinylpyrrolidone few or no rouleaux form. The measured interaction coefficient is less than unity but the magnitude is of doubtful significance because colliding cells do not always adhere. Above 1.0 g/l rouleaux form and the interaction coefficients are greater than one (Sewchand et al. 1982b). The coefficients do not rise linearly with increasing concentration to 6.0 g/l. Rather the Ξ versus concentration relation may be a step function but more work is needed to settle this point.

With Dextran 70, at low concentration, again there are few rouleaux and no interaction ($\Xi < 1$). A concentration of 20 g/l (i.e. twenty times that of PVP) is needed for rouleaux and an interaction to arise. Again there is a suggestion of a step function (Sewchand et al. 1982b). Clearly the minimum concentration necessary for an interaction is not inversely proportional to the molecular weight.

In the blood experiments, sticking of the cells to the floor of the chamber was avoided by silicone treatment, but this did not entirely overcome the problem when "artificial" polymer solutions were used. Stuck cells had to be identified and excluded from the count, which made the analysis tedious. These experiments should be repeated and extended but it would be far more satisfactory to use the method of O.G. Fritz [(1984) see later: Sect. 5].

3.3 Albumin (MW ~60,000)

Serum will sometimes form rouleaux, but a solution of serum's predominant constituent, albumin, will not. The interaction coefficients are always less than unity even at concentrations of albumin higher than that in blood (Sewchand et al. 1982b). Albumin is a globular protein, a macromolecule folded in on itself into a pseudosphere. On the contrary, PVP and Dextran have an extended structure. So, apparently, does fibrinogen, which seems from the blood and serum experiments to be the molecule in blood which gives rise to rouleaux and to the ultra-long-range interaction.

Albumin can, however, be unfolded from a globule into an extended structure by heating the solution at 62 °C for 20 min and then allowing it to cool. When this was done, rouleaux formed and the interaction coefficients became substantially greater than unity (Rowlands et al. 1983b). Moreover, when serum was heated in the same way the magnitude of the interaction was doubled (Rowlands et al. 1983b).

3.4 Fibrinogen (MW ?340,000, ?680,000)

Our first experiments were done with the partially purified fibrinogen which is used clinically to treat patients who have dangerously low concentrations of fibrinogen in their blood. At concentrations of 1.0 g/l or less there were neither rouleaux nor an interaction. Above this concentration, rouleaux and a large interaction appeared, with again a suggestion of a step function for Ξ versus concentration (Sewchand et al. 1982b). To be quite sure that it was fibrinogen and not an impurity which was producing the interaction, we asked our biochemistry colleagues to provide highly purified samples (Masri et al. 1983). To our surprise, purified fibrinogen gave no rouleaux and no interaction (Sewchand et al. 1984). So we added to pure fibrinogen the few "clotting factors" which we were able to obtain; which were those used in the treatment of haemophilia. Addition of the preparation used in treating classical haemophilia (Factor VIII, antihaemophilia A) did nothing. On the contrary, the one used in Christmas disease (antihaemophilia B) restored rouleaux and an interaction in the presence of an adequate concentration of pure fibrinogen (Sewchand et al. 1984). In the publication cited (Sewchand et al. 1984) there is an extensive discussion of these experiments, which leads to the suggestion that pure fibrinogen is a globular structure and that one or more of the four factors (II, VII, IX and X) in antihaemophilia B extends it, allowing it to produce rouleaux and an ultra-long-range interaction. Regrettably, the team which did this work dispersed, so preventing further investigation of these interesting findings.

4 The Specificity of the Interaction

A Fröhlich interaction should be specific; that is, attractive, repulsive or null, depending on the relative frequencies or phases of the vibrations in the interacting entities, and on the dispersion of the dielectric constant of the surroundings. The experiments of Sewchand and Canham (1976) showed that mixtures of red cells from different mammalian species (human, dog, cat, rat etc.) would form rouleaux in PVP 360. Plasma cannot be used because of immunological incompatibility analogous to a mismatched blood transfusion.

The cells from the different species can be recognized within a rouleau from differences in size. Sewchand and Canham noticed that the ordering of the cells from the different species in a rouleau was not random. Rather, the cells from the same species had a greater affinity for each other than for cells of the second species. Overbeek (1952) has shown that in a colloid consisting of particles of varying size (polydispersed colloid) the *least* likely collision in Brownian motion is between particles of the *same* size. The rouleau experiments show the reverse behaviour, which suggests that cells from the same species attract each other more strongly than do cells from different species. But non-random rouleaux could also result if collisions between cells of different species did not result invariably in adherence. To settle this point, we filmed four experiments with mixtures of dog and human cells and ten experiments with cat and human cells. In each of the 14 experiments about 40 pairs would collide; there was no case of cells, once in contact, coming apart again (Sewchand and Rowlands 1983).

We also measured interaction coefficients for mixtures of cat and human cells and compared them with the coefficients for cat cells only and for human cells only. The results for the cat-human mixture were significantly lower than for cat or human cells alone (Sewchand and Rowlands 1983). There were technical difficulties in these experiments in that the same suspending fluid was not ideal for both species. Nevertheless they show that the attractive force between cells of the same species is greater than that between cells from different species. The results do not exclude a zero interaction between cat and human cells.

5 A Different Method: Quasi-Elastic Laser Light Scattering (QELS)

The method used in the experiments of the previous sections measured deviations from Brownian motion and showed that the rate of aggregation of living red cells is greater than if they are dead or inactive. Living cells therefore move faster on the average, which implies that their diffusion coefficient is higher. Quasi-elastic light scattering (QELS) is a well-established method for measuring the diffusion coefficients of large molecules (Berne and Pecora 1976). In recent years the method has been used to study the motility of bacteria and of spermatozoa, which are not much smaller than human red cells. Fritz has shown that, with care, QELS gives the correct diffusion coefficient for particles of the same size as human red cells (Fritz 1984).

In brief the method is as follows: A narrow beam of light from a laser is incident on the cell suspension. Light scattered from the cells is collimated at a measured angle and falls onto a photoelectric cell (photomultiplier) from which the photoelectric pulses are fed to an autocorrelator and computer.

The effect of the Brownian motion of the cells is to broaden a very sharp spectral line of the laser by the Doppler effect. The frequency shift to either side of the line is a measure of the velocity of the cells. Unfortunately, no spectroscope can measure this very small change in the frequency of the light, so another method has to be used. If the scattered light is viewed by eye it is seen to "twinkle"; that is the amplitude varies. If there was no variation in amplitude, the photomultiplier pulses would be random in time of arrival and would exhibit a Poisson distribution. The autocorrelator records the pulse-time distribution and analyzes it for deviations from a Poisson distribution. From this altered distribution the computer calculates the diffusion coefficient of the cells. Mathematically, the intensity-frequency profile of the scattered light and the so-called autocorrelation-time function of the scattered optical electric field form a Fourier transform pair and the one can be derived from the other.

Fritz has made an exhaustive analysis of the application of this method to measuring the diffusion coefficient of mammalian red cells (Fritz 1984) and his experiments are well controlled. He has repeated a number of the experiments described above and performed some others. The two approaches, QELS and the haemacytometer method, differ in only two significant ways: (a) In QELS the cells are in suspension and the problem of the drag of the cells against the floor of the haemacytometer is eliminated, (b) the number/volume concentration of cells is more than an order of magnitude smaller in Fritz's experiments. This restriction is imposed by the necessity for avoiding multiple scattering of the laser light; if the number of cells in the light beam is increased, then more and more of the light scattered from one cell may be scattered by another cell before reaching the photomultiplier. This invalidates the analysis. On the other hand, it is gratifying that the similarity of the results given by the two methods occurs in spite of the great difference in cell concentration.

Fritz could not experiment with cells in plasma because of the strong absorption of light in this coloured liquid. In the experiments published up to now (Fritz 1984; Paul et al. 1983) he has limited himself to cells suspended in solutions of polyvinylpyrrolidone (PVP) of molecular weights 360,000, 40,000 and 10,000 (PVP 360, PVP 40 and PVP 10), in phosphate-buffered saline plus a low concentration of albumin to maintain cell shape (see Sect. 3.2 above). In summary, Fritz finds a significantly increased diffusion coefficient for human red cells suspended in PVP 360. The controls are glutaraldehyde-fixed cells, living cells with zero membrane potential, cells which are depleted of their stores of energy, microspheres, and living cells in PBS only; all of which give the diffusion coefficient expected for unperturbed Brownian motion. His findings are therefore compatible with the presence of an ultra-long-range interaction which depends on the presence of an intact membrane, a membrane potential and on a supply of metabolic energy in accordance with Fröhlich's theory. Again the abolition of the interaction by reduction in membrane potential and in the metabolic stores is reversible.

In all major respects, then, the QELS method confirms the findings with the haemacytometer method. The theory of Fritz's method, which incorporates Fröhlich's long-range potential instead of Smoluchowski's "zone of attraction", has been published (Paul et al. 1983). When the quantitative results of the two methods are compared, the agreement is slightly better than 4% (Paul et al. 1983).

In connection with a later part of this chapter (Sect. 7.2) it is significant that Fritz found no interaction using polyvinylpyrrolidone of molecular weight 40,000 nor 10,000.

6 Contractile Fibrils Between Red Cells: Contractils

A solution of red cell rouleaux on a microscope slide can be perturbed by micromanipulation. If the manipulation is done by micropipettes in which the fluid pressure can be measured, it is found that in pipettes of internal diameter greater than 3 μm the rouleaux will slip in and out of the tubes at pressures of ~1 mm H_2O. If, however, the diameter is less than 3 μm a rouleau will not enter, even under pressures orders of magnitude greater (Canham and Burton 1968). Rather part of the end cell of a rouleau is drawn in leaving a now spherical part of the cell outside (Fig. 1). If the two end cells of a linear rouleau are sucked into two oppositely directed micropipettes of suitable diameter, the rouleau is held firmly. When the tips of the micropipettes are separated, the rouleau does not break up. The cells elongate in the direction of traction and then separate, but remain attached to each other by fibrils (Rowlands et al. 1983a) which may reach a length of 100 μm. Moreover, the fibrils are stable over periods of 30 or more minutes. If the negative pressure in the micropipettes is removed, the fibrils will rapidly pull the cells together at velocities of up to 180 μm/s until the original rouleau is restored. The proceedure can be repeated indefinitely. We call these contractile fibrils "contractils".

To see if this phenomenon was dependent on the requirements of the Fröhlich theory, the cells were modified by the techniques used in the departure from Brownian motion experiments described above (Sect. 2.5). Glutaraldehyde-fixed cells, which stick to each other, were quite rigid and could not be aspirated into pipettes. They could, however, be separated by shear with the tips of the pipettes. There were no contractils. When the membrane potential was abolished by the ionophore (A23187), rouleaux of normal appearance formed. The end cell would enter a pipette under negative pressure but immediately became detached from the rouleau with no contractil formation. The cells could be picked off one by one from a rouleau with a single pipette. When the membrane potential was abolished by change in pH, some rouleaux fell apart on micromanipulation, as was the case with the ionophore but others formed short contractils which would break easily and would not remain for more than a few seconds. Restoring the pH and so the membrane potential gave contractils more like normal ones, but they never entirely recovered their properties. When the cells' metabolism was depleted, the results were the same as when the membrane potential was abolished by change in pH. Restoring the metabolism with adenosine improved the contractils significantly, but once again their properties were never fully restored.

A mixture of cells from man and rabbit forms rouleaux in which the order of the cells is not random (Sect. 4; Sewchand and Canham 1976; Sewchand and Rowlands 1983). This finding is evidence for the specificity of the Fröhlich interaction (Sewchand and Rowlands 1983). On micromanipulation, a mixed rouleau always comes apart at the junction between a human cell and a non-human cell, sometimes, but not usually, with the formation of a short contractil. However, if a mixed rouleau is stripped until all that is left is a rouleau of the cells of one species it produces normal contractils on applying tension (Rowlands et al. 1983a).

The interaction-at-a-distance demonstrated by both the haemacytometer method (Sects. 2, 3) and QELS (Sect. 5) are dependent on an adequate concentration of extended macromolecules, which infers that the interaction is transmitted by such macromolecules. Contractils are also dependent on the presence of extended macromolecules but strong solutions of such macromolecules can be pulled out into thin threads [e.g. poly(ethylene oxide)]. If two objects are coated with extended macromolecules and pulled apart, then a thin thread will remain between them. Contractils, however, do not form with all "objects": Glutaraldehyde- and ionophore-treated cells do not form contractils at all and in the other cases the properties of the contractils vary significantly when metabolism and membrane potential are changed. When the tension on an extended rouleau is released, the contractils shorten rapidly to zero length. A viscoelastic thread would shorten but not down to zero length. Moreover a passive thread would remain attached at the point on the surface of the cell from which it originated. In Fig. 1 it can be seen that the line of attachment of the contractil to the spherical part of the end cell held in the micropipette is directed towards the centre of the sphere. If the pipettes are moved in and out of alignment with each other, the point of attachment moves freely over the spherical surface and the line of the contractil always points to the centre of the sphere, however fast this manipulation is performed.

There is additional evidence that the contractils are not just a "chemical" phenomenon. We have manipulated a branched rouleau with three micropipettes arranged in a Y-formation to produce three contractils meeting at a point (Rowlands et al. 1984). The triple contractil is as stable as a single one. By appropriate movement of the pipettes the junction can be moved back and forth and even run onto one of the cells so that three contractils are attached to one of the three cells. This manoeuvre is reversible. If we assume that the junction of the three contractils is simple, it will be the smoothed junction of three coplanar cylinders: that is, at the junction there will be surfaces of both positive and negative curvature. Such a composite junction is incompatible with an interfacial tension, because the pressure difference associated with surface tension would have to be oppositely directed across the oppositely directed curvatures.

Young observers can just see contractils by light microscope, but no unequivocal photograph has been produced. Contractils can be photographed by scanning electron microscopy, but the technique demands considerable patience (Rowlands et al. 1983a). Work had to be abandoned before we achieved a picture of a triple contractil but a good photograph of a single contractil is seen in Fig. 2. There is nearly always an artefactual break along a contractil, and in the case of Fig. 2, such a break is in the middle. More consistent are the discontinuities where the contractil meets the drawn-out cell membranes at either end. On the negative there is seen to be some fainter structure within the discontinuities. Del Giudice et al. (1985) have suggested that contractils are formed and maintained by forces arising from self-focussing of Fröhlich waves by analogy with the forces of self-focussing in non-linear optics. If that is so the gap between the drawn-out membrane and contractil is the distance required for self-focussing to become established (Del Giudice et al. 1985). The gaps may, however, be artefactual, since they are not so obvious when platinum rather than gold is used in the shadowing process.

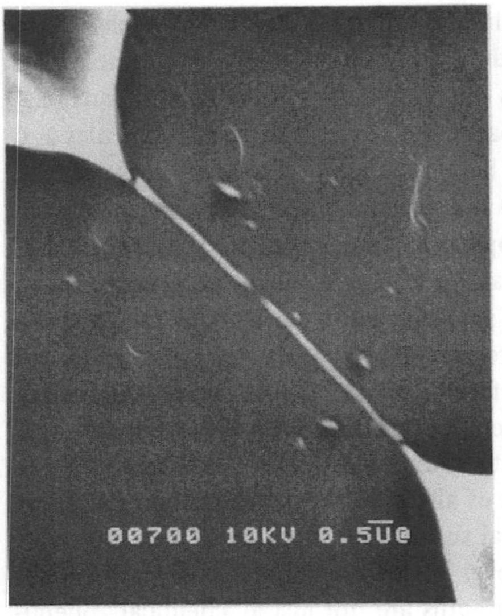

Fig. 2. Scanning electron micrograph of part of a rouleau extended as in Fig. 1. The macromolecular solution was slowly replaced with isotonic glutaraldehyde while tension was maintained by the pipettes. After fixing by glutaraldehyde, the specimen was washed in distilled water, dried overnight and then gold-shadowed. Scale line at lower right is 0.5 μm long

7 Future Work

The work described in this chapter started in the summer of 1980 and finished in the summer of 1984 with the dispersal of the team. We explored a new field of research widely, leaving a host of questions unanswered. The results are intriguing, even startling. What is needed is consolidation.

7.1 Macromolecules

Perhaps the most pressing questions surround the nature of the macromolecules needed to transmit the ultra-long-range interaction. This necessity was not explicit in the Fröhlich theory (Fröhlich 1980). Pohl (1980, 1981) has detected an attraction between cells, other than erythrocytes, and inorganic particles without macromolecules being present. Pohl's experiments were done in artificial media of very low electrical conductivity, whereas blood and the solutions in PBS we used are highly conducting. If a non-conducting medium could be found in which the red cells would not deteriorate within an hour it would be worth establishing whether extended macromolecules are a sine qua non for an interaction or whether they, by forming insulating "bridges", avoid the high conductivity of the medium.

7.2 What Macromolecules at What Concentration?

We believe that fibrinogen is the carrier in blood but not pure fibrinogen. Pure fibrinogen is likely to be a globular protein which extends into a pseudo-linear structure in the presence of one or some of the many factors associated with the "extrinsic" pathway of the cascade of reactions which lead to the coagulation of blood (Sewchand et al. 1984). Clearly there are some interesting experiments to be done with mixtures of fibrinogen and the dozen "factors" of blood coagulation. Albumin is not a carrier of the interaction, but it can be transformed if its globular structure is unravelled into an extended form (Rowlands et al. 1983b). We have found interactions with polyvinylpyrrolidone of molecular weight 360,000 dextran of molecular weight 70,000 (Sewchand et al. 1982b) and poly(ethylene oxide) of molecular weight 300,000 (unpublished), all of which have an extended structure. Fritz (1984) found, however, that there was no interaction with PVP of molecular weight 40,000 or 10,000, and our unpublished observations confirm these findings. The same is true for low molecular weights of dextran and of poly(ethylene oxide).

So one would like to know the minimum molecular weight of PVP to give an interaction and its minimum concentration; then the minimum concentration of higher molecular weight PVPs to give an interaction. These observations should be compared with structural observations to see if an interacting network of molecules is needed to sustain an interaction.

There was also a suggestion from some experiments (Sewchand et al. 1982b) that the interaction coefficient-concentration relation might be non-linear, perhaps even a step function. Further experiments would be easier and much quicker using QELS (Fritz 1984) rather than the haemacytometer method (Rowlands et al. 1982a,b). It might turn out that once a network formed, an increase in its density made no difference to the magnitude of the interaction.

Finally, in connection with macromolecules, there are numerous others as yet untried, and many globular forms which might be converted into extended structures. But there are snags. We tried to unravel the tertiary and secondary structures of haemoglobin with sodium dodecyl sulphate and urea respectively, but both these substances killed the cells!

7.3 Vicinal (Ordered) Water

For 20 years or more a controversy has raged over whether the structure of water was modified or not in the immediate neighbourhood of intracellular structures composed of extended macromolecules. An extended protein chain with alternating negative (CO) and positive (NH) chemical groups should polarize the neighbouring water and reduce its translational and rotational motion. This would modify the properties of intracellular water to such an extent that intracellular reactions might be different from the same reactions studied in vitro in dilute aqueous solution (Clegg 1983).

A comparable controversy has raged over how the marked difference in the concentration of ions across a cell membrane is maintained against the "leakiness" of the membrane. For instance, sodium has a low concentration inside compared with outside and the reverse is so for potassium. Most biologists believe that the concentrations are maintained by "pumps", but Ling (Ling et al. 1973) has shown that no cell has anything like the supply of energy necessary to maintain all the postulated "pumps"! Ling's alternative hypothesis, in brief, is that potassium inside the cell is almost entirely bound to protein and that the sodium ion, which has a large hydration volume, is much less soluble in the "ordered" water within a cell than it is in normal water. Hence potassium is bound and sodium is excluded by processes which do not require a constant supply of energy.

Recently, Ling and his colleagues have completed a major study on the physical state of water in living cells and model systems (Ling and Ochsenfeld 1983; Ling and Murphy 1983; Ling 1983; Ling and Zhang 1983; Zhang and Ling 1983). Very briefly: they show that the presence of gelatin, polyvinylpyrrolidone, poly(ethylene oxide) and other extended macromolecules reduces the solubility of sodium ion in water (Ling and Ochsenfeld 1983) and alters the NMR relaxation times of water protons in a manner consistent with the reduced solubility for sodium (Ling and Murphy 1983). They also show highly abnormal osmotic activities in their macromolecular solutions and hence a reduction in the chemical activity of water which parallels the reduction in sodium solubility (Ling 1983). The two other papers (Ling and Zhang 1983; Zhang and Ling 1983) concern some interesting details of the state of their "ordered" water during freezing and thawing. The relevance to the present work is that perhaps the property of the macromolecule which mediates the transmission of the Fröhlich interaction is the molecule's ability to polarize the surrounding water into oriented, ordered multilayers, along which Fröhlich polar waves could be transmitted. Ling demonstrates ordered water with PVP and poly(ethylene oxide), both of which transmit the ultra-long-range interaction. Denaturing albumin with sodium dodecyl sulphate, which unravels the tertiary structure, did not produce ordered water. On the contrary, treatment with urea, which affects the secondary structure and exposes NH and CO groups, did result in the ordering of the water. This parallels our conversion of albumin from a non-transmitter to a transmitter of the Fröhlich interaction by unravelling its secondary structure by heating. So there would be good sense in seeking a closer correlation between the induction of ordered water and the transmission of the Fröhlich interaction.

7.4 The Cells

The work has been done mainly on human red cells (erythrocytes). Cells from other mammalian species were employed only in establishing specificity for the interaction (Sewchand and Rowlands 1983). The species which we studied all have biconcave rouleau-forming erythrocytes although their sizes vary significantly. Size itself is unlikely to make much difference, although a study might be interesting, but some mammals (e.g. camel) have ovoid cells which differ in important respects (e.g. resistance to osmotic haemolysis) from the cells which we have studied. Mammalian cells have no nucleus; amphibian cells are nucleated and might show quite different properties.

It was pointed out above that a sample of mammalian cells consists of cells of all ages, so some are senescent with perhaps abnormal properties. There are also considerable well-documented variations in the properties of the cells in a single sample from a single individual. There is, for instance, a very wide span (factor of 3) in the permeability of single cells to various non-electrolytes (Jay and Rowlands 1975). Human red cells can be separated into young and old by sedimentation in a density gradient, but our experiments using this technique showed no significant differences (unpublished). Reticulocytes, immediate precursors of mature red cells, and earlier precursors such as the normoblasts which have nuclei can be separated. These categories might show interesting differences.

Returning to human erythrocytes: In establishing accord between the ultra-long-range interaction and Fröhlich's theory, the membrane potential and the metabolic stores were changed from normal to zero. In some unpublished experiments we reduced the membrane potential step by step to zero. We saw signs of a reduction of the interaction coefficient, but this work needs repeating. Reducing metabolism other than from normal to zero necessitates measuring metabolism and this we did not attempt. Our disorganization of the cell membrane with glutaraldehyde was a crude method, although the cells retained their shape and size and formed rouleaux. The surface charge was retained, but it is probable that the membrane potential and the metabolic pathways were affected. Reference to the extensive literature on membrane function will provide many subtle ways of perturbing the membrane and the cells' metabolism. Finally red cells can be depleted of nearly all their haemoglobin by haemolysis. With care, the resultant "red cell ghosts" will reform as biconcave discs which retain many membrane functions. We have not tried "ghosts".

7.5 Cells Other Than Erythrocytes

The erythrocyte is the simplest of mammalian cells, although still very complicated. It has no intrinsic motility, which makes it ideal for these studies because only the physical, well-studied and mathematically tractable Brownian motion has to be subtracted to reveal an interaction. The erythrocyte is likely to remain the ideal biophysical system for studying Fröhlich's interaction for years, not only because of its simplicity, but also because a great deal is known about the structure and function of this cell. Controls on experiments can be firmly and acceptably established.

The next most profitable cell seems certain to be the platelet (thrombocyte). Studies on its interaction at a distance may be of great significance in studies of the major killer diseases of the human cardiovascular system, atherosclerosis and coronary heart disease (Haemostasis, Br. Med. Bull. 1977). Platelets are the smallest cells in human blood, of average diameter 2–4 μm. They are oval or disc-shaped but not biconcave. By light- or electron-microscopy they have have a complex structure but, like red cells, they have no nucleus. Unlike red cells they can quickly change their properties when stimulated with a variety of agents, including agents such as collagen, which are released when blood vessels are cut or injured. They become sticky, the membrane wrinkles and they adhere strongly to each other. They form a plug which plays a major role in sealing up blood vessels damaged by trauma (Haemostasis 1977; Born et al. 1981). The rate of aggregation of platelets, in vitro, plays an important part in the investigation of disorders of haemostasis. Instrumentation ("aggregometer") is commercially available. In such an instrument separated platelets, when stimulated, aggregate in a few seconds. Born (1977), however, has shown that, in vivo, aggregation can occur a thousand times faster. If a small blood vessel is cleanly punctured, a platelet in the blood spurting through the wound spends no more than 2 ms in the neighbourhood of the hole, yet the platelets aggregate on the cut edge and seal it in a few minutes! Is this very fast interaction between the damaged tissue and the platelets a Fröhlich interaction?

There is some indication that platelet aggregation in vivo is quicker in the presence of red cells than in their absence. I have looked for a physical interaction between unstimulated platelets and red cells but seen nothing obvious. I have seen a spectacular interaction between platelets and small lymphocytes (members of the family of white cells in blood). Small lymphocytes are spheres of about the same diameter as erythrocytes. They repel each other and a preparation photographed by time-lapse cinematography resembles a preparation of microspheres. Occasionally a lymphocyte will transform into an amoeboid form and when the film is run at 24 frames per second this transformed cell has all the appearance of a Walt Disney ghost! With platelets in the lymphocyte preparation one in about a hundred lymphocytes will be "attacked" by platelets. They stick to the surface of the lymphocyte which throws out pseudopods wherever platelets are attached. The appearance is quite different from the rather purposeful ghost-like dance of the amoeboid cell. The significance of the platelet-lymphocyte interaction is quite unknown. It is clear from these few observations that the platelet is well worth investigating, particularly by QELS, with which it should be possible to stimulate the cells while under measurement. The behaviour of the white cells is even more complex in their function of seeking out and destroying "enemies" of the host organism and it may be years before their properties can be related to Fröhlich's coherent excitations.

7.6 Contractils

Experiments on contractils are difficult, particularly to make them quantitative. Some who have tried the technique abandoned it without making even one satisfactory pipette! To make, fill and mount two or three pipettes, to be able to manipulate them and their pressure taps for hours without breaking them or letting them clog is a skill of the few. Experiments on contractils should not be embarked on lightly.

Nevertheless these quantum-mechanically maintained fibrils may well be the ideal model system for studying the network of macromolecular fibres which are increasingly being shown to occupy what was previously thought to be the liquid compartments of cells (Clegg 1983) and of nerve axons (Nagele and Roisen 1982). It is rapidly becoming apparent that few, if any, large molecules are in simple solution in living matter. Rather they are associated with or part of a complex and ordered mesh which has been described as "a vectorial framework upon which some other force-generating component is organised (Hyams 1981)". These networks are dynamic. They form and reform in apparent fulfillment of the needs and function of the tissue.

Up to now we have only experimented on contractil formation in blood (in which we presume fibrinogen forms the fibrils), and in solutions of PVP 360 and poly(ethylene oxide) of molecular weight 400,000. In cells, the so-called cytoskeleton consists of actin, spectrin, tubulin and other less well-defined proteins (Clegg 1983; Del Giudice et al. 1985). If contractils are of biological significance they should form from these substances. If contractils form from tubulin they should be reversibly disorganized by Concanavalin A, and this drug is unlikely to affect the red cells which contain no tubulin. Cytochalasin B affects actin fibrils, but unfortunately it also affects spectrin, which is a component of erythrocytes. There are other agents which might be worth a trial.

We have made preliminary studies of two means of measuring the tension in contractils. If the pressure in one pipette is released, the stretched rouleau will recoil. Its velocity can be measured with a high-speed ciné camera. Using the results of our studies on the drag coefficients of red cells (Sewchand et al. 1982a), we estimated a rough upper limit for the tension as 10^{-6} N. Assuming the diameter of the contractil to be 0.2 μm the force per unit area is of the order of 10^7 Nm^{-2} (Rowlands et al. 1983a). The other method consists of measuring the minimum pressure in the pipette to maintain a rouleau in a state of stretch. Unfortunately the held cell does not seem to make a perfect seal of the end of the pipette and in any case red cells are very permeable to water. So there is a flow of water in the pipette and therefore a pressure gradient. This vitiates the measurement of the pressure which is made, unavoidably, downstream of the narrow end of the pipette. The flow can be measured with an oil-water interface in the pipette and corrected for, but making the experiments is a frustrating experience. Preliminary results give tensions of the same order as does the recoil method (K. Kaler, unpublished).

7.7 Imposed Electromagnetic Fields

In view of the interest in, and the practical importance of interactions between electromagnetic fields and cells (Chiabrera et al. 1985) the Fröhlich attraction demonstrated in mammalian erythrocytes should be examined in electric and magnetic fields, steady, alternating and pulsed.

8 Significance

The significance of rouleau formation is unknown and though it must play at least a small part in the early stages of the stemming of haemorrhage, this has never been stated. Our knowledge of platelet aggregation and of fibrin formation in clotting blood is largely arrived at from observations made in vitro, when red cells are usually excluded. There is some indication that red cells enhance platelet aggregation in vivo, although this could be caused merely by axial streaming of red cells in flow (Rowlands 1966) which displaces platelets to the periphery of the lumen of blood vessels and so increases their concentration at the wall of the damaged vessel. There seems to be little function for an ultra-long-range interaction between erythrocytes in flowing blood in which the cell concentration is so high (~50%) that the cells are normally in contact with one another. The interaction will, of course, hold the cells more tightly together in shed, virtually stationary blood.

Possibly the red cell interaction is vestigial. It has been suggested that there is a "second nervous system" (Rowlands 1983a) much more subtle than the "action-potential" system as we know it at present. This idea is an elaboration of "neurotrophism" which may be defined as interactions between nerves and other cells which initiate and control molecular processes. Red cells are part of a complex organ called the reticulo-endothelial system and one part of this is the "erythron", which denotes the red cell from its birth as a precursor cell in the bone marrow to its destruction in the spleen. The reticulo-endothelial system and so the erythron are under control by the nervous system and perhaps the erythrocyte interaction which we have studied is a vestige of the "receiver" mechanism which is essential for control of function of cells from their birth to their death. The "transmitter" would be the second nervous system.

The innumerable chemical reactions studied by biochemists occur in tissue in an ordered fashion with little waste of energy. It is now inconceivable that, as was earlier thought, molecules wander around inside the cell by quite random motion until they happen to meet at a suitable time and orientation for a useful reaction. The conceivable explanations are coming from non-linear (Scott 1981) and coherent phenomena (Fröhlich 1980) and there is no lack of speculation (Rowlands 1983a,b, 1985). The erythrocyte interaction provides a mathematically tractable non-linear coherent system. It is a simple biological system suitable for the study of vectorial interactions in life and the effects on them of electromagnetic fields.

References

Berne BJ, Pecora R (1976) Dynamic light scattering. John Wiley, New York

Born GVR (1977) Fluid-mechanical and biochemical interactions. In: Thomas D (ed) Haemostasis. Br Med Bull 33:193–197

Born GVR, Görög P, Kratzer MAA (1981) Aggregation of platelets in damaged vessels. Phil Trans R Soc Lond B294:241–250

Canham PB, Burton AC (1968) Distribution of size and shape in populations of normal human red cells. Circ Res 22:405–422

Chandrasekhar S (1943) Stochastic problems in physics and astronomy. Rev Mod Phys 15:1–89

Chiabrera A, Nicolini C, Schwan HP (1985) Interactions between electromagnetic fields and cells. Plenum, New York

Clegg JS (1983) Intracellular water, metabolism and cell architecture. In: Fröhlich H, Kremer F (eds) Coherent excitations in biological systems. Springer, Berlin Heidelberg New York, pp 162–177

Del Giudice E, Doglia S, Milani M (1985) Ordered structures as a result of the propagation of coherent electric waves in living systems. In: Chiabrera A, Nicolini C, Schwan HP (eds) Interactions between electromagnetic fields and cells. Plenum, New York, pp 157–169

Einstein A (1926) Investigations on the theory of the Brownian movement. Transl. and ed. by Fürth R, Dover Publications, New York

Enns EG, Fung TS, Rowlands S, Sewchand LS (1983) Planar Brownian motion in the presence of an attractive force. Cell Biophys 5:189–195

Fåhraeus R (1929) The suspension stability of blood. Physiol Rev 9:241–274

Fritz OG (1984) Anomalous diffusion of erythrocytes in the presence of polyvinylpyrrolidone. Biophys J 46:219–228

Fröhlich H (1980) The biological effects of microwaves and related questions. Adv Electron. Electron Phys 53:85–152

Hyams J (1981) Hooked on microtubules. Nature 291:107–108

Jay AWL, Rowlands S (1975) The stages of osmotic haemolysis. J Physiol (Lond) 252:817–832

Ling GN (1983) Studies on the physical state of water in living cells and model systems. III. The high osmotic pressure of aqueous solutions of gelatin, polyvinylpyrrolidone and poly(ethylene oxide) and their relation to the reduced solubility for Na, sugars and free amino acids. Physiol Chem Phys 15:155–165

Ling GN, Murphy RC (1983) Studies on the physical state of water in living cells and model systems. II. NMR relaxation times of water protons in aqueous solutions of gelatin and oxygen-containing polymers which reduce the solvency of water for Na, sugars and free amino acids. Physiol Chem Phys 15:137–154

Ling GN, Ochsenfeld MM (1983) Studies on the physical state of water in living cells and model systems. I. The quantitative relationship between the concentration of gelatin and certain oxygen-containing polymers and their influence upon the solubility of water for Na salts. Physiol Chem Phys 15:127–136

Ling GN, Zhang ZL (1983) Studies on the physical state of water in living cells and model systems. IV. Freezing and thawing point depression of water by gelatin, oxygen-containing polymers and urea-denatured proteins. Physiol Chem Phys 15:391–406

Ling GN, Miller C, Ochsenfeld MM (1973) The physical state of solutes and water in living cells according to the association-induction hypothesis. Ann NY Acad Sci 204:6–50

Masri MA, Masri SA, Boyd NG (1983) Isolation of human fibrinogen of high purity and in high yield using polyethylene glycol 1000. Thromb Haemostas 49:116–119

Nagele RG, Roisen FJ (1982) Ultrastructure of a new microtubule-neurofilament coupler in nerves. Brain Res 253:31–37

Overbeek JTG (1952) Colloid Science Vol 1. Kruyt HR (ed) Elsevier, Amsterdam

Paul R, Chatterjee R, Tuszyński JA, Fritz OG (1983) Theory of long-range coherence in biological systems. I. The anomalous behaviour of human erythrocytes. J Theor Biol 104:169–185

Pohl HA (1980) Oscillating fields about growing cells. Int J Quantum Chem 7:411–431

Pohl HA (1981) Natural electrical RF oscillation from cells. J Bioenerg Biomembr 13:149–169

Rowlands S (1966) In: Rotblat J (ed) Aspects of medical physics. Taylor and Francis, London

Rowlands S (1983a) In: Fröhlich H, Kremer F (eds) Coherent excitations in biological systems. Springer, Berlin Heidelberg New York

Rowlands S (1983b) Some physics aspects for 21st century biologists. J Biol Phys 11:117–122

Rowlands S (1985) Quantum mechanical coherence in human red blood cells. J Statis Phys 39: 543–549

Rowlands S, Skibo L (1972) The morphology of red cell aggregates. Thromb Res 1:47–58

Rowlands S, Sewchand LS, Lovlin RE, Beck JS, Enns EG (1981) A Fröhlich interaction of human erythrocytes. Phys Lett 82A:436–438

Rowlands S, Sewchand LS, Enns EG (1982a) A quantum mechanical interaction of human erythrocytes. Can J Physiol Pharmacol 60:52–59

Rowlands S, Sewchand LS, Enns EG (1982b) Further evidence for a Fröhlich interaction of erythrocytes. Phys Lett 87A:256–260

Rowlands S, Eisenberg CP, Sewchand LS (1983a) Contractils: quantum mechanical fibrils. J Biol Phys 11:1–4

Rowlands S, Sewchand LS, Skibo L (1983b) Conversion of albumin into a transmitter of the ultra long-range interaction of human erythrocytes. Cell Biophys 5:197–203

Rowlands S, Skibo L, Eisenberg CP, Sewchand LS (1984) Triple contractils. J Biol Phys 12:31–32

Scott AC (1981) In: Enns RH, Jones BL, Miura RM, Rangneker SS (eds) Nonlinear phenomena in physics and biology. Plenum, New York

Sewchand LS, Bruckschwaiger D (1980) Observed differences in dextran and polyvinylpyrrolidone as rouleaux-inducing agents. Can J Physiol Pharmacol 58:271–274

Sewchand LS, Canham PB (1976) Induced rouleaux formation in interspecies populations of red cells. Can J Physiol Pharmacol 54:437–442

Sewchand LS, Rowlands S (1983) Specificity of the Fröhlich interaction of erythrocytes. Phys Lett 93A:363–364

Sewchand LS, Rowlands S, Lovlin RE (1982a) Resistance to the Brownian movement of red blood cells on flat horizontal surfaces. Cell Biophys 4:41–46

Sewchand LS, Roberts D, Rowlands S (1982b) Transmission of the quantum mechanical interaction of erythrocytes. Cell Biophys 4:253–258

Sewchand LS, Masri MA, Fritz OG, Boyd NG, Rowlands S (1984) Fibrinogen and the ultra long-range interaction of human erythrocytes. Cell Biophys 6:215–221

Zhang ZL, Ling GN (1983) Studies on the physical state of water in living cells and model systems. V. The warming exothermic reaction of frozen aqueous solution of polyvinylpyrrolidone, poly(ethylene oxide) and urea-denatured proteins. Physiol Chem Phys 15:407–415

Note added in proof with H. Fröhlich's permission

The criticism in a paper by Müller-Herold, Lutz and Kedem (J Theor Biol (1987) 126:251–252) is incorrect, as will be shown on another occasion.

The Genetic Code as Language

F. Fröhlich[1]

There is a concrete paradox between the fundamental law, or profound assumption, of the survival of the fittest and the newly realized fact that all DNA does not code for proteins – it is "selfish" (Dawkins 1976) or to put it bluntly, as Orgel and Crick (1980) have done, is "junk".

In the unravelling of the genetic code only a single celled body having only one chromosome, *E. coli*, was used. It was assumed that since the code applied universally in so far as the codons, although degenerate, designated universally one and only one of 20 amino acids, one could extrapolate from the mechanism of *E. coli* to all beings, especially to multi-celled organisms. But once one had began to investigate many-celled organisms, this was found not to be true; in fact, they contained a predominating quantity of copy DNA or sheer nonsense in imprecise sequences which made no contribution to the phenotype. What could be the survival value of having such quantities of mostly functionless DNA? This is the concrete paradox.

Its paradoxical status is somewhat diminished by the fact that there are also non-selective mutations in, for instance, the colour of flowers in one ecological niche. But this does not involve the negative form of the survival of the fittest. So we must modify the survival of the fittest, saying that if it has negative value to the organism, it will be selected against. Thus modified, the assumption remains valid.

In contrast to the survival of the fittest, or not relevantly at a disadvantage among the phenotypes in which the actual members must struggle for survival, (or more cogently formulated, must fit into a given niche), the selfish genes must struggle for survival within the chromosome as a whole. It is this struggle for survival within the genotype itself which forms the hypothesis of selfish DNA; the junk riding as a passenger on functional DNA. Orgel and Crick (1980) say "The conviction has been growing that much of this extra DNA is 'junk'", in other words that it has little specificity and conveys little or no selective advantage to the organism.

However, Crick was acutely aware of the lurking paradox, and explored possibilities for how the excessive DNA might function in providing central mechanisms for the function of the genes. It is perhaps in developing new controls that further evolution might take place. He cites Calavier-Smith's hypothesis that the sheer quantity of random DNA might be used for the cell's own purposes – for instance to slow up the cell's development or to make larger cells. Thus what begins as a parasite might develop into symbiosis.

1 Dept. of Physics, The University of Liverpool, Oliver Hodge Laboratory, Liverpool, L693BX, U.K.

He minimizes the difficulty involved in carrying random DNA by saying that even for a relatively long section of selfish DNA the energy requirement would be of the order of 10^{-6}, in fact, very small. But this takes no account of its three-dimensional nature – of the unknown way in which a chromosome must be rolled up at each cell division and its complex way of being partially unrolled during protein synthesis.

However, he rejects this idea, saying "It is difficult to accept the idea that all human parasites have been selected by human beings for their own advantage." And although admitting that an excessive baggage of functionless DNA might be seen as a cancer of the genome and that one does not know whether extinction of genotypes occurs in nature for this reason, he prefers to think that the cell only tolerates the excessive baggage of selfish DNA. Concluding he says "The main facts are, at first sight, so puzzling that only a somewhat unconventional idea is likely to explain them".

The third unconventional hypothesis suggested here is to broaden the "code" to include more of the functions of language – in a restricted sense to be elaborated below. This will be illuminating in bringing many questions and some not-even posed questions in biology together in one nexus, but, of course, if pursued too far it will be misleading. One could protest that taking this view was too anthropomorphic. However, the philosophic problems posed by the relation of code to the world of biology are similar to those posed by the relation of language to this same world – to say nothing of the even much more anthropocentred idea of "selfish genes" of "junk", or even the notion of "translator RNA". They are all, to a certain extent, anthropomorphic as is, even in physics, the idea of attraction at a distance, which was accepted.

This hypothesis might account in a physical way for all the "junk" or "selfish DNA" which occurs in the chromosomes by indicating how it might function in higher levels of complexity, which are not present in the bacterium (except in reproduction). This quantity of "junk" might have a real physical interpretation, giving bulk and a specific shape to the chromosome, transforming it into a three-dimensional coherent phenomenon. Thus, giving some functional meaning to this "junk" as not just a linear code but as a modified form of language might solve the paradox of its reconciliation with the modified form of the survival of the fittest. Of course, this language would not be exactly a human language, but while differing in some respects, it would have certain analogies.

The restrictions on the genetic language are outlined below.

1. The genetic language would provide no ambiguity, not give rise to jeux de mots or to poetry.
2. It would be imperative and not indicative. It might be modelled on Wittgenstein's (1953) elementary not-quite language from the "Investigations"; his language for the builder who says "Brick" when he means "Bring me a brick," "Trowel" when he means "Bring me a trowel." But in the genetic language it would not be "bring" but rather "create" (a protein).
3. Meaning. There is an arbitrary element in how a human word refers to things. If you do not know the word for "chaise" you will not understand "La chaise est rouge". It is learned in a social context. In fact, there is a tower of Babel erected among human languages. The chromosome, however, is not taught meaning although

degenerate. The codon specifies one and only one amino acid. It is not species-dependent – it is universal. One might naively ask why this universality in the chromosomal language, when in human language there is Babel?, and be answered that it is a question of physical fit; the three codons have a geometrical form which through the complication of the RNA fits uniquely to one amino acid.

But one can pose the question in a more sophisticated way by asking how it has come about in terms of evolution. It is a likely hypothesis that life commenced to emerge from the pre-biotic soup in several places in the world more or less simultaneously. If such were the case, it would be more probable that the genetic language would be different in different life forms. The amino acids might themselves be different and so there would be competing languages. It does not seem clear on a simple inspection of the formula for the 20 amino acids why there should be just these and not more or varying. But there are not. Why not? This is an unposed problem for which there is no simply conceivable answer. It seems to be a chance which has developed into a necessity.

There is thus this fit between codons and amino acids; not arbitrary as European letters are arbitrary, but intrinsic as parts of the Chinese and Japanese ideogram fit together to make up the whole idiogram or Kangi. This would then be the linguistic analogue.

4. The chromosomal language is concretely embedded within the cell along with the mitochondria, the substrates, the cell membrane as part of the cell. It functions on a level with the other cell components in biological, chemical and physical ways. It is embedded, one might say, as are the arithmetic functions in the wider Gödel function.
5. It is not a spoken language, but a "written" language.
6. Who uses the chromosomal language and to whom? The chromosome is used to communicate within the cell and in higher-celled organisms, between cells in a multitude of different situations. It replies by making proteins of a certain type or refraining from making certain proteins or by redoubling itself. There are cases of short-range communication creating transfer RNA which fits into the amino acids ingested through food, they already containing information, the information which enables them to fit into the right transfer RNA. Or there may be in the interstices of a cell certain substrates, a protein, which requires a specific enzyme to digest it. This substrate communicates over fairly long distances with the chromosome giving the message "Create a specific enzyme" or it could be the presence of antigens which communicates in an elaborate way but ultimately *through* the chromosome saying "Create a specific antibody". The second type of communication with the one-celled organism is that which tells the chromosome to divide as a whole (see 3). Beyond the single-celled organism, where the case of selfish DNA or junk first occurs, there arises the question of differentiation. Here there must be a series of long-distance communications from the first-ordered differentiation, through the cell's migration to the target cells, to its subsequently becoming the right sort of cell for the organ to which it has migrated. Beyond the chromosome, the genome as a whole must contain some sort of long-distance communication in order not to produce on one section antigens produced on other parts. So there is a complexity of entities which use genetic language in different ways. There might be said to be a logic of cells.

The replies would be fairly simple, either in producing a certain protein, in refraining from producing certain proteins or in reproducing itself. This is a hypothetical sketch of who or what might be said to use the chromosomal language.

It might be asked whether these constituents using the chromosomal language possess any type of mind. Obviously they do not have mind as self-consciousness, but a more illuminating answer might be prised from Whitehead (1929). He has a holistic metaphysics in which a multitude of actual entities have both a physical pole, in which they collect their reaction from the given world from their perspective on it, and a mental pole, in which they react, "feel and organize themselves to have an influence on the future". This he calls the concrescence of the world. In his sense, a certain low type of mind might be said to be in operation in the cells, a mind of a primitive enough nature to communicate in the ways sketched above.

Although the restrictions on the concept of language might make one wonder why it is introduced at all, the code model leads one to expect that all that is necessary for a complete understanding of the cell's functions is the genes coding for proteins, whereas there are many more collective aspects of the functioning of a chromosome which might have clear analogies with language.

The investigation here will proceed in a linear way from the shortest segments of the chromosome to progressively longer segments, finally including the chromosome as a whole and indeed the whole group of the organism's chromosomes, the genome, planetary genome. In the following schematism these chromosome segments will be indicated under "I". The biological entities and functions controlled by these will be indicated under "II". The physical correlates will be suggested under "III". The units of language suggested as analogous to the functioning of the segments of the chromosomes are indicated under A.B.C. etc.

A. I. Nucleotides
   II. None
   IV. Dot-dash-dot of Morse code

B. I. Codons
   II. Amino acids
   III. Atomic repulsion and attraction
   IV. Letters

C. I. Genes
   II. Inactive proteins: Active proteins
   III. Coherent phenomena
   IV. Words (nouns): sentences

D. I. Longer sections of chromosome
   II. Operon or super-gene
   IV. Logical relations of implication or contradiction or relevance

E. I. Chromosome as a whole in communicating with others with protein
   II. Differentiation
   IV. Language as usage

F. I. Chromosome as communicating with itself
   II. Switch between proteins production and DNA production
   IV. Object language meta-language distinction

G. I. Genome
II. Self-recognition
IV. Consistency

H. I. Planetary genome
II. Modified self-recognition on one planet as opposed to other planets
IV. Modified consistency of planetary genome

A. The smallest element of the chromosome – the nucleotides – have individually no product within the cell. They might be said to correspond to the dot-dash-dot of the Morse code.

B. The next element, the codon, consisting of four nucleotides, has chemically a "fit" and through an elaborate mechanism involving several types of RNA, fits ultimately into an amino acid. This chemical "fit" comes about through physical means – through the attraction and repulsion of atoms. The amino acids also have a precise three-dimensional form (again created by forces of short-range interaction) which enables them, through the elaborate processes of RNA, to fit into the correct transfer RNA. By means of this fit the amino acids are recognized, but since there are only 20 amino acids and 64 possibilities of combination of nucleotides in a codon, their fit is rather degenerate, several codons coding for one amino acid and several being there to start and stop the process of coding. IV. These codons might be correlated with letters, with one provision. While the letters of our European languages are arbitrary – having no intrinsic relation to their meaning – these letters would be more like the components of Chinese ideograms or Japanese Kangi in that they have the complex geometrical shapes which enable them to fit into the definite shapes of the receptors, the amino acids.

C. I. The next-long functionally distinguished sequence is the gene, which contains information for a protein (or one of the protein's constituents – a polypeptide chain).

II. The long sequence of the gene codes indirectly for the long sequence of amino acids making up the proteins – a string of unrolled inactive amino acids. This is what is meant by decoding the genetic code – identifying the genes with their protein correlates in the linear inactive form. This would reflect the bio-chemist's method of searching for an isolatable enzyme corresponding to the one specific gene. This identification might correspond to the semantics of the genetic language.

However, there is an essential difference, which might find a linguistic analogue, between the linear inactive one-dimensional protein and its active three-dimensional form. The rolled-up active protein can find sites, enter into reactions, by virtue of its three-dimensional fit. This fit endows proteins with their remarkable ability to identify the specific substrates they should digest or the anti-genes they should attack. "Self-organization" is the word for this, but it is a word which, seemingly elucidating, masks difficulties. Thus the linguistic analogue would be "Digest this (and no other) substrate" or "Attack this anti-gene". So it would seem that the gene corresponds, not to a word – a noun, as it would with the unrolled protein, but to a simple imperative sentence. This would reflect the biologist's method of identifying a gene by what it does – what processes are hindered by its absence and hastened by its presence. Thus the unrolled inactive protein would have analogues with a word – a noun – to semantics, where as the three-dimensional active protein would have analogies with a single imperative sentence, under the limitations of its grammatical correctness and with syntax. So we have a switch from semantics to syntax in the process of self-organization within the genes product, the protein.

In human language, once words are given by convention, all combinations of words are no longer possible in a correct sentence. They are limited by the restriction on grammatical correctness. According to Noam Chomsky (1968), underlying the surface structure of sentences in various languages there is a universal deep grammatical structure common to all possible human languages.

Now does an analogue for his universal deep grammar exist for the chromosome as some sort of general restriction common to the chromosome's language? Its existence would mean that not every possible sequence of codons is possible within a gene – that given a certain sequence there would be limitations on those that could follow. But are there any such restrictions? Mathematical analysis of known codons has revealed no recognizable regularities in their occurrence. Reflecting this lack of order, the newly assembled polypeptide chain or protein has no restriction on its amino acid sequences.

But it is not yet an active protein. In order to become active, it must assume its correct three-dimensional shape. One might expect that it could roll up in an almost infinite variety of ways, but in fact it (nearly) always assumes only one definite configuration. This fact imposes severe physical conditions such that the free energy of this, the preferred configuration, must be sufficiently lower than that of any other possible configuration and possible conditions on the pathways to reach this. They should be miminized to meet the dynamical physical requirements. Such physical restrictions on the possible codon sequence might be the biological equivalent of universal grammar.

IV. In linguistics the possibility of certain words combining in meaningful sentences is manifested within the wider context of the activities of a linguistically competent person (his brain or his mind). This wider context might be said to correspond in biology to the cytoplasmic medium of the whole cell within which only certain sequences of amino acids can assume the unique configuration and thus become functional proteins which have, so to speak, meaning in the life of the organism, while others which do not fulfil these dynamic physical conditions cannot do so but remain as non-functioning nonsense. (Such protein nonsense might constitute a disease in which there might be protein chains floating inactively in the cytoplasm unable to take on their active configuration, but evolutionary pressures would limit their spread).

E. I. Longer regions of chromosomes: Here we have reached what would be the case biologically if the genes were floating separately randomly in the cytoplasm, but actually they are parts of chromosomes hundreds of times the gene's length. This would, in human terms, correspond to the situation in which a person would have the competence to make well-formed sentences but could not connect several sentences together in relations of implication, contradiction or even of relevance. (Chomsky does not concern himself with relations *between* sentences). One can see the human difficulties in being unable to connect sentences relevantly, but one might also ask biologically which kind of information beyond single gene specification for protein manufacture could be contained in this long series. This question has been answered by Jacob and Monod (1961). Within the bacterial chromosome the next-long operative segment is the operon or supergene. This could be, for example, a group of genes which code for several enzymes involved in the breakdown of a given substrate whose presence activates them. Or it might be a self-regulatory feed-back inhibition group for which the presence of the end product switches off the whole group. These short sequences linking the production of certain proteins, which function together, might parallel logical connections between several sentences. The substrate-induced production of the group of enzymes capable of digesting it would provide an analogue for the mutual relevance of a group of sentences to each other and to the occasion of their utterance. Thus these short gene sequences, the operon, suggest an analogue with the logical relations among sentences. The breakdown of operon functioning linguistically – the production of non-sequitors or meaningless repetition – might correspond to diseases of lack of control of protein functioning.

E. I. Beyond these short sequences of logically connected gene structure, one might expect some complex information conveying function in this large molecule, the chromosome. If this were not the case, then the very size of the chromosome, involving as it does complicated operations of unhelixing and separating during mitosis, finding the appropriate (sexual) partner with which to pair and thereafter being folded again without being tangled up in a space many times less than its own length would work against its survival. (Think of the complex operations of trying to fold up several very long hoses in a small garden). By this time, having been through countless efficiency tests according to the survival of the fittest, the chromosome would be no longer than necessary – perhaps operon-sized fragments. But it is not.

What sort of information would be lost if the entire chromosome were broken down into operon-sized fragments? Of course, that could not be done in a living cell – it would cease to be alive. It is more of a thought experiment. The answer would depend on the type of biologist to whom you posed the question. The molecular biologist would reply "None", maintaining that the rest of the components were mere "selfish DNA", going along for the ride or to put it bluntly "junk". The embryologist, on the other hand, might reply "A great deal" ... its ability to differentiate in a coherent way, not to reproduce itself in an uncontrolled manner, the ability to change its destiny if placed by the biologist in the wrong organ in the course of experiments.

On the side of the molecular biologist it has been found that beyond the sequence of several genes the gene order is not vital, whole genes being moved by translocation from one place to another, to say nothing of the movability of section of selfish DNA. So it would appear that great length of chromosome is not being used. What then of its survival value?

Since this is the point where our paradox first becomes apparent – the change from one-celled organisms coding for proteins to many-celled organisms which also code for proteins, but also carry with them a large load of selfish DNA or junk, we might here begin to notice the essential difference between the concept of "code" and the concept of "language." The code was decoded using a bacterium, *E. coli*, a one-celled organism having only one chromosome and without a nuclear membrane, and it was assumed that this would apply universally. When it was found, on investigating higher organisms, that this was no longer the case, there being quantities of repetitious DNA or junk, the very anthropomorphic concept of selfish DNA was introduced to side-step the paradox. But perhaps instead of denying that the junk has major survival value, one should have looked for more complicated methods of control using this for the cell's purpose.

Let us return to the embryologists. There are two aspects of differentiation which must be dealt with – the one natural, the other induced by the embryologist himself to see how the cell changes if forcefully transplanted into another organ.

The first is the differentiation of the fertilized egg cell into a many-celled organism. This begins, obviously, with cell multiplication, but not random multiplication, as with the bacterium, but in a highly ordered way. Indeed, these changes in the embryologist organization were formulated mathematically in terms of "catastrophes" by Rene Thom (1974), the catastrophes corresponding to fundamental changes and discontinuities of shape (leading to changes of function). From a very early stage the cells are predestined to become the sort of cells they will become, even sometimes moving over large distances to join their potential neighbours. During differentiation, cells become gradually fixed from their originally labile state into specialized protein production by their surrounding cells. So the cells must communicate during development with their neighbouring cells. One might suspect that the excessive amount of DNA, in the chromosome since it arises at this stage, has a part to play in this fixing.

One might suspect this even more strongly in considering the experiments performed to see how the implanted embryological cell develops differently, depending on the organ into which it is transplanted and the time scale of its gradually moving from lability to fixity. For instance, a young liver cell transplanted to a kidney cell would become a kidney cell, while a more mature one would be perplexed. This is referred to as embryological induction. So the environment, its neighbours, and the degree of its maturity determine how such a transplanted cell will develop.

Does this induction depend on some collective property of the chromosome as a whole, including both the coding DNA and the non-coding selfish DNA? It might be that the apparently excessive length of DNA is necessary for such a collective functioning. This could represent its apparently absent survival value.

E. III. Physical explanation in terms of excessively long DNA. Arising from their excessive length, chromosomes might have a collective property – namely coherence.

Two hypothesis about how the chromosome might function collectively in the control and maintenance of differentiation have been suggested. One by R.R. Cook (1973) in Oxford assumes a model with large control bands of nucleotides whose order is not significant (the junk which is not transcribed) and with small significant interbands (which is transcribed). The large bands would function collectively in some sort of communication with the cell and its neighbours and fix the particular chromosomal configuration so that selected sections would be exposed for the production of the relevant proteins. How could they do this? The second hypothesis, by H. Fröhlich, indicates a way. It suggests that long-range coherent vibrations will lead to resonance between a differentiated cell with its own characteristic vibrations and the chromosome such that the chromosome-particular region responding to this characteristic frequency will be activated or opened up so it can produce the appropriate proteins. Such a resonance could transport the embryological, already partially induced cells to their target cells and there they would be further fixed into producing the correct proteins for this organ by their neighbours. It might, furthermore, be used to explain how a fairly young cell – say, a kidney cell – transplanted to a liver organ would adapt to that organ through superimposed resonance and why if it were too mature, it would not so adapt. There might be degrees of resonance.

E. IV. Here, beyond the gene and the supergene, there might be analogies with communication with neighbouring cells – the precise specification of a chromosome-possible range of utterances through the activation by its neighbours. For instance, there might be unspoken conversations between neighbouring cells and an embryological cell giving it directions about what sort of cell it becomes – step by step – culminating in the embryological cell itself becoming a mature member of the community and joining the others in giving further directions to the young. When the organ was large enough, there would have to be some intra-cellular form of communication to end further proliferation. Without this, a sort of cancer would develop. It would, however, be a limited language – strictly imperative, leaving no room for wit or poetic ambiguities.

The element of black wit might enter artificially with the embryologist who, for his own scientific purposes, transplants a young kidney cell into a developed liver organ. Here the question would be "What will the neighbours say?", and "What would the cell itself respond?". This depends on its state of development. What would you say if you were a young kidney cell and began to get messages from a liver? Unlike Austin's ordinary language example where the response is "I would not know what to say," the young kidney cell does know exactly what to say, depending on the stage of development.

Thus the precise specification of a small part of the chromosome's repertoire of possible utterances by its neighbours displays analogies with language as usage.

F. Not only must the chromosome in some way communicate with its own and neighbouring cells, but in some circumstances it must also communicate directly with itself. In linguistic terms it must include the function of a meta-language.

II. There are three cases in which the chromosome must regulate its own functioning and give directions for the way in which it is to be read.

1. The messenger RNA must be produced on only *one* strand of the DNA, although both strands will necessarily be exposed simultaneously. This asymmetry, maintained only for relatively long strands of DNA, suggests that some collective property is here at work.

2. Switching for producing, very indirectly, proteins to reproducing DNA before cell division. It takes place via an enzyme which instructs the chromosome to replicate itself. But *why*, or rather *how*, does the chromosome produce such an enyzme? We know in the case of substrate or an antigen's presence in the cytoplasm that the chromosome must react in a definite way, but here the chromosome as a whole must react.

3. After replication, the chromosome must again begin to produce the necessary proteins. This takes place in an amazingly intricate way through the interaction of three types of RNA molecules with chromosomes and with amino acids floating in the cytoplasm.

Three types of RNA molecules

a) Most frequent (making up 80% of RNA) is ribosomal RNA, which has no specific relation to the order on the DNA. It attaches to the DNA at sites of the genes and provides energy for the assembly of the amino acids.

b) Messenger RNA reads the opened sections of the DNA – a gene or supergene – and it is attached to ribosomal RNA. It is finely specific.
c) Transfer RNA has one site which recognizes a codon in the messenger RNA and another which recognizes a site on the amino acid, through molecular fit. It goes out to search for the appropriate amino acids in the cytoplasm and returns to attach them to the messenger RNA.

These three ways in which the chromosome's information can be used must be contained in the chromosome itself. It would thus seem to be in the very centre of the situation in which self-referring paradoxes arise.

So far the chromosomal language has been modelled on the normal human language, with the modifications and limitations suggested, which might still suggest unwished-for anthropomorphic analogies, but now consider extending this to the meta-mathematical language which underlies elementary arithmetic and hence all mathematical, physical and biological extensions based on it. This would remove the reproach of anthropomorphism, except insofar as our understanding of the world is conditioned by possible mathematics, certain branches of physics. This would imply certain mathematical limits on what could be said, but mathematics, unlike the categories underlying language, can grow.

Frege formulated this concept of whole numbers, for instance a 5 as all classes of 5 entities. Russell found therein a paradox. Does the class of all classes contain itself? If it does, it does not. If it does not, it does. In ordinary language this would be formulated for a particular case. Thus; "I am a liar". If it holds, it does not. If it does not, it does. Russell avoided this in his theory of types by specifying that a statement *about* a statement is of a higher logical type. Thus from distinguishing logical types arose the distinction between object-language and meta-language.

Applying this object-language meta-language distinction to the biological language will be both illuminating and misleading. It will be perplexing when applied to the reading of the chromosome insofar as we have therein a direct method applying to the replication of the chromosome itself and a highly indirect method through many types of RNA molecules regulating the production of protein. It would a priori seem that the contrary should be the case, that the complexity of elaborate RNA mechanisms should correspond to the meta-language.

However, biology provides surprises. It is not to the replication of the DNA on itself, but, on the contrary, to the production of the protein that this complexity belongs. But the difference in the method of referring is here the essential feature, corresponding to the meta-language, object-language distinction.

In this paper so far, diseases which *might* arise from the faulty functioning of the genetic language have been mentioned hypothetically. However, in this case, a mechanism has been worked out biologically for a mix-up in function which has real consequences. It occurs in some rare cases that from RNA DNA is produced, this DNA then being projected into the cells's chromosome as a virus-type infection. There are four diseases in which this unlawful DNA enters the chromosome – the first three lead to different types of leukemia and the fourth leads to Aids (Gallo 1986). This contravenes in a way the basic dogma of molecular biology that states that DNA functions through RNA to produce proteins and that the proteins have no effect on the DNA and that RNA never produces independently DNA.

G. I. Genome II. Self-Recognition III. Coherence IV Consistency.
One might even look beyond the possible long-range properties of the single chromosome and ask whether *any* chromosomes whatever could combine in some viable organism or whether there are limitations of those which can be combined – whether there exist conditions for consistency for the whole genome. The genome must somehow contain within itself the information to make anti-bodies, not only against currently circulating diseases or possible new diseases which might arise, but also against any foreign proteins which would act as anti-bodies provoking a new and very specific production of anti-genes.

A well-functioning organism must somehow contain this genetic consistency. There might be thought to be four stages of the self-recognition. The drama begins within one cell because there are histicompatibility proteins on the cell membrane which effectively (or ineffectively) hinder the entrance of foreign proteins. So the first act plays among the chromosome of one cell, the germ cell or gamete and the chromosome or another such cell which have mated. During the complicated process of reproduction when the two gametes have lined up and are fusing into a zygote, analogous chromosomes would line up, crossing over to combine parts of the female and male chromosome. If the female chromosomes were inconsistent with the male's, did not at least partially agree in not producing antigens for the other anti-bodies, this might be one of the causes for infertility. In a later stage of embryological development when it became apparent that the agreement had not been adequately profound, the genome of the resultant cells might continue to produce anti-bodies to its own antigenes. This might be one of the causes of miscarriages.

The third and most harmless – not necessarily lethal – stage of this conflict within the genome might be the case of the individual who had survived the hazards of birth, developing an auto-immune disease, more or less serious. This, however, would occur later in life by random mutations in an adult. That auto-immune diseases are relatively infrequent and arise only with age indicates that the initial compatibility system worked fairly well.

The next stage is the healthy individual who is invaded by foreign proteins, either genuine antigenes or foreign proteins or cells introduced by a scientist or by a doctor. The bacteria having an antigenic character are attacked by antibodies created specifically from a somewhat more generalized protein. If a foreign protein is introduced for research or a whole organ is introduced medically, the genome might react by rejecting it in different degrees. This arises in medical cases, for instance when a kidney has been transplanted, when one does not know quite when or whether it will be rejected.

This poses the problem of self-recognition. What basically differentiates a self-cell from a foreign cell which would be attacked? The molecular answer to this is that there is a self-marker on all of the cells being coded for by the genetic code, but the efficient cause of this, how it comes about, is a question that is not even asked. It must be a collective property of the genome as a whole, since all of the markers are the same. This is a coherent phenomenon since knowing one such marker, one knows all of the others.

III. So must we consider some kind of pre-established harmony among the chromosomes or some chess-like game in which short-term survival would be balanced against long-term benefits (as Waddington does in the *The Strategy of the Genes*) or the hypothesis of language analogies?

The physical explanation is relevant in terms of the excessively long chromosomes might be some form of coherent excitation among members of the genome which would place some constraints upon them.

This has been done more specifically in the case of cell division during mitosis, where it has been hypothesized that corresponding chromosomes line up through resonance having the same frequency, finding each other from the apparent confusion of the genome by long-range communication (Holland 1972). Resonance oscillations draw like to like. More generally there might be some kind of coherent long-range interaction among the members of the genome creating the self-marking. If this broke down through subsequent mutations, auto-immune diseases would arise.

IV. Language: In the complete language there is no assurance that inconsistent statements will not arise; rather the contrary – they will. But it is only a part of the language possibilities which might be said to make up the genome of an individual. So in this limited language of an individual, such inconsistencies might be excluded.

G. I. Planetary genome
II. Translocation by viral infection
IV. Some degree of consistency

One might take this question of consistency even further in the concept of the "planetary genome" – evolved from translations of segments of chromosomes from one species to another through viral infections, as suggested by Dainelli. This might result in some degree of mutual tolerance – consistency – among all the organisms which have developed in each other's environment on one planet, as opposed to those which have developed on another planet.

Summary

Noticing the paradoxical relationship between the assumption of the survival of the fittest or not, and the excessive amount of DNA which does not code, this chapter has tried to resolve this by indicating other ways in which this excessive "junk" DNA might function, having more analogies with this complex functioning of language than with code. Such a language should be restricted and modified in the following ways

1. Unambiguous.
2. Imperative not indicative.
3. Meaning by molecular fit.
4. Embedded with the cell.
5. Not spoken but "written".
6. Used within the cell and between cells by entities which might be said, in a Whitheadian sense, to have a mental pole.

Investigation proceeds in a linear way from the shortest segments of the chromosome to the genome and finally to the planetary genome. Progressively longer segments of the chromosome will be indicated under A, B, C, etc.

A. I. Nucleotides
   II. None
   IV. Dot-dash-dot of Morse code

B. I. Codons
   II. Amino acids
   III. Atomic repulsion and attraction
   IV. Letters (as formed components of idieograms or Kangi)

C. I. Genes
   II. Inactive proteins, active proteins
   III. Coherence
   IV. Word (nouns) sentences

D. I. Longer sections of chromosome
   II. Operon or supergene
   IV. Logical relations of implication, contradiction and relevance

E. I. Chromosome making proteins
   II. Differentiation
   III. Coherence
   IV. Language as usage

F. I. Chromosome as communicating with itself
   II. Switch between protein production and DNA production.
   III. Object-language meta language distinction

References

Chomsky N (1968) Language and mind. Harcourt Brace & World, New York
Cook RR (1973) Nature 245:23
Dawkins R (1976) The selfish gene. Oxford Univ Press
Gallo RC (1986) Sci Am Dec 1986:78
Holland BW (1972) J Theor Biol 35:395
Jacob F, Monod J (1961) J Mol Biol 3:318
Orgel L, Crick F (1980) Nature 288:645
Thom Rene (1974) Modeles mathematics de la morphogenèse. Union Generale D'éditions
Whitehead AN (1929) Process and reality. Macmillan
Wittgenstein L (1953) Philosophical investigations. Basil Blackwell

Electromagnetic Effects in Humans

C. W. Smith [1]

1 Electromagnetic Effects and Humans in History

The recognition of electromagnetic effects by and in humans must begin with man's conscious perception of the light from the sun and other heavenly bodies, thunderstorms, geomagnetic fields and through magnetite (lodestone) to the navigators' compass. Gilbert (1600) published his treatise, *De Magnete*, in which he included an account of the "medicinal virtues" of lodestone.

Man's need for clocks and calendars presupposes the existence of a coherence in Nature. Both Kepler (1571–1630) and Newton (1642–1727) were receptive to ideas of planetary influences on humans. This interest has revived through the work of Gauquelin, particularly following his recent collaborations with Eysenck.

Unfortunately, electrical phenomena in biomedical studies have continually been the centre of much, often acrimonious, controversy. Mesmer (1733–1815) has been described as the author of the concept of "animal magnetism" (Didot 1861) although both he and a Fr. Hell S.J. hotly disputed priority for the therapeutic use of magnets. There was also much money involved, the King of France under-bid at least sevenfold for the "secrets" in his approaches to Mesmer through the French Academy of Sciences.

Galvani, an obstetrician and Professor of Anatomy at the University of Bologna, concluded from his many years of experiments on the relation between biology and electricity, that "animal electricity" was the long-sought-after "vital force". He announced this to the Bologna Academy of Science in 1791. Injury potentials, which he also discovered, were similarly interpreted. However, within a couple of years, Volta, a physicist of the University of Padua, had violently disagreed and concluded that the electricity was not animal, but electrochemical in origin (Geddes and Hoff 1971; Becker and Marino 1982; Becker and Selden 1985). The basic phenomena of bioelectricity had been demonstrated by the 18th century. Samuel Hahnemann (1755–1843) was clearly aware of the therapeutic possibilities of electricity and magnetism which he described in his *Organon of Medicine* (Hahnemann 1982).

Tesla (1856–1943), a prolific inventor and a visionary in many fields of electrical engineering and electronics, suffered an illness which from the account given by his biographer (O'Neill 1968) must have been an electromagnetically triggered hypersensitivity. The person who can justly be called the Father of Electrical Engineering must,

1 Electronic and Electrical Engineering Department, University of Salford, Salford M5 4WT, England

through his obsessional habits of working, have become the first well-documented case of this condition.

Lakhovsky (1939) investigated electromagnetic and other environmental factors in cancers and was the first to make systematic use of high-frequency electromagnetic fields (150 MHz) in biological experiments on cancerous plants. He regarded health as equivalent to the oscillatory equilibrium of living cells and illness as oscillatory disequilibrium.

Piccardi (1895–1972) particularly concerned himself during his last 20 years with the investigation of non-reproducible or fluctuating phenomena and their correlation with extraterrestrial and cosmic influences (Piccardi 1962).

Gauquelin (1973) has outlined the history of research into biological rhythms and biological clocks. These can be remarkably resistant to the effects of all environmental factors, except the particular one that functions as that clock's Zeitgeber. The existence of such clocks implies a high degree of coherence within living systems.

2 Electromagnetic Fields and Coherence

Superimposed upon the order of living systems is the fundamental randomness associated with the thermal energy of any system above absolute zero of temperature. Ambient temperature (300 K, 27 °C) corresponds to a mean energy of 4.142×10^{-21} Joules, about 1/40 eV, a mean frequency of about 6×10^{12} Hz, a wavelength of 50 microns (μm), a wave number of 200/cm, or a chemical energy of about 2 kilojoules per mole (0.5 kcal/mol).

Frequencies greater than 6×10^{12} Hz are associated with quanta which are, on the average, more energetic than those corresponding to the environmental thermal background; hence single quanta should be detectable as such and this includes visible light, green light has a frequency of 6×10^{14} Hz. At frequencies below 6×10^{12} Hz, the wave properties can be expected to predominate over the quantum properties and large numbers of quanta will be needed to have enough energy to exceed the thermal background or "noise" level and give a signal-to-noise ratio (coherence-to-incoherence ratio) greater than unity. Here, one is usually concerned with the signal power in a communications channel of specific bandwidth rather than the energies involved, although the spectral power density expressed in watts per cycle of bandwidth is numerically the same as the energy if a bandwidth of 1 Hz is taken (Bell 1960).

However, it is necessary to consider the number of "degrees of freedom" involved. The pendulum of a clock has only two degrees of freedom for its energy – gravitational potential energy and kinetic energy. The forces of gravity and inertia act over the whole pendulum, which responds in a coherent manner at a specific frequency. In any practical situation, the pendulum cannot have infinite coherence (zero spectral line width), if only because it must start and stop within the confines of finite time and because of small changes in its resonance frequency arising from random thermal vibrations affecting, for example, its length.

For an oscillating electrical circuit (Bell 1960), there would appear to be two degrees of freedom represented by the energy stored as an electric field in the capacitor and the energy stored in the magnetic field of the inductance. In each case, an energy of 1/2 kT would be apportioned to each of the two degrees of freedom of a system at a temperature of T K (k is Boltzmann's constant). However, the magnetic interaction between the conduction electrons gives coherence to the whole circuit, so that even though the conduction electrons are vibrating at random in the available six directions, if one measures the electric and magnetic fields over the whole of the circuit, inductance and capacitance, the circuit behaves as if it only has a single degree of freedom, to oscillate or not.

"Les électrons indépendants sont une fiction. Il va falloir maintenant tenir compte de leurs couplages magnétiques ..." (Brillouin 1934).

The living world has evolved in an environment flooded with most electromagnetic frequencies. Optical radiation is only coherent over time intervals less than about 10^{-8} s, which corresponds to the propagation time for radiation through larger organisms and hence spatial and temporal coherences are defined. Throughout evolution, Nature would have been able to make use of high coherence, narrow spectral bandwidth channels for parallel data processing channels without risk of interference from the coherence present in sunlight.

Living systems may be able to optimize the degree of coherence they use for any given biocommunication channel, choosing between one broad-band high carrier frequency, high data rate, serial communication channel; or the corresponding number of narrow band, low data rate communication channels capable of parallel data processing, each coherent enough to overcome the ambient noise but together having the same overall data capacity.

There is a fundamental limitation to the degree of coherence which can be achieved simultaneously in both time and space (Popp 1979) and E. del Giudice (pers. commun. 1986). If the power supply frequency has a long-term stability of a fraction of a second per day, then radiation at this frequency is coherent enough to exceed the Uncertainty Principle condition for spatial coherence throughout an organism of the order of a metre in size. There certainly appear to be clinical problems associated with living in proximity to power lines (Smith and Baker 1982).

The lowest frequency that could be meaningful to a living organism is the frequency corresponding to the reciprocal of its life span. Wever (1985) has shown that diurnal circadian rhythms can be entrained using ELF fields acting as a Zeitgeber. The involvement of highly coherent frequencies throughout the life-span of a living system carries with it the risk of sensitivity to and disruption by coherent electromagnetic fields in the environment just as there is a risk of chemical disruption of the genetic information by antagonistic chemicals in the environment. Electrical and chemical signals may be regarded as interchangeable in respect of living systems. If this fundamental duality between chemical structure and coherent oscillations did not exist, chemical analysis by spectroscopy would be impossible.

In the predator-prey situation, the ultimate survivors will be those that have sensors limited only by the fundamental laws of physics. Such sensitivity to electromagnetic radiation makes it unlikely that the responses will be linear, but will manifest thresholds of sensitivity and an "all-or-nothing" response. Weak stimuli, below thresholds, will evoke no response; stimuli above threshold will evoke a maximum or "panic response".

It is important to realize that any measurements expressed as a time-averaged frequency spectrum have serious limitations. The time sequence of the signals from which the spectrum was generated has been lost in the averaging process. It is not possible to recover the original speech, music or computer programme from its time-averaged spectrum. If Nature is producing a series of time-sequential electrical control signals, like a computer programme, there is no way of recovering them from the time-averaged frequency spectrum common for the presentation of biological data, although they remain in the original data; it seems that biological signals should be analyzed by "code-breaking" computer programmes.

3 Coherent Environmental Electromagnetic Fields and Humans

The first extensive review of the biological effects of the geomagnetic field (GMF) was published by Dubrov, in Leningrad, in 1974 (Dubrov 1978). The GMF at the surface of the Earth has a steady component about 0.5 Gauss [1 Gauss = 10^5 gamma (γ) = 100 microtesla (μT)]. The actual value varies between 0.35 and 0.70 Gauss over the surface of the earth. The direction of the local maximum field is inclined to the horizontal at the so-called angle of dip of the compass needle. The GMF approximates to the field that would be given by a powerful bar magnet about 400 km from the earth's centre and inclined at about 11.5 degrees to the axis of rotation. Measurements on rocks show that the polarity of the GMF reverses every 10^4 to 10^5 years and has smaller variations with a periodicity of 10^2 to 10^3 years.

Superimposed on this steady field is a variable field amounting to less than 2%, but very important biologically. Under "quiet" conditions, there are variations synchronized with the solar day and the lunar day and month; there are also annual variations. Disturbances occur in synchronism with the 11-year sunspot cycle, and the 27-day period of rotation of the sun about its axis. The earth's orbit is within the outer atmosphere of the sun. It experiences a "solar wind" of radiation and charged particles which become trapped in the earth's (Van Allen) radiation belts, giving rise to aurorae at the poles and to currents in the upper atmosphere which can oscillate between the poles along the magnetic field lines. Magnetic storms are classed according to the magnetic field strengths involved – very strong are over 200 gammas, weak are over 50 gammas (1 gamma = 1 nT). Such fluctuations are well above the theoretical threshold sensitivity of man and animals.

Experiments involving the shielding of living systems from the GMF have also been carried out (Dubrov 1978), but they are difficult and involve expensive shielding materials or techniques. In view of the extreme sensitivities which now seem to be achievable by living systems, it is doubtful whether many such experiments achieved the necessary degree of shielding for effective isolation. Magnetic fields are very difficult to screen. A 1-mm-thick shield of a high permeability magnetic alloy will only reduce the steady field within its enclosure to about 50 nT. The shielding of ELF magnetic fields presents an even greater problem.

A number of experiments have been conducted involving persons living for prolonged periods in underground dwellings, shielded by steel with the moist soil adding to the ELF shielding. The control subjects lived in similar rooms but with ordinary

GMF conditions. Dubrov refers to some Russian work as well as to the experiments on human circadian rhythms commenced by Professor Wever at the Max Planck Institute for Psychiatry more than a quarter of a century ago.

In the course of these investigations involving human circadian rhythms, Wever (1973, 1985) found the remarkable sensitivity of the circadian system to weak ELF fields rather as a by-product. Early on, he was able to describe the circadian system in terms of a mathematical model. To check this out experimentally he needed a suitable stimulus to control the circadian rhythms. In various experiments with animal species, he found that regular changes in the intensity of light were the most effective external stimulus. However, the human circadian rhythm was remarkably resistant to this as a timing stimulus (Zeitgeber). He took this to imply that another stimulus was operative. This he found to be weak ELF fields. He used as his "standard" field, a vertical electric field of 2.5 V/m (peak-to-peak) at a frequency of 10 Hz and having a square waveform. The rise time was less than 1 μs, so that enough current would be flowing through the capacitance of the electrodes to generate a small magnetic pulse; the possibility that this was having an effect could not be excluded. Natural environmental fields of corresponding frequencies are in the range 2.5 mV/m, although man-made environmental fields can be much greater.

All the experiments were carried out in one of two underground isolation units. These comprised a living room with bed, a kitchen and a bathroom. A double-door lock connected and isolated the control room from each unit. The units were constructed with double walls giving sound insulation; one unit also had five layers of continuous soft iron shielding to reduce the GMF. Electrodes and coils, invisible to the subjects, enabled any electric field or magnetic field, steady or alternating, to be applied. Over nearly 20 years they tested 325 subjects; most subjects were isolated singly, but there were some group experiments. The period of an experiment was usually 4 to 6 weeks, but some lasted up to 3 months. Nearly all subjects felt positively well during the isolation and 80% spontaneously asked for a repeat. The most relevant effect was the field-induced shortening of the circadian period and systematic alterations in all the rhythm parameters.

In a constant environment without clues as to the local time, the sleep/wake activity and the deep body temperature periodicities were liable to undergo a sudden change, for example from a 24-h day to a 25-h day. It is remarkable that the human circadian rhythm was found to be insensitive to changes in the level of illumination, from total darkness to 1500 lx and whether under voluntary or involuntary control. The dark/light cycle does not have the ability to function as a human Zeitgeber.

There have now been 52 experiments in which internal desynchronization has occurred (pers. commun. Prof. Dr. R.A. Wever, 1987). In this case, the circadian rhythm abruptly alters its periodicity with the deep body (rectal) temperature retaining the former periodicity. Changes in the sleep-wake cycle of exactly half or twice the temperature cycle have been called "apparent desynchronization" because the internal coupling is, in these cases, as great as it is in the case of full internal synchronization (with a 25-h sleep-wake cycle). Only in the case of real internal desynchronization, with the different cycles not being in an integral relationship, is the age of the subjects relevant. The electric 10 Hz field influences this state and the tendency is greater in subjects older than 40 to 45 years. The continuously operating weak 10 Hz square-wave electric field was able to prevent, or reduce, internal desynchronization and affect the autonomous periods. When it was operated periodically, it could exert a Zeitgeber influence and entrain a free-running rhythm into synchronism. The sleep/wake cycles and the deep body temperature cycles could even be synchronized simultaneously to the weak 10 Hz field at a periodicity of 23.5 h per day, for which there are no known natural or artificial Zeitgebers in the environment.

Wever also reports that preliminary animal experiments showed that the animal circadian rhythm had a similar sensitivity to weak ELF fields, his special birds could differentiate between fields of 9 Hz and 10 Hz.

Jacobi (cited in Wever 1975) found an effect on thrombocytes which could only be observed with 10 kHz "sferics" after the 10 kHz had been modulated with 10 Hz. Previously drawn blood samples which were subsequently exposed to the same field did not show any effect, so it was the living subject that was responding. Furthermore, when the heads of the subjects were shielded by a cover of copper gauze during exposure to the field no effect could be observed. The conclusion is that the head contains the sensitive region.

In mammals, the pineal gland (attached to the posterior part of the 3rd ventricle of the brain) is a light-sensitive time-keeping organ in which the cell activity is affected by magnetic field pulses of the order of the strength of the geomagnetic field (Semm et al. 1980). For the 2-g human pineal gland (Barr 1979) to be able to respond cooperatively to changes in magnetic field, the energy of the magnetic field within the volume of the pineal gland must be not less than thermal energy (Smith 1986), that is the magnetic field must be greater than 0.24 nT (2.4 μG).

The other possible fundamental limitation to magnetic sensitivity is set by the size of the quantum of magnetic flux (see H. Fröhlich, this Vol.). The value of magnetic field which would give only a single magnetic flux quantum through a human pineal gland is 75 pT (0.75 μG). In this case, the thermal energy or signal-to-noise limit would be reached before the quantum limit.

Honeybees *(Apis)* and birds are reported to be sensitive to magnetic changes probably less than 1 nT (10 μG) (Keeton 1979). The pigeon has a 1.5-g pineal gland, large enough to react to such magnetic field changes. For *Apis* to be able to sense fields of the order of 1 nT, the whole of the insect must be cooperatively involved. All such creatures should be able to sense and distinguish the "quiet" diurnal and the "disturbed" variations of the GMF.

In experiments to try to cancel out weak magnetic fields and field gradients at levels which dowsers could detect, C. Brooker (Smith 1985) found that although the magnetic fields and field gradients could be reduced with suitable coils and currents so as to be unmeasurable with a magnetometer, dowsers could still detect the objects producing the original magnetic field perturbation. The only magnetic field which the coils would not have cancelled would be a uniform field travelling at a constant velocity. Brooker constructed a modified "Earth Inductor" which had a second commutator so that the voltage generated in the part of the coil travelling in the direction of the earth's rotation was measured separately from that generated in the part of the coil moving against the motion of the earth. These two voltages differed by an amount corresponding to almost the full velocity of the earth's rotation at the latitude of the experiment; appropriately, the voltages became zero when the axis of rotation was along the angle of dip. Faraday had recorded in his Diaries (Faraday 1838) experiments which demonstrated that a magnetic field does not rotate when the magnet producing it is rotated about its axis.

A steady magnetic field moving at the peripheral velocity of the earth will generate an electric field vector directed relative to geographical north and of magnitude 15 mV/m in the U.K. This is well within the theoretical sensing capabilities of the larger organisms and could give a navigational "fix", anywhere on the surface of the earth when combined with the "local time" information derived from the off-set geomagnetic pole component.

Dubrov (1978) lists a wide range of effects obtained with in vitro cultures shielded from the geomagnetic field (GMF), some of which did not appear until after many sub-cultures. These include histological and physiological disturbances as well as peculiar tumours; also there were definite alterations in biochemistry, physiology and behaviour through the range of creatures from insects to higher animals and man following prolonged total shielding from the GMF. Of particular note were the physiological changes to submarine crews spending long periods shielded from the normal GMF by a thick steel hull and from much of the ELF by salt water. There are also physiological changes in earth-orbiting astronauts who experience the complete cycle of variations in the GMF at each orbit of the earth.

The spectrum of the earth-ionosphere cavity resonance (Schumann Resonance) covers the 1 Hz to 30 Hz region (König 1979), the 8 Hz component being thought to be particularly important and generally beneficial to living systems, it also coincides with the brain's alpha-rhythm (8–12 Hz). The equivalent power flux density of the order of 10^{-10} W/m^2/Hz in the Schumann radiation (Gendrin and Stefant 1964) is greater than the theoretical threshold sensitivity for a 70 kg man (see Table 1). Ludwig (1987) has measured and compared a large number of the ELF rhythms in human subjects with resonant frequencies in homoeopathic remedies. To find these measurements, it is necessary to use a spectrum analyzer able to resolve frequencies with an accuracy of a millihertz. A number of frequencies have been found to be common to all the subjects and to relate to the specific physiological functions.

For example, the frequency 0.1 Hz relates to the circulatory system. 7.8 Hz relates to the hippocampus, 10 Hz relates to the circadian rhythms, 33 Hz relates to the lymphatic system and 65 Hz relates to toxin elimination and the lymphatic system (pers. commun. Dr. W. Ludwig 1987).

Baker (1984, 1985a,b) has investigated the sensing of magnetic fields by man and other primates and the possibilities for navigation and direction finding which this

Table 1. Electric and magnetic fields within the volume of the biological system which represent an energy of kT at 310 K

	Cell membrane	Prokaryotes	Eukaryotes		Hen's egg	Man
Size of system	10 nm x 1 μm	1 μm	10 μm	100 μm	70 g	70 kg
E (min)	70 kV/m	3 kV/m	100 V/m	3 V/m	250 μV/m	8 μV/m
B (min)	500 μT	100 μT	3 μT	100 nT	9 pT	300 fT
Flux quanta Φ_0	$\Phi_0/300$	$\Phi_0/300$	$\Phi_0/10$	$\Phi_0/3$	8 Φ_0	25 Φ_0
Poynting vector	30 MW/m^2	80 kW/m^2	80 W/m^2	80 mW/m^2	600 pW/m^2	1 pW/m^2

sensing ability opens up. He recounts how the idea of animals having an ability to sense magnetic fields and navigate by the earth's magnetic field was also ridiculed when this was suggested in 1855 by von Middendorf. The location of magnetite deposits in bacteria, and in the bone of birds, mammals and man suggests that an electromechanical torque sensor is involved, a combination which might form a magnetic compass.

4 Coherent Electromagnetic Fields, Enzymes and Cells

Although this chapter is primarily concerned with electromagnetic field effects in humans, even the contents of the human alimentary tract represents an appreciable fraction of the total body weight and contains cells which can all too readily function in relative isolation from the body's detailed control.

A single living cell has about 3000 enzymes to effect and regulate its chemical reactions, yet there is nothing in the chemical structure of an enzyme to explain its enormous catalytic power. In terms of electronics an enzyme-substrate system can be considered as an "amplifier", if one regards the input signal as the amount of enzyme present and the output signal as the amount of reaction product formed per minute; the substrate consumption would be analogous to the current drain from the battery and its voltage analogous to the chemical potential of the particular reaction.

The simplest and most general negative-feedback analogue-control system in electronics is achieved with negative feedback around a high-gain amplifier. In the case of enzyme catalyzed reactions, the gain of the enzyme as an amplifier has been determined, although it is not usually described in these terms by chemists. If the gain is taken as the ratio of the rate of reaction – molecular turnover – when catalyzed by the enzyme to the uncatalyzed reaction rate (Callinan 1985), then the gains can be 3.6×10^6 for carbonic anhydrase to 3.2×10^{10} for kinases.

It can be shown that so long as the open-loop gain of the amplifier is very high, its precise value does not matter, the characteristics are determined by the feedback path. Any fault which reduces the amount of negative feedback will make the gain tend towards the open-loop gain. Then, either the control system will fail because it saturates in the "on" or "off" state, or if the necessary gain-phase criteria are satisfied, it will oscillate; it can even do both. In either case it will cease to exert a controlling function.

Living systems can maintain homeostasis over a wide dynamic range of perturbations. They usually show a logarithmic response in respect of a stimulus. These properties imply not only the presence of such high gains but non-linear elements like diodes in the feedback circuit to generate the necessary logarithmic characteristic.

The specificity of enzyme action, which is not confined to living systems, has often been modelled on a mechanical "lock and key" principle, but this still leaves many features unexplained. It is still necessary to consider how the enzyme "key" can find the substrate "lock" and fit into it without jamming. This requires an information exchange at a distance. The "key" must "see" the light coming through the "keyhole". Fröhlich (1969, 1975, 1978, 1980) represented such a system by two highly polarizable dielectrics capable of giant dipole oscillations and separated by a distance greater than their dimensions. He showed that there would be a strong interaction if both oscillated at precisely the same frequency. For elastically bound particles of molecular dimensions the frequency was estimated to be of the order of 10^{13} Hz. This is close to the frequency corresponding to the mean thermal energy at an ambient temperature of 27 °C, 300 K (6.25×10^{12} Hz). The biological cell has the possibility of resonating at such frequencies as shown in an electrical engineer's model of a biological cell, Fig. 1 (Smith 1986).

At distances up to the thickness of a typical biological membrane (10 nm) from a single electronic charge the electrostatic field will be of the same order as the membrane field. The 10 MV/m membrane field provides a bias field ensuring that charges oscillating on, or in, the membrane will do so at the same frequency and not at twice the frequency of the membrane field. Two opposite electronic charges on opposite sides of a biological membrane have an electrostatic energy of attraction greater than thermal energy at 310 K (37 °C), and thus will remain stable against dissipation by thermal diffusion. This will apply down to the size of the smallest living cells, as shown in Table 1.

Preliminary dielectric measurements (Ahmed et al. 1976) did little more than conform that moist proteins behaved like moist ferroelectrics. The choice of method used for introducing the moisture into the enzyme powder appeared to be important. Samples which had been humidified by exposing the enzyme to water vapour in a refrigerator at 6 °C for more than 12 h gave greater dielectric effects than specimens which had been humidified by the addition of an equivalent amount of liquid water. In

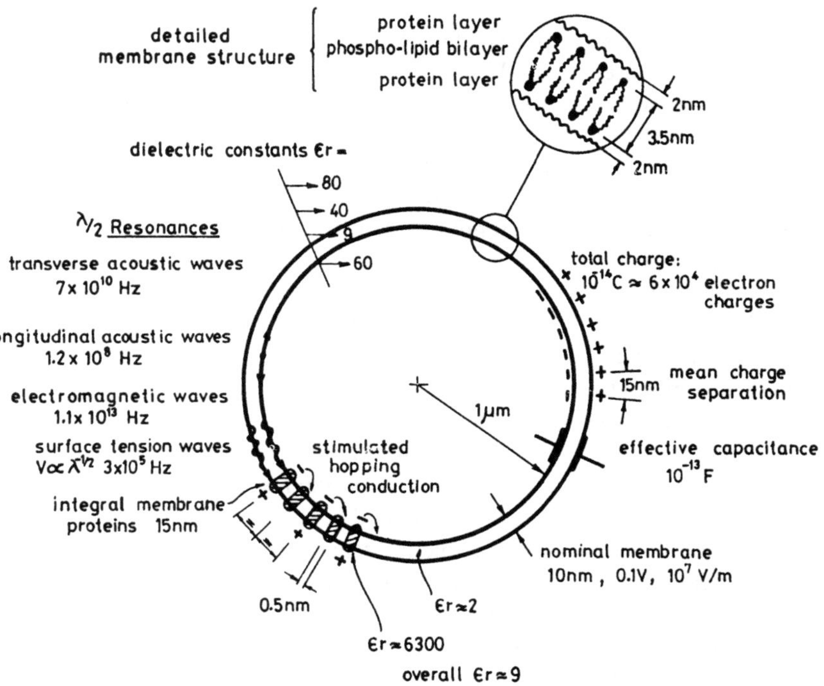

Fig. 1. A simple electrical model for a nominal biological cell. (Smith 1986)

retrospect, this was the first indication of memory and structural effects in water; these reappeared later in connection with allergy therapy.

There should not be any appreciable magnetic field interaction involving biological dielectrics and water. However, it was found that the magnetic field from a permanent magnet gave a reduction in both the permittivity and loss by about 40% for humidified powders of lysozyme, ribonuclease and ovalbumin. Similar results were obtained with an alternating (50 Hz) magnetic field for lysozyme and trypsin, and subsequently with various concentrations of lysozyme solutions.

Experiments confirmed that diamagnetism was definitely involved. This can only come from the equivalent of a short-circuited current loop. Further measurements using magnetic fields of the order of 0.06 T (600 Gauss), showed that dilute solutions of lysozyme in water had an increment in the diamagnetic susceptibility 10^4 times higher than expected. This anomaly disappeared above certain critical magnetic field strengths (Ahmed et al. 1975; Smith 1985).

It was found to be essential to use newly made quartz test cells for each magnetic susceptibility experiment and to test them individually prior to each experiment. The reason for this is that in the early work, attempts to repeat measurements using fresh lysozyme in a previously used glass or quartz cell all gave a null effect. It was eventually found that lysozyme forms a film which adheres tenaciously to glass and quartz; this gave such an overwhelmingly large effect in both the calibration and test measurements that it masked all other effects and differences. This emphasizes the care necessary in such work. The only technique that was successful in removing the lysozyme deposit from glass and its associated magnetic effects was a prolonged bake in air, above 200 °C. Ordinary detergents and ultrasonic cleaning bath were ineffective.

Other laboratories did attempt experiments related to this work, but without any published success (Sorensen et al. 1976; Chu et al. 1976; Careri et al. 1977). The anomalous effects reported persisted until sterile laboratory facilities were installed – flame sterilization of the quartz cells, autoclave sterilization of all liquids, ultraviolet (TUV) sterilization of the lyophilized lysozyme powder, and all handling and flame sealing carried out in a TUV flooded enclosure. Then, magnetization measurements for water and lysozyme were linear, reproducible within experimental error and without the anomalies.

The enzymatic activity of lysozyme was also found to be affected by electromagnetic fields (Shaya and Smith 1977).The substrate used for assaying lysozyme was *Micrococcus lysodeikticus*, obtained as a freeze-dried powder; it was found to be still alive and the saline meant to lyze and kill it to leave just the cell membrane fragments, did not do this. Further investigations showed that the enzymatic activity of the lysozyme was affected by the nutrition available to the substrate and by the phase of the micrococcus cell division cycle. The activity of lysozyme was also photosensitive, even to the light beam of a spectrophotometer and to the geomagnetic field. Furthermore, lysozyme solutions seemed to be able to "remember" for long periods, the frequency of any magnetic fields to which they had been exposed.

The widespread use of magnetic stirrers, which subject the solutions being stirred to magnetic fields of the order of 0.01 T (100 Gauss) at a frequency of a few Hertz, ensures that many magnetic anomalies are saturated and remain unobserved.

A delay time, dependent on the substrate concentration, was found to occur before the lysozyme reaction (lysis) got going. If this delay is the time needed for coherent oscillations to build up (the reciprocal of the band-width) in the enzyme-substrate system, then it demonstrates one of the consequences arising from Fröhlich's (1980) theoretical treatment of the excitation of coherent electric vibrations by random metabolic energy.

The lysozyme reaction was also found to be sensitive to proton magnetic resonance conditions. If the lysozyme assay was run in the usual way but when the substrate was only partly lyzed, the cuvette was exposed to the combination of a steady magnetic field and a radiofrequency field which exactly satisfied the proton magnetic resonance condition – the lysis ceased completely. It resumed when the cuvette was returned to ordinary laboratory ambient magnetic field within the spectrophotometer. It was necessary to expose the cuvette to the steady magnetic field (split the energy levels) before the radiofrequency field (Jafary-Asl and Smith 1983; Jafary-Asl et al. 1983).

As commonly occurs in living systems under good homeostatic control, larger, often bi-phasic effects are obtained when the biological system is stressed. A convenient way of doing this for the lysozyme – *Micrococcus lysodeikticus*, enzyme-substrate system – is by use of a competitive inhibitor for the reaction. When N-acetyl-D-glucosamine was used to give partial inhibition, a complicated pattern of frequency-dependent effects involving the enzyme activity was found between 50 kHz and 300 MHz. The effect of the inhibitor could be enhanced or cancelled depending on the frequency applied (Shaya and Smith 1977).

When the anomalous effects were found to be associated with live *Micrococcus lysodeikticus* substrate, investigations were commenced on another living system, the bacterium *Escherichia coli*, which originates from the colon.

To be quite certain that nothing would be missed between spot measurement intervals, batches of 18 cuvettes were situated in a gradient of magnetic field so that no value of field went untested, although this resulted in a loss in the accuracy with which the magnetic field could be specified. The magnetic field at each cuvette was an average over its 1-cm dimension in the field gradient. More than 1000 cultures were grown and the mean generation times measured under carefully controlled conditions. To eliminate the possibility of merely observing thermal effects, the cells were cultured at the temperature which gave maximum growth rate, so that any change in temperature, whether an increase or decrease, would slow down the cell growth. Yet, at certain values of magnetic field strength, an increase in the rate of cell growth was observed. The standard deviation in the control measurements was reduced to 0.5% by attention to detail; this enabled effects at a 4% level to be determined.

The mean generation times for *E. coli* grown in various strengths of magnetic fields of square waveform showed a marked threshold effect and strong indications of a periodicity with field strengths above this threshold (Fig. 2). The application of the statistical F-ratio test indicated a probability of less than one in two million that the effects were due to chance (Aarholt et al. 1981).

These results demonstrate an important feature of the response of biological systems to an external stress when they are under good homeostatic control. The effects do not become larger as the electromagnetic stress is raised, they become more com-

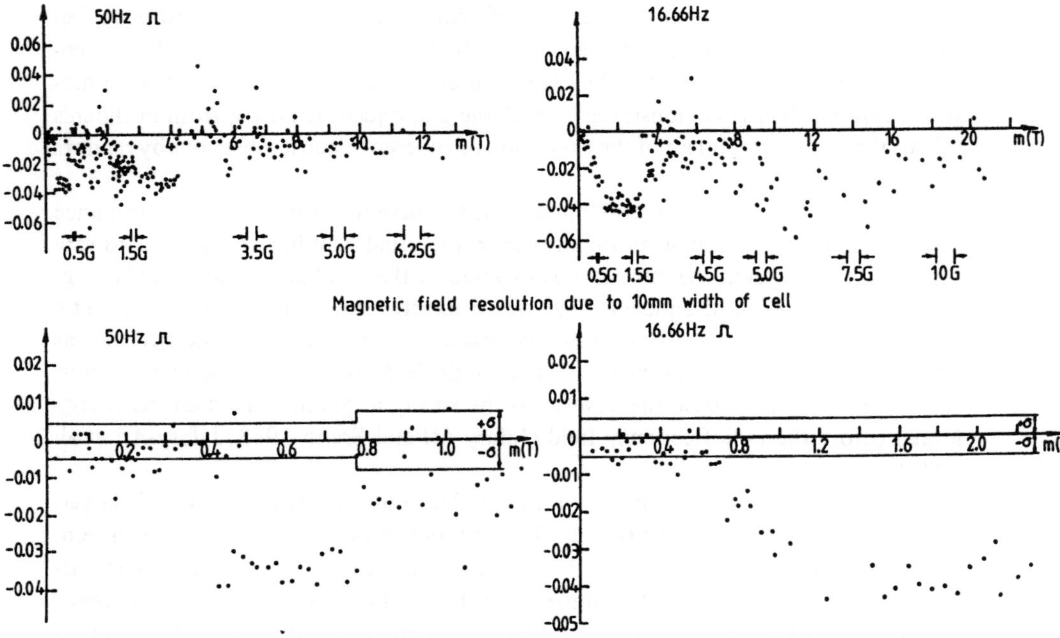

Fig. 2. Variation in mean generation time (MGT) with peak value of applied magnetic field. σ is the standard deviation of the MGT for control cultures from the same batch as the experimental cultures. Note the onset of the effects, the small values (4%) reached, the complication and non-proportionality at higher field strengths. (Aarholt et al. 1981)

plicated and require greater precision for their detection. Biological systems are non-linear and often discontinuous in their response to stress. The argument that a null effect at a strong stress implies an even smaller effect at a weaker stress, is not valid.

To obtain effects larger than the 4% obtained, the system needed to be biologically stressed. When *E. coli* is stressed by feeding it lactose which it cannot use directly, it transcribes instructions from the DNA to enable it to manufacture a protein, the enzyme β-galactosidase, which enables it to use the lactose. This gene-controlled process is normally repressed by a protein which binds very strongly to the specific site on the DNA, located just outside the structural gene. The repressor protein comes off the DNA in the presence of lactose (Davies and Walker 1979). The relative rates of β-galactosidase synthesis were measured as a function of the strength of a 50 Hz square wave magnetic field using the field-gradient magnet as previously. The results (Fig. 3) showed a fivefold reduction at 0.3 mT ± 5% (3 gauss) and a two-and-a-half-fold increase at 0.6 mT ± 10% (6 gauss). There was no effect at cell concentrations above 10^8 cells/ml (Aarholt et al. 1982).

If spot readings had been taken at 0.1 mT (1 gauss) intervals, instead of using the gradient of magnetic field to test at all values of magnetic field within the range, only *one* non-zero measurement would have been obtained. It is essential to carry out measurements on living systems with sufficient physical precision. This principle of applying a steady stressor to take bioregulatory systems away from homeostasis ap-

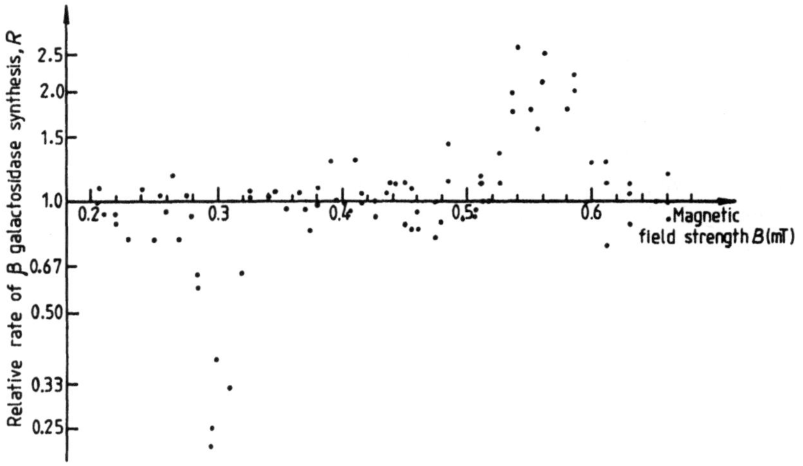

Fig. 3. The relative rate of synthesis of β-galactosidase with magnetic field strenght. Each *point* represents one test culture. Only points for weighted mean concentrations between 1.5×10^7 to 5×10^7 cells per ml are shown. A value of unity for the relative rate of β-galactosidase synthesis represents a zero magnetic field effect. Note the large changes in a biologically stressed system compared with Fig. 2. (Aarholt et al. 1982)

pears to be of general application in the investigation of bio-responses to a given perturbation (Popp 1986b).

It was by then becoming apparent from experiments on a number of different living cells of various sizes, that the critical magnetic field strength for the onset of the various magnetic effects corresponded to that at which a single quantum of magnetic flux would be linking a single cell or cell pair. The periodicity, observed as the field strength was increased, corresponded to integer increments in the numbers of magnetic flux quanta linking the cells (Jafary-Asl et al. 1983).

Since 1980, there have been a number of instances where the Josephson equation and magnetic flux quantization conditions seemed to be applicable to the results of the experiments (Ahmed and Smith 1978; Jafary-Asl et al. 1983; Smith 1984; Smith 1986). If nature is able to make use of magnetic flux quanta, then in principle it also has the Josephson effect available for its use, since both derive from the quantized nature of a magnetic field (see H. Fröhlich, this Vol.).

Liboff and co-workers (1984) have reported that time-varying magnetic fields affect DNA synthesis. From the data published, it seems likely that the onset of these effects occurs when flux quantization conditions are satisfied.

The clinical implication of the effects of coherent frequencies on *E. coli* and yeasts is that their proliferation may be triggered electromagnetically. This may contribute to the *Candida* problem. One allergy patient has reported attacks of colitis when working next to a computer which had an 8 MHz clock frequency, a frequency that has been shown to be associated with effects in actively growing yeasts (Fröhlich and Kremer 1983; Gründler 1985; Smith 1984; Smith 1986). The frequency of 16 MHz is also in use for computer clocks.

In experiments on bovine eye lenses, it was found that sub-capsular cataracts located in the posterior cortex of the lens developed particularly well when the microwaves were modulated so that the modulation frequency satisfied proton NMR conditions in the ambient magnetic field (Marsh 1986; Aarholt et al. 1987). In this case, the microwaves would be merely acting as a carrier capable of inducing currents at the NMR frequency through the demodulation of the microwave signal within the tissue by its electrical non-linearities. Since many microwave oscillators contain magnets, a technician moving about in the vicinity of such an oscillator would regularly satisfy the necessary field and frequency resonance conditions within the body tissues, thus increasing the risk of cataract formation.

Once the possibility of magnetic resonance in living systems is appreciated, it is clear that it could also be a source of highly coherent radiation. If a magnetic field pulse is applied to a living system (as in stimulated bone healing), the electric field induced by the changing magnetic field could excite the protons (or other particles having spin) and stimulate an increase in the proton population of a higher energy state. If the magnetic field is then reduced relatively slowly to zero but in a time that is short compared with the spin relaxation time, the protons will not be stimulated to make the downwards transition back to the ground state because of the smaller induced EMF, the excited spin states will relax in their own time and in the original ambient magnetic field, thereby subjecting all the surrounding tissues to highly coherent spin relaxation frequencies, containing all that information relating to the body chemistry and its state of health or disease, which would be contained in an NMR spectrum (Aarholt et al. 1987).

Cancers are also tissues which function outside the control of the body's regulatory systems. Nordenström has simulated the injury potentials which failed to appear in the vicinity of tumour tissues, by inserting electrodes – one into the tumour and the other into surrounding tissue. The application of appropriate voltages to simulate the injury potentials from an external source gives a current flowing from one electrode, through the tissue to the capillaries where it passes through ion-permeable pores and thence within the electrically insulating veins and arteries to capillaries in the vicinity of the counter electrode. This technique has had some remarkable successes in the dissolution of inoperable tumours (Nordenström 1983, 1985).

The number of photon counts registered with a photomultiplier viewing cell suspensions decreases with cell concentration in the case of healthy cells, but increases with cell concentration in the case of tumour cells (Popp 1986a). This suggests that the changes necessary to turn a healthy cell into a cancer cell may be mediated by coherent radiation and that a function of cyto-chemotoxic agents may be to act as highly specific optical absorbing filters, possibly with fluorescent re-emission at benign frequencies.

5 Electromagnetic Hypersensitivities, Allergic Responses and Coherence

The involvement of electromagnetic fields in the human organism and its regulatory systems can be demonstrated through the disturbance of regulatory systems with extremely weak electromagnetic fields in very allergic subjects. Allergy used to be concerned solely with skins and respiration, but in recent years allergic responses have been found to occur so widely that allergy is now defined as "the failure of a regulatory system" (W. J Rea, 4th Ann Conf Man and his environment in health and disease, Dallas, 1986). The more severely allergic patients have acquired allergic responses to many chemical, environmental and nutritional substances; these may be counted in tens and even exceed a hundred in extreme cases. It appears that about 15% of a given population function to some extent below their best performance capability due to some degree of allergy; that is, one or more of their regulatory systems functions inadequately.

One is not here concerned with an L.D. 50 situation, but with effects arising from chronic long-term exposures to very low environmental levels, whether of chemicals or coherent electromagnetic fields. The significant blood levels of chemicals may be in the range nanograms to femtograms per millilitre. The effects saturate at higher levels (Rea et al. 1984). The significant electromagnetic field levels are likely to be near the theoretical limits set by the physics of the situation.

Endogenous opiates have been linked with allergies as well as electromagnetic fields (Smith and Aarholt 1982). This led Dr. Jean A. Monro (Allergy Unit, The Lister Hospital, London, SW1W 8RH), to contact the writer in 1982, to seek help with the treatment of her electrically sensitive multiple-allergy patients. Since then, more than 100 such patients have been tested in London and in Dallas in cooperation with Dr. W.J. Rea. Most of the patients tested were found to have electromagnetic hypersensitivities critically dependent on frequency over a range extending from at least millihertz to gigahertz, but less dependent upon the field strengths so long as this exceeded a certain threshold value specific for the individual patient, and the allergic state at the time of testing.

New allergic responses can be acquired or transferred. If the patient receives a chronic exposure to some hitherto innocuous substance while reacting strongly to a triggering allergen, this may subsequently trigger the same set of symptoms. In such circumstances, it seems that a coherent electromagnetic frequency in the environment can also become an allergen and in hypersensitive patients trigger their specific pattern of allergic responses on a subsequent encounter of that particular frequency. In general the pattern of allergic responses is the same whether the trigger is chemical, environmental, nutritional or electrical. In principle, an accurate "memory" for frequency is no different to the "absolute pitch" facility that many musicians possess.

A therapy for chemical and nutritional allergic responses was originally described by Carlton Lee and modified by Miller (1972), who showed that the diameter of the wheal which developed on the skin of a patient following a "skin-prick" (intradermal injection) test with an allergen depended upon the dilution of the allergen used. Furthermore, if a whole sequence of serially diluted allergens was applied successively, certain dilutions gave large wheals, but after further serial dilutions, one would eventually attain a dilution of the allergen at which no wheal was produced. Still further serial dilutions would recycle through similar patterns of response and no-response. The dilution at which no wheal resulted from the "prick-test" was termed that patient's "neutralizing dilution". It could be injected to give neutralization of the symptoms provoked by that particular allergen and could provide prophylactic protection against subsequent environmental or nutritional exposure to it.

There is evidence from four double-blind studies that the therapy is satisfactory in the treatment of food allergy and from a further double-blind study in respect of food allergy in migraine (Monro 1983; Monro et al. 1984).

In the course of treating many extremely sensitive allergic patients, Dr. Jean Monro and her colleagues had found that such patients merely needed to hold a glass tube containing a dilution of the allergen for the symptoms or neutralizing effects to become manifest, even if the tube contents were still frozen solid following removal from deep-freeze storage. The most sensitive patients could distinguish, double-blind, tubes of allergen from tubes of placebo if these were merely brought into the room. This led to the development of the technique of desensitization using surface application to the skin of antigens in serial dilution. It paralleled the intradermal technique in provocation and neutralization of symptoms. This technique is much quicker and easier to apply, because the drop of antigen can be merely wiped off the skin when the full allergic response has been observed; the patient is then immediately ready for testing with the next serial dilution.

Based on these techniques, a method of testing and therapy has been devised for the treatment of electrically sensitive, multiple-allergy patients. Those who experience symptoms which they can describe within a few seconds of being exposed to the allergen or electrical frequency (Smith 1984; Choy et al. 1987) can be tested on this basis. In cases where there are no describable symptoms, the techniques of kinesiology may be successful. This is based on an abrupt decrease in muscle strength occurring within seconds of the patient being exposed to an allergen, and can be tested on the fingers or arm.

The testing of electrically hypersensitive allergy patients must be regarded as a clinical procedure and not be attempted without the immediate availability of facilities and staff medically competent to treat the remote, but serious risk of anaphylaxis, a serious shock reaction arising from a hypersensitive condition of the body. The most effective and rapid therapy for this is to place a drop of the patient's neutralizing dilution of chemical allergen on the skin.

It appears that increasing frequency has the same clinical effect as increasing allergen dilution. It is usually possible to find particular frequencies at which the allergic reactions cease, just as one can usually find a serial dilution of allergen (the "neutralizing dilution") which results in the cessation of the allergic reactions ("neutralization"). The procedure for testing and treating electrically hypersensitive patients is therefore based on the provokation-neutralization therapies of Monro, Miller and earlier workers. In the electromagnetic case, it is the frequency and its coherence which is important, the field strength is less important so long as it is above a certain threshold, particular to the individual patient. When one is near to the neutralizing frequency, the field strength may be raised to give increased sensitivity. To use too strong a field at frequencies where a strong allergic reaction is encountered would overstimulate and saturate the patient's responses so that no further testing would be possible for some hours or days. Thus, great care is needed at the start of testing. It is the writer's practice to commence with the oscillator in the next room and to determine whether the patient can tell whether it is on, or off.

One is seeking in this clinical situation to produce clinical effects by merely having the patient in an environment containing the sort of electromagnetic fields that leak from ordinary electronic equipment, such as laboratory oscillators or signal generators, televisions and computers. It is sufficient to use an ordinary laboratory oscillator such as might be borrowed from a school or university teaching laboratory. It is possible to do much useful testing over the frequency range 0.1 Hz to 1 MHz. The output level of such oscillators is a few volts. With the most sensitive patients, it is sufficient to use the radiation which leaks from the case and down the power cord. For the less sensitive, it is sufficient to trail a metre length of wire to the floor. The tests can be carried out double-blind to the patient and the clinician, if the dials are only visible to the tester. No electrical connection to the patient is necessary, nor should be contemplated for reasons of electrical safety and the risk of overstimulation of the patient's responses, which would prevent further testing through saturating the weak-field effects being sought. All that needs to be done is to supplement the environmental background electromagnetic radiation with the addition of a weak but highly coherent signal, well within the allowed limits for non-ionizing radiation and comparable to the radiation leakage from a domestic television set, home computer, or other piece of electronic equipment. If the patient does not respond to this level of signal, then the patient has no problem with electromagnetic allergic hypersensitivity, or does not have a problem unless triggered by some other allergen. A caveat: delayed responses are also a possibility, even 24 h after the original challenge. In general, the symptoms provoked on electrical testing are the same as those provoked by chemical or nutrition tests on that same patient.

However, not only are these reacting allergic patients extremely electrically sensitive, they can also emit electromagnetic signals when reacting allergically, rather like an electric fish or dividing yeast cells (Smith 1986). The signals from such patients can be large enough to interfere with electronic apparatus, as many clinical case histories testify, as well as the testing of other patients. The electrical leakage signals which make it possible to eavesdrop on computers are sufficiently strong to trigger allergic reactions in sensitive patients. They are then liable to feed "garbage" back into the computer or other equipment. One needs to give serious consideration to making electronic apparatus compatible with people. The problems which allergy patients have described are very wide-ranging; besides upsetting computers, one patient made a robotic system in a factory completely malfunction each time he stood near it; another has had the electronic ignition system on successive new cars keep failing as soon as the allergic reaction was triggered by fumes from a diesel truck ahead; yet another only sends "garbage" over a facsimile link.

Electromagnetic emissions in the audio-frequency part of the spectrum from reacting allergic subjects may be readily demonstrated by getting the subject to hold a plastic-cased tape recorder with the tape running and the recorder in the "record" mode, but with no microphone connected. If the subject is reacting strongly enough, there will be sufficient interference passing through the plastic case to be picked up by the amplifier circuits. A wide variety of signals are obtained on replay, these vary not only from patient to patient but on different occasions with the same patient. The control in this case is a section of tape recorded with the patient out of the room. Sometimes the patient emits a continuous sinusoidal oscillation, perhaps with distinct sidebands, on other occasions there will be a series of clicks; both these types of electrical waveform have also been observed by us with species of electrical fish. Fish can also have such electrical sensitivities approaching the theoretical limits (Bullock 1977).

The spectrum of the signals obtained from a reacting allergic subject holding a tape recorder can be subsequently analyzed. A spectrum analyzer can be zeroed on the background noise level of clean section of the tape recorded with the patient out of the room. This checks for the absence of environmental electrical interference. It has not yet been possible to determine whether the frequencies seen in the recorded spectrum correlate with those noted during the testing of electrically sensitive allergy patients or with those found using other techniques.

There is equipment originating from West Germany (Brügemann 1984) which makes use of the patient's own oscillations for testing and/or therapeutic purposes; in one case these are picked up, filtered, phase-inverted and then fed back to the patient. The patient's emissions are likely to be highly coherent so that quite coarse electrical filtering suffices to select the frequency band. The oscillations corresponding to the diseased state appear to be much more coherent than those corresponding to the healthy state. The healthy state oscillations can be reinforced at the expense of the disease state oscillations if the feedback to the patient is applied as pulses repeating at 10-s intervals (Ludwig 1987). This is a particularly important development, because it indicates how to enter organ-specific regulatory feedback loops of the human control systems as determined by the placement of the probe electrode on the acupuncture point, as well as providing further evidence of the importance of coherence.

On the one occasion that it was possible to make a direct measurement on an allergic subject using a spectrum analyzer (Hewlett-Packard, HP-8553B), electrical signals superimposed on the spectrum of the medium waveband radio transmissions were observed; they ceased when the subject left the room. The measurements were made with the subject holding the 50-ohm impedance input lead of the spectrum analyzer. A signal of 800 μV was obtained. The subject's measured dc skin resistance was 50 kohm so that there should have been an open circuit voltage of 800 mV available at frequencies where the skin reactance was negligible, enough to trigger the less sophisticated TTL-logic in computer circuits and thus to feed in false signals ("garbage"). The observed "comb" spectrum of spikes extended to at least 2 MHz. In this case, ambient electrical signals were probably needed to trigger the allergic reaction in the subject because no emission was obtained with the subject and apparatus in an electrically screened laboratory (Smith 1986).

For a simple test to estimate the region of the frequency spectrum which is triggering an allergic reaction, a set of metal sieves or meshes as used for grading powders and granules can be used as a set of high-pass microwave filters. It should first be demonstrated that solid metal will screen off the interaction; then one should progress from the smallest mesh size to larger mesh sizes until an effect is observed. Only wavelenghts less than twice the mesh aperture can pass through each sieve. Thus, the first onset of the interaction can be bracketed between two mesh sizes, and hence two frequencies.

Brown and Behrens (1985) have used dowsing pendulums comprising copper rods cut to resonant lengths. They show how to determine the resonant wavelengths effective in foods that cause allergic responses, as well as the frequencies of radio and radar (speed trap) transmitters. Following the writer's suggestion that they should use metal mesh filters to check that the mesh size and the resonant length of the metal pendulum bob were consistent in giving the same wavelength, they were able to confirm that this was indeed the case. This experiment with metal meshes is also effective using the muscle-test technique (kinesiology) for the detection of an allergic response.

The microwave cooker was invented by Professor H.P. Schwan specifically for the purpose of cooking meals in submerged submarines. It is interesting that many allergic patients tolerate microwave heated or cooked food and water better than that heated with the 50/60 Hz power frequency current in a kettle or cooker; using microwaves reduces the cooking time and these patients will have had less exposure to the less coherent 2450 MHz of the microwave cooker in their environment. The 50/60 Hz power frequencies are highly coherent; sideband peaks in the spectrum have been reported at 0.24 Hz (-50 dB) and 0.75 Hz (-72 dB) (Pallas-Areny 1987); the long-term coherence is as good as the timekeeping of electric clocks.

While it is possible for patients to gain relief from their allergic responses by having an oscillator set to a neutralizing frequency left on in the room, this is not a satisfactory or cost-effective solution, particularly in the microwave region. Furthermore, one patient's neutralizing frequency is another's allergic trigger and in a crowded hospital or clinic, a multiplicity of personal oscillators would be as destructive to allergy testing as the presence of perfumes or other odours.

When faced with the problem of a patient who could only be neutralized at a frequency in the microwave part of the electromagnetic spectrum, it was remembered that the homoeopathic "Materia Medica" (pharmacopoeia) includes potentized water which has been exposed to electric currents, magnetic fields or X-rays. Dr. Monro has used serial dilutions of water exposed to ultraviolet radiation as a therapy for hypersensitivity to light. She has also confirmed that allergens specially prepared as homoeopathic potencies have the same clinical effectiveness as the same allergens correspondingly serially diluted with a syringe according to the practice in allergy therapy.

A mineral water, known to be tolerated by the above patient, was exposed to a magnetic field at the frequency which neutralized the allergic responses. This then became clinically as effective when held by the patient as having the microwave oscillator switched on in the room. Water exposed to magnetic fields at a neutralizing frequency can be used as a neutralizing dilution of allergen. It resembles in its effects a "potentized" homoeopathic preparation. It is easier to keep a tube of such potentized water away from other allergy patients than to provide an electrically screened room to shield an oscillator from affecting them. It seems that such water retains this effectiveness for at least 1 or 2 months (Smith et al. 1985). When patients use such a tube of water to neutralize a strong allergic reaction, it seems subsequently to have become ineffective. Superimposing an allergy-triggering frequency and a neutralizing frequency in the same tube produces a tube of water which triggers allergic reactions rather than neutralizing them. It is possible that the strongly reacting patients overwrite the water in their tube with their own electromagnetic emissions. An extended metal surface near a reacting allergic patient seems to act as a mirror for the patient's own electromagnetic emissions and also makes their reactions worse. A large glass container of salt water (about sea water concentration) near the patient helps to damp these reactions and make the effects of the oscillator easier to determine during testing. This may also explain the problems that some patients experience in cars and other vehicles having a metal box construction and near regular metal structures such as railings. It was observed that some allergy patients reacted to passive microwave resonators tuned to an appropriate frequency, in this case the energy source must have been the patient.

6 Coherence and Structured Water

Some years previously, it had been noticed that lysozyme solutions seemed to "remember" frequencies and fields to which they had been exposed for long enough to allow measurements to be made on them in a different building. Homoeopathic remedies based on electric currents and magnetic fields, as well as X-ray and ultra-violet and coloured light radiation are well known; "potentized" preparations from water or lactose exposed to these radiations appear in homoeopathic "Materia Medica" (pharmacopoeia). Homoeopathy is clinically effective at potencies or dilutions where no chemistry should remain. Potencies produced by serial dilutions beyond 10^{24} (24X or 12C homoeopathic potencies) should have none of the original tincture molecules remaining. In allergy therapies using fivefold serial dilutions of (1 +4), this is equivalent to 5^{34}. That is, 34 serial dilutions are needed to reach the dilution corresponding to Avogadro's Number (Loschmidt's Number).

It is necessary to investigate how water can take up a "structuring" having the properties of an electrical resonator or delay line, since such a mechanism would provide a physical basis for homoeopathy. It has been objected that the structure of water has been very thoroughly investigated by neutron diffraction and nothing of this nature has been reported. But such water was not "proved" as if it were a homoeopathic remedy and demonstrated to have been unaltered by the neutrons.

The required "water structuring" needs only to take the form of coherent, cooperative, metastable, interactions between the protons in specific hydrogen bonds to give an electromagnetic resonator which can be sensed by allergic subjects. Smith et al. (1985) proposed a helical conducting path structured in water. This is obtained by taking the pentagonal ice structure and instead of bond-bending to close the ring, continuing to add water molecules to develop a helix (Fig. 4). A helix would be equivalent to a solenoidal current sheet and if located around a quantum of magnetic flux then, from classical physics, the induced emf could be large enough to give rise to a current which would in turn regenerate the line of magnetic flux. Once some such metastable state is established, the current paths and resonances must persist in the water after the alternating magnetic flux had been removed. A helical structure in water is the only magnetic field-compatable structure having solenoidal symmetry.

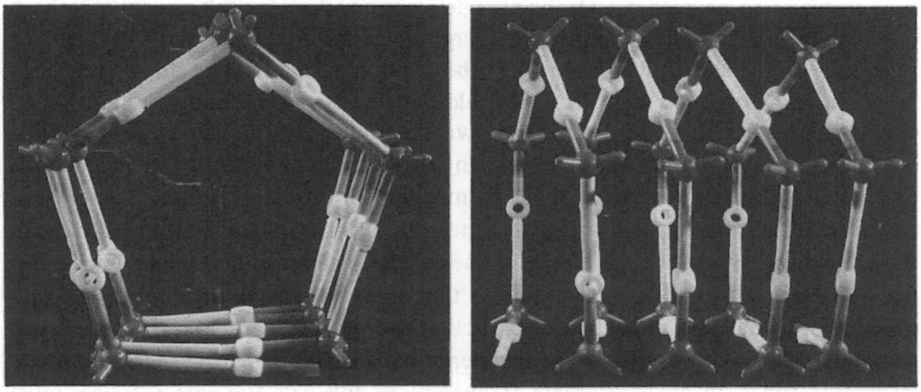

Fig. 4. Photographs of a molecular model of the proposed helical structure in water

Dr. H. Hayashi (pers. commun. 1987) in cooperation with Jhon (Moon and Jhon 1986) makes clinical use of a controlled water environment for his patients. The water used for all drinking and food preparation purposes is taken from one of a number of available devices which comprise an electrolytic cell with the anode and cathode compartments separated by an ion permeable membrane. The electrodes are (non-rusting) ferrite magnets giving a field greater than 1.2 T. The cathodic water is alkaline, has enhanced calcium ion content and is used for ingestion, the anodic water is acid and used for external purposes only. From the nature of a magnetic field, it is likely that these devices structure the ions into the "electrolytically reformed water" by threading them within a water helix generated by the magnetic field at the electrodes.

Clinically, water seems to have a memory for all exposures to coherent frequencies which have taken place since it was last distilled. In biological and medical work, all such exposures ought to be documented throughout the course of an experiment. Apart from the clinical effects, there is a simple experiment to demonstrate a coherent memory effect in water.

A capacitance bridge is balanced with a test cell containing water. A bridge frequency of 50,000 Hz is convenient. Chlorided silver electrodes are used to minimize electrode polarization effects. Then, without disturbing the cell, the output of the same oscillator is transferred to a coil which generates in the water at the same 50,000 Hz, and alternating magnetic field of about 100 μT (1 gauss). This is applied for a few minutes. The 50,000 Hz oscillator is then re-connected to the capacitance bridge. It will be found that the bridge has gone "off-balance". The original capacitance and loss values are recovered if the frequency is slightly changed: 50,020 Hz or 49,980 Hz suffice. The bridge will still be off-balance at 50,010 Hz to 49,990 Hz. To take this experiment further, it would be necessary to have the use of an oscillator which can reproduce frequencies to parts-per-million. It has been suggested that X-ray diffraction might show an ordering in the ice formed by freezing water in an alternating magnetic field. It is known that a frozen allergen dilution is clinically as effective as when it is a liquid and also that the process of crystallization is sensitive to homoeopathic potencies in the mother-liquor (Williams and Sabarth 1987). Freezing drops of water on a microscope slide in a magnetic field results in an asymmetry in the ice crystals respectively growing parallel and perpendicular to the field direction. Work on this is continuing in the writer's laboratory.

Havelock Fidler (1983) describes how memory effects imparted into stone by hammering (succussion), which are presumably structured into the absorbed water, can be rendered permanent by heat treatment. This may be the result of incorporating of the water ordering into stable hydrates.

7 Organ-Specific Coherent Effects

Acupuncture is a very ancient healing art; in China it dates back to 3000 B.C. It was said that soldiers wounded with arrows obtained relief from diseases at bony sites remote from the injury and that this led to the discovery of what appears to be some form of connection between internal organs and precise locations at the body's skin surface. About 1000 of these connecting lines, "meridians", are described in classical

Chinese acupuncture. Modern electroacupuncture techniques are finding still more. However, acupuncture is not an exercise in meridian "plumbing"; it is an philosophy of life with the checks and balances which determine health or disease. The life forces, body energy or Chi, are supposed to circulate through the meridians, 25 times each day and each night. This is yet another of the circadian rhythms in the body. Wever (1985) has shown how the synchronism of some of these rhythms can be affected by environmental parameters and particularly by the very difficult to shield, environmental ELF fields. These include the naturally occurring Schumann Radiation, that is itself subject to the effects of tides and extraterrestrial radiation in the upper atmosphere, as well as all the man-made sources of ELF.

Recently, Darras and De Vernejoule (World Research Foundation Congress, Los Angeles, 1986) reported success in visualising the acupuncture meridians by injecting a radioactive tracer (containing one of the radioactive isotopes of technetium, mercury or xenon) at an acupuncture point. Using a gamma-ray camera, they found that the radioactivity travelled along the acupuncture meridian with a velocity of 3-5 cm/min, the right order of magntidue to give the 25 circulations, day or night. The velocity was found to be less in the case of diseased organs. They confirmed that the radio isotope did not diffuse appreciably if injected other than at an acupuncture point, and that it did not enter into the lymphatic system or the blood circulation. It penetrated a tournequet preventing blood circulation. It only diffused towards the target organ if injected at an acupuncture point a part way along a meridian. The diffusion was not due to electrophoretic forces because a non-ionic radio-tracer (xenon) was equally affected. The rate of diffusion along the meridian was increased when the acupuncture point was stimulated, whether by a needle, electrically or by a helium-neon laser. M.A. Trelles (ZDN Congress, Essen, 1985) has also used low-power laser beams applied to acupuncture points to stimulate effects which occur in the appropriate remote organ. Andreev and co-workers (1984) have similarly applied highly coherent millimetre wave radiation and reported clinical successes in the treatment of duodenal ulcers.

The electrical characteristic of acupuncture points and their meridians is a decrease in the resistance measured on the skin extending for a few millimetres around an acupuncture point. Modern acupuncture (Kenyon 1983a,b, 1985) now makes extensive use of electrical devices in many ways both for diagnosis and therapy.

The clear impression gained from watching demonstrations of electrical diagnosis and therapy apparatus, is that the operator and the patient together form an essential part of such apparatus and the measurement procedures. The significant biocommunication signals may not actually travel in apparently obvious directions. For example, an electrical output from the apparatus may trigger allergic responses in hypersensitive subjects and their subsequent electromagnetic reactions may affect the electrical resistance measured at the acupuncture points of the patient, or the person testing the patient.

If it is remembered that a reacting allergic subject can identify a tube of an allergen merely brought into the room, then the ability of these pieces of electronically simple apparatus to indicate the correct potency of a homoeopathic preparation when used by a sensitive operator should not be cause for surprise. Although in many cases they are versions of Wheatstone's Bridge or are high-gain low-frequency amplifiers and filters, this need not imply that they are actually responding to the low frequencies for which they were designed. Most commercial chart recorders will produce all sorts of complicated traces on the chart paper, with no connections at all made to the input, if an oscillator on the other side of the laboratory is merely tuned across a band of frequencies from 10 MHz to 1000 MHz. These frequencies travel around the laboratory and along power wiring, they are converted to small steady voltages by nonlinearities in high-gain amplifiers. It should not be forgotten that living systems too, can emit radio frequencies of sufficient intensity to affect electronic equipment.

8 Conclusions

The lowest electric field that has been observed to evoke a response from certain fish is 1 μV/m (Bullock 1977), which corresponds closely with the most extreme sensitivities of allergic subjects. Comparing this to the observed radiofrequency emission levels from yeast cells (0.1 μV/μm) (Smith 1984) and assuming that there is nothing particularly unique about yeasts in this respect, then it is clear that fish will have no problem in locating food electrically, particularly if it happens to be some electrically "noisy" plankton. The same ability might have applied to man in primitive conditions and may still be possessed by aboriginal peoples; scouts and hunters in the bush and jungle may also unwittingly make use of it.

The response of living systems to an electromagnetic stress is exemplified by the responses of *E. coli* to alternating magnetic fields which show:

1. a critical threshold for the onset of any response,
2. a response which does not increase proportionately to the field, but becomes more complicated and may require high resolution and sensitivity for its detection,
3. a magnitude of response which remains small unless the homeostatic system is under a steady biological stress (Popp 1986b).

The theoretical limits to the electromagnetic sensitivities of living systems can be estimated by the following method:

a) take the equation for the energy per unit volume of the appropriate field,

b) multiply this by the volume of the biological system presumed to be cooperatively involved in detecting the radiation; assume this is the apparent volume of the organ sensing the field or the volume of the entire organism,

c) equate the energy calculated above to the ambient thermal energy (kT). This is equivalent to assuming that the system is operating with a signal-to-noise ratio of unity, or that the signal is only just as strong as the background thermal noise (Bell 1960; Smith 1986),

d) the only unknown in the equation is the field strength satisfying the initial conditions. Table 1 lists these for a range of biological structures.

If the whole of a 70-kg man is, in effect, the antenna for the detection of an electric field, the above criteria would be met with a threshold electric field of 8 μV/m. This is of the order of the sensitivity of the most sensitive to the multiple-allergy patients tested, but it only occurred while the patients were in a reacting allergic condition. It corresponds to an incident power density of less than a picowatt per square metre.

Enzymes enable and control the biochemistry of living systems. There are about 3000 different enzymes in each cell. There is evidence that electromagnetic phenomena are important even at this basic level when one considers the biological cell. The electric field across a cell membrane is a good indicator that the cell is alive. It is likely that all electromagnetic frequencies from the ultraviolet to the reciprocal of the life-time of the cell are involved in its life processes. Because many of the biomolecules are themselves electrically charged, acoustic vibrations and electrical vibrations are equivalent and interchangeable. There is evidence for highly coherent microwave acoustic modes in DNA (Edwards et al. 1985).

The biochemical oxidation reactions by which an organism gets its energy involve free radicals (Levine and Parris 1985), if electromagnetic fields and frequencies are able to alter free radical kinetics, the stability of the regulatory system could be affected.

It seems that just as living systems are able to respond to near single quanta at optical frequencies, they are also able to respond to single quanta of flux in a magnetic field. If this is the case, they also have the 500 MHz/μV frequency/voltage Josephson interconversion, available (H. Fröhlich, this Vol.).

In the past, most bioelectric measurements have been concerned with the measurement of small, low frequency, or steady currents and voltages. Work on the radiofrequency and microwave effects in yeast cells strongly suggests that the Josephson relation (500 MHz/μV) should be applied to all experiments with living systems involving small steady voltages; where voltage steps occur, coherent oscillations should be sought at the frequencies indicated. This is how the writer adjusted the experimental conditions before detecting radiofrequency emissions from dividing yeast cells.

Any biological membrane is more likely to behave as a diode rectifier and Zener diode than as an ohmic conductor. This implies that the Nernst equation is being causally interpreted the wrong way round, the membrane potentials arise from the rectification of oscillations on the membrane, the ion concentrations are consequent upon this.

In any case, the interpretation of the results of bioelectrical experiments limited by microelectrode time-constants to d.c. or low frequencies, is as difficult and as restricted as would be the interpretation of a television transmission or a computer programme with nothing more than a "cat's-whisker" crystal set. Biological systems have so far been viewed with only one eye open. The coherent electromagnetic oscillations which are the dual of the chemical bond are as fundamental and interconvertable.

A question frequently asked, is whether one should choose to live under power lines. The writer's reply has been, only if:

1. there is no history of allergy in the family,
2. the medical adviser can recognize and treat environmentally triggered illnesses,
3. there is no history of past exposure to toxic chemicals or risk of future exposure and dietary habits avoid the consumption of food additives where possible.

One is also frequently asked about the possibilities of significant effects from electromagnetic fields on large populations. From what has been written in this chapter, it should now be clear:

1. that homoeopathic remedies stimulate in healthy persons the symptoms of the disease that they seek to cure in the sick,
2. that coherent oscillations are able to reproduce homoeopathic potencies,
3. that the highly coherent microwave oscillators now available could be modulated with homoeopathic frequencies,
4. that the non-linear conductivity of living tissues would demodulate such absorbed microwave radiation to produce currents at homoeopathic frequencies and give rise to "proving" symptoms in healthy persons,
5. that prolonged exposure to such homoeopathic "proving" symptoms may result in persons acquiring hypersensitivities to them, particularly in the presence of chemical or nutritional "triggers".

This has been written so that the possibilities may be duly recognized, in case this should ever be attempted in a clandestine manner.

Unfortunately, there may be accidents which will require mass medication from a distance and in this connection it should be noted that homoeopathic potencies of radium bromide have been used for the treatment of radiation sickness. There are cases where allergic responses appear to have been triggered by radioactive "fall-out".

However, one must hope that World Peace does not depend upon the allergy-free status of strategic computer operators throughout the world.

Acknowledgments. The writer gratefully acknowledges the many years of advice and encouragement from Professor H. Fröhlich, F.R.S.

Thanks are also due for the clinical cooperation of Dr. Jean Monro, Dr. Bill Rea and Dr. Ray Choy which has made it possible to relate the theoretical to the clinical and the environment, and also to colleagues and the many students whose research reported in theses and publications has contributed so much to the progress.

References

Aarholt E, Flinn EA, Smith CW (1981) Effects of low-frequency magnetic fields on bacterial growth rate. Phys Med Biol 26:613–621

Aarholt E, Flinn EA, Smith CW (1982) Magnetic fields affect the lac operon system. Phys Med Biol 27:603–610

Aarholt E, Jaberansari M, Jafary-Asl AH, Marsh PN, Smith CW (1987) NMR conditions and biological systems. In: Marino AA (ed) Handbook of bioelectricity. Marcel Dekker, New York, Chap 26

Ahmed NAG, Smith CW (1978) Further investigations of anomalous effects in lysozyme. Collect Phenom 3:25–33

Ahmed NAG, Calderwood JH, Fröhlich H, Smith CW (1975) Evidence for collective magnetic effects in an enzyme: likelihood of room temperature superconductive regions. Phys Lett 53A: 129–130

Ahmed NAG, Smith CW, Calderwood JH, Fröhlich H (1976) Electric and magnetic properties of lysozyme and other biomolecules. Collect Phenom 2:155–166

Andreev EA, Beliy MU, Sitko SP (1984) The appearance of characteristic frequencies of the human body. Dokl Akad Nauk Ukr SSR No 10 Ser B Geol Khim Biol Nauki, pp 60–63 (in Russian)

Baker RR (1984) Signal magnetite and direction finding. Phys Technol 15:30–36

Baker RR (1985a) Magnetoreception by man and other primates. In: Kirschvink JL, Jones DS, McFadden BJ (eds) Magnetite biomineralisation and magnetoreception in organisms: a new magnetism. Plenum, New York, Chap 26

Baker RR (1985b) Human navigation: A summary of American data and interpretations. In: Kirschvink JL, Jones DS, McFadden BJ (eds) Magnetite biomineralisation and magnetoreception in organisms: a new magnetism. Plenum, New York, Chap 34

Barr ML (1979) The human nervous system. Harper & Row, Hagerstown

Becker RO, Marino AA (1982) Electromagnetism and life. SUNY, Albany

Becker RO, Selden G (1985) The body electric. Morrow, New York

Bell DA (1960) Electrical noise. Fundamentals and physical mechanism. Van Nostrand, London

Brillouin L (1934) Fluctuations de courant dans un conducteur. Helv Phys Acta 7 (suppl 2):47–67

Brown E, Behrens K (1985) Your body's responses. Madison Ave, Dallas

Brügemann H (1984) Diagnose- und Therapieverfahren im ultrafeinen Bioenergie-Bereich. Haug Verlag, Heidelberg

Bullock TH (1977) Electromagnetic sensing in fish. Neurosci Res Program Bull 15(1):17–22

Callinan P (1985) The mechanism of action of homoeopathic remedies. Complementary Med 3(1):35–56

Careri G, de Angelis L, Gratton E, Messana C (1977) Magnetic susceptibility of lysozyme. Phys Lett 60A:490–491

Choy RYS, Monro JA, Smith CW (1987) Electrical sensitivities in allergy patients. Clin Ecol 4(3): 93–102

Chu CW, Chen VKH, Sugawara K, Huang CY (1976) Search for magnetic field induced aggregations of lysozyme molecules in dilute aqueous solutions. Solid State Comm 19:357–359

Davies KE, Walker IO (1979) The structure and function of chromatin in lower eukaryotes. In: Nicolini CA (ed) Chromatin structure and function. Plenum, New York

Delgado JMR, Leal J, Monteagudo JL, Gracia MG (1982) Embryological changes induced by weak, extremely lowfrequency electromagnetic fields. J Anat 134:533–551

Didot F (ed) (1861) Nouvelle biographie generale. Firmin Didot, Paris, Col 147–162

Dubrov AP (1978) The geomagnetic field and life: Geomagnetobiology. Plenum, New York

Edwards GS, Davis CC, Saffer JD, Swicord ML (1985) Microwave-field-driven acoustic modes in DNA. Biophys J 47:799–807

Faraday M (1838) Experimental researches in electricity. Taylor and Francis, London, (reprint (1965) Dover)

Fidler JH (1983) Ley lines: their nature and properties. Turnstone, Wellingborough

Fröhlich H (1969) Quantum mechanical concepts in biology. In: Marois M (ed) Theoretical physics and biology. North Holland, Amsterdam

Fröhlich H (1975) The extraordinary dielectric properties of biological molecules and the action of enzymes. Proc Natl Acad Sci USA 72:4211–4215

Fröhlich H (1978) Coherent electric vibrations in biological systems. IEEE Trans MTT 26:613–617

Fröhlich H (1980) Biological effects of microwaves and related questions. Adv Electronics Electron Phys 53:85–152

Fröhlich H, Kremer F (eds) (1983) Coherent excitations in biological systems. Springer, Berlin Heidelberg New York

Gauquelin M (1973) The cosmic clocks: from astrology to a modern science. Paladin, St. Albans

Geddes LA, Hoff HE (1971) The discovery of bioelectricity and current electricity – the Galvani – Volta controversy. IEEE Spectrum 8:38–46

Gendrin R, Stefant R (1964) Magnetic records between 0.2–30 c/s. In: Blackman WT (ed) Propagation of radio waves at frequencies below 300 kc/s. Pergamon, London

Gilbert G (1600) De magnete, magnetisque corporibus, et de magneto magnete tellure; physiologia nuoa, plurimis & argumentis & experimentis demonstrata. Short, Londini (Reprint (1958) Dover)

Gründler W (1985) Frequency-dependent biological effects of low intensity microwaves. In: Chiabrera A, Nicolini C, Schwan HP (eds) Interactions between electromagnetic fields and cells. NATO ASI Ser 97A. Plenum, New York, pp 458–481

Jacobi E, Kruskemper G (1975) Wirkungen simulierter sferics (wetterbedingte, elektromagnetische Strahlungen) auf die Thrombozytenadhäsivität. Inn Med 2:73–81

Jafarl-Asl AH, Smith CW (1983) Biological dielectrics in electric and magnetic fields. Annu Rep Conf Electrical Insulation & Dielectric Phenom. IEEE Publ 83 CH 1902-6, pp 350–355

Jafary-Asl AH, Solanki SN, Aarholt E, Smith CW (1983) Dielectric measurements on live biological materials under magnetic resonance conditions. J Biol Phys 11:15–22
Keeton WT (1979) Avian orientation and navigation. Brit Birds 72:451–470
Kenyon JN (1983a,b, 1985) Modern techniques of acupuncture. Thorsons, Wellingborough, 3 vols
König HL (1979) Bioinformation – electrophysical aspects. In: Popp F-A, Becker G (eds) Electromagnetic bio-information. Urban and Schwarzenberg, Munich, pp 25–54
Lakhovsky G (1939) The secret of life. Heinemann Medical, London
Levine SJ, Parris MK (1985) Antioxidant adaptation, its role in free radical pathology. Allergy Res Gp, San Leandro CA
Liboff AR, Williams Jr T, Strong DM, Wistar Jr R (1984) Time-varying magnetic fields: effect on DNA synthesis. Science 223:818–820
Ludwig HW (1987) Electromagnetic multiresonance – the base of homeopathy and biophysical therapy. In: Proc 42nd Congr Int Homeopathic Med League, 29 Mar–2 Apr 1987, Arlington, Am Inst Homeopathy, Washington DC
Marsh PN (1986) The biological and biochemical effects of microwave and radiofrequency radiation on the bovine eye lens in vitro. Thesis, University of Salford
Miller JB (1972) Food allergy, provocative testing and injection therapy. C.C. Thomas, Springfield IL
Monro J (1983) Food allergy in migrane. Proc Nutr Soc 42:241–246
Monro J, Carini C, Brostoff J (1984) Migrane is a food-allergic disease. Lancet 2:719–721
Moon MJ, Jhon MS (1986) The studies on the hydration energy and water structures in dilute aqueous solution. Bull Chem Soc Jpn 59:1215–1222
Nordenström BEW (1983) Biologically closed electric circuits: clinical, experimental and theoretical evidence for an additional circulatory system. Nordic Medical, Stockholm
Nordenström BEW (1985) Biokinetic impacts on structure and imaging of the lung: the concept of biologically closed electric circuits. Am J Roentgenol 145:447–467
O'Neill JJ (1968) The life of Nikola Tesla: a prodigal genius. Nevil Spearman, London
Pallas-Areny R (1987) On the simulation of real 50/60 Hz electrical fields. IEEE Eng Med & Biol Mag 6:58
Piccardi G (1962) The chemical basis of medical climatology. C.C. Thomas, Springfield IL
Popp F-A (1979) Photon storage in biological systems. In: Popp F-A, Becker G (eds) Electromagnetic bio-information. Urban and Schwarzenberg, Munich, pp 123–149
Popp F-A (1986a) On the coherence of ultraweak photon emission from living tissues. In: Kilmister CW (ed) Disequilibrium and self-organisation. Reidel, Hingham MA, pp 207–230
Popp F-A (1986b) Bericht an Bonn. VGM Verlag, Essen, p 85
Rea WJ, Butler JR, Laseter JL, DeLeon IR (1984) Pesticides and brain-function changes in a controlled environment. Clin Ecol 2(3):145–150
Semm P, Schneider T, Vollrath L (1980) Effects of an earth-strength magnetic field on electrical activity of pineal cells. Nature 288:607–608
Shaya SY, Smith CW (1977) The effects of magnetic and radiofrequency fields on the activity of lysozyme. Collect Phenom 2:215–218
Smith CW (1984) Proc 6th Annu Conf IEEE Eng Med and Biol Soc. IEEE Publ No CH2058, pp 176–180
Smith CW (1985) Superconducting areas in living systems. In: Mishra RK (ed) The living state II. World Scientific, Singapore, pp 404–420
Smith CW (1986) High sensitivity biosensors and weak environmental stimuli. Proc Colloq Bioelectronics and biosensors. UCNW Bangor 17–19 April 1985. In: Industrial Biotechnology Wales, April/May 1986, Art 4:2–85
Smith CW, Aarholt E (1982) Possible effects of environmentally stimulated endogenous opiates. Health Phys 43(6):929–930
Smith CW, Baker RD (1982) Comments on the paper "Environmental Power-Frequency Magnetic Fields and Suicide". Health Phys 43:439–441
Smith CW, Choy R, Monro JA (1985) Water – friend or foe? Lab Pract 34(10):29–34
Sorensen CM, Fickett FR, Mockler RC, O'Sullivan WJ, Scott JF (1976) On lysozyme as a possible high temperature superconductor. J Phys C:Solid State Phys 9:L251

Wever R (1973) Human circadian rhythms under the influence of weak electric fields and the different aspects of these studies. Int J Biometeorol 17(3):227–232

Wever RA (1985) The electromagnetic environment and the circadian rhythms of human subjects". In: Grandolfo M, Michaelson SM, Rindi A (eds) Static and ELF electromagnetic fields: Biological effects and dosimetry. Plenum, New York

Williams HN, Sabarth E (1987) In vitro demonstration of homeopathic effects. In: Proc 42nd Congr Int Homeopathic Med League, 29 Mar–2 Apr 1987, Arlington, Am Inst Homeopathy, Washington DC

Coherent Properties of Energy-Coupling Membrane Systems

D. B. KELL [1]

1 Introduction and Scope

An understanding of the organization of biological systems is predicated upon an understanding of their energetics. Thus the present chapter will give a very broad overview of some of the current thinking in bioenergetics, with especial reference to the formation of ATP linked to the transport of electrons down their electrochemical potential gradient, as catalyzed by biomembranes containing mobile protein complexes participating in the two half-reactions (Fig. 1). Here we see how the downhill reactions of electron transport are coupled to the otherwise endergonic ATP synthase reaction through the transfer of one or more quanta of free energy (Fig. 1A). Arguably, the major problem of bioenergetics concerns the nature of this free-energy-transducing quantum, and Fig. 1B shows some of the salient possibilities under discussion (Kell and Harris 1985a). Only the celebrated chemiosmotic model may be regarded as reasonably well developed (Nicholls 1982; Harold 1986), but since its perceived shortcomings have been discussed elsewhere in extenso (e.g. Ferguson and Sorgato 1982; Kell 1979, 1986a, 1987a, 1988; Ferguson 1985; Kell and Hitchens 1983; Westerhoff et al. 1984a; Kell and Westerhoff 1985), I shall not concentrate on it in detail here, where a more heuristic overview is appropriate.

One distinction to be made between various approaches to describing energy coupling in electron transport-linked phosphorylation (ETP) is whether the intermediates they postulate for such a process are or are not thermally activated (Blumenfeld 1983; Welch and Kell 1986; Kell 1987b, 1988). For this reason, and in view of the subject matter of the present volume, I shall devote my space to describing how in principle we should best seek to treat an energy-coupling system, and to reviewing some recent studies which have sought to estimate the degree of coherence involved in such processes. As before (Kell and Hitchens 1983; Kell and Westerhoff 1985; Kell 1988), I shall take "coherence" to cover the idea that the motions of one or more parts of a system are directly and functionally linked to those in a spatially separate part of the system, so that free energy may be transferred between them in an essentially dispersionless fashion.

1 Department of Botany and Microbiology, University College of Wales, ABERYSTWYTH, Dyfed SY23 3DA, U.K.

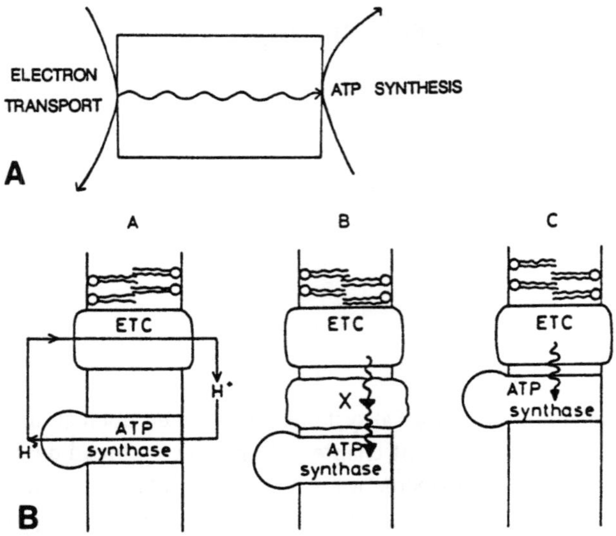

Fig. 1. A The central problem of energy coupling in electron transport-linked phosphorylation. The system is modelled as a black box in which the input reaction is electron transport catalyzed by a particular enzyme complex whilst the output reaction is ATP synthesis catalyzed by a separate enzyme complex. Free energy *(wiggly line)* is conserved by the first enzyme and passed to the second enzyme. Both enzymes are embedded in a so-called energy coupling membrane. **B** Three possible models for this process: in the chemiosmotic model *A*, the electron transfer complex *(ETC)* pumps protons across the membrane; these protons may return to the aqueous phase in which they originated by means of a proton-translocation ATP synthase. The lateral mobility of these complexes is irrelevant. By contrast (*B*, *C*) a special arrangement of the ETC and ATP synthase complexes may be necessary to effect free energy transfer by means of an unknown energised state, involving electrical and acoustic modes of the membrane (proteins). There may (*B*) or may not (*C*) be a requirement for other proteins (*X*) to participate in this energy coupling

2 Conformational States of Protein Molecules

A typical protein of molecular weight 20 kD may possess or explore some 10^{80} conformational substates (e.g. Jaenicke 1984). However, even if we let it explore them at a rate of 10^{15} per second, it cannot possibly explore all of those available in passing from a given conformational "state" to another, since the Universe is "only" some 10^{17} s old (Barrow and Silk 1983). Thus even an isolated protein, ostensibly in equilibrium with a heat bath, is not an ergodic system. Nonetheless, it is usual to speak of "the" structure of a protein, and the changes in angles and lengths of non-covalent intramolecular bonds caused by thermal energy (kT) are both relatively small and are observable in electron density maps obtainable by X-ray diffraction (e.g. Ringe and Petsko 1985). This allows us to make progress, but does not tell us if anything about whether something qualitatively different takes place when we allow our protein to do something, such as to catalyze the approach to equilibrium of a chemical reaction involving small molecules.

A consideration of typical free energy diagrams as drawn for enzymatic processes (e.g. Fersht 1985), in which the free energy of the system is plotted against the reaction coordinate, indicates that enzymes are equilibrium thermochemical machines which may exploit (at least part of) the energy of ligand binding to lower, and facilitate passage over, the energy barrier separating the enzyme-substrate and enzyme-product complexes. But whilst these typical aqueous globular enzymes do transduce free energy in the sense that they attain (pass through) states with a higher free energy than the ground state, their macroscopic effect is, as stated, simply to catalyze the approach of a reaction to equilibrium. There is then no thermodynamic paradox here, despite the fact that our enzyme is acting isothermally, since it is only required that the overall ΔG for the system $\leqslant 0$ (Kemeny 1974; Lumry 1980; Cooper 1984; Somogyi et al. 1984). Indeed, the view that an enzyme acts at least in part as an "equilibrium chemodynamical machine" (Somogyi et al. 1984) has been significantly strengthened recently by the demonstration that an important partial reaction catalyzed by engineered derivatives of the *Bacillus stearothermophilus* tyrosyl-tRNA synthetase exhibit linear free energy relationships (Fersht et al. 1986). By contrast, studies by Frauenfelder and coworkers (Ansari et al. 1985) and by Isied and colleagues (Bechtold et al. 1986) have shown that even aqueous, globular proteins can apparently exhibit out-of-equilibrium states for considerable periods.

The problem then is (Welch and Kell 1986): how does an enzyme which, whilst it is fluctuating considerably about its "mean" conformation(s) [and hence free energy (Welch et al. 1982; Welch 1986; Englander and Kallenbach 1984)], succeed in distinguishing a conformational substate which is arrived at by means of a "favourable" set of thermally-induced fluctuations (and which it is not allowed to use to conserve free energy) from one which arises by virtue of a (macroscopically) thermodynamically favourable reaction such as ligand-binding (McClare 1971), since at the submolecular level they are in principle indistinguishable? Evidently, and whilst this is to an extent self-defining, the latter case is associated with regions of conformational phase space which are only rarely if ever encountered by the enzyme in its "ground state" ($\pm$ kT), and it must be taken that these regions of conformation space are characterized by the fact that collective and microscopically "irreversible" motions are required to attain them. It is the "rarely if ever" phrase in the previous section which gives us a further clue, since it immediately introduces the idea of time into thermodynamics, as required by our earlier simpler analysis indicating the impossibility that an ensemble of protein molecules could be an ergodic system even when at equilibrium in a heat bath. Thus our enzyme, by catalyzing a reaction in a "forward" direction, exhibits irreversibility de facto. Similarly, the fact that an enzyme obeys the Haldane relationship (see Fersht 1985) does not of itself prove that the forward and back reactions are mechanistically identical, merely that the free energies (chemical potentials) of the intermediates are. In this regard, it is worth remarking that an imperfectly coupled molecular energy machine catalyzing a reaction against the chemical potential of its ligands may be identified on the basis that it will *not* obey the Haldane relationship.

At the current state of knowledge we can say little more about this at the fundamental or even mechanistic levels, beyond indicating that the acceptance of this fact alone shows that protein molecules must exhibit collective or coherent *intra*molecular motions. A recent series of calculations of relevance to the present problem (Astumian et al. 1987; Westerhoff et al. 1986, 1987) shows that the electrical noise spectrum of a prototypical membranous energy-transducing enzyme is far from white, and that such an energy transducer can harvest exogenous electrical energy (both sinusoidal and "noise") of mean free energy < kT, provided that the noise spectrum of the exogenous field is not complementary to that of the enzyme. This provides a reasonably simple and well-defined type of mechanism by which a membrane-located energy converter may apparently disobey the second law in its usual formulations (i.e. act non-thermally sensu lato), although other thermally based but non-linear mechanisms such as field-induced "lateral electrophoresis" (see Harris and Kell 1985; Kell and Harris 1985b; Kell 1987b; Pething and Kell 1987) might appear to possess similar properties under appropriate conditions. In a similar vein, a recent study by Careri and colleagues (1985) of the dielectric behaviour of lysozyme has shown a highly cooperative channelling of protons from sites along the enzyme's entire surface (with a 7th order dependence on the number of bound protons) towards the enzyme's active site.

Since macroscopically observable conformational states ("primary macroergs" in the terminology of Blumenfeld 1983) may persist at least for seconds in bioenergetics, some mechanisms must be operative to restrict, in a kinetic sense, the decay of these conformational states to equilibrium (with consequent heat evolution) whilst permitting, in appropriate cases, their conservation as "high-energy" conformational states within the same molecule or by transfer to another protein molecule. The latter is what bioenergeticists refer to as energy coupling, or, in Fröhlich's (1969) phrase, the ability of a system "very strongly to excite a few modes of motion".

Two major classes of mechanistic proposal for the existence in living systems of the types of behaviour alluded to above include a variety of the possible solitary excitations or solitons (see e.g. Webb 1980; Bilz et al. 1981; Blumenfeld 1983; Davydov 1983; Jardetzky and King 1983; Scott 1983; Careri and Wyman 1984; Del Giudice et al. 1984; Lomdahl 1984; Lomdahl et al. 1984; Somogyi et al. 1984) and Fröhlich's (1968, 1969, 1980, 1986; Fröhlich and Kremer 1983) theory of collective excitations in biological systems. All of these theoretical formalisms share the idea that particular metastable, high-energy (out-of-equilibrium) states of proteins in vivo may be created as intermediates in and by natural processes and thus may be created or destroyed by very weak (non-thermal) exogenous stimuli. Based upon evidence from inhibitor titrations inter alia (see Kell and Hitchens 1983; Kell and Westerhoff 1985; Herweijer et al. 1986; Kell 1986, 1987a; Pietrobon and Caplan 1986a,b; Petronilli et al. 1987; Westerhoff and Kell 1987), it appears that the membranous systems of electron transport phosphorylation possess precisely this property of a coherent or dispersionless transfer of free energy between the spatially separate but mobile complexes catalyzing the appropriate half-reactions of electron transport and of phosphorylation. The question then arises as to whether they might also be expected to respond to weak exogenous stimuli.

3 Response of Membrane Proteins to Exogenous Electrical Fields

The imposition of an exogenous, sinusoidal electrical field of peak-to-peak field strength E_0 V/m and frequency ω radians/s via two macroscopic electrodes between which are held a suspension of (spherical) membrane vesicles of radius r induces a transmembrane electrical potential $\Delta\psi$ given approximately (Tsong 1983) by:

$$\Delta\psi = (1.5\ E_0 r \cos\theta)/[1 + (\omega\tau)^2]^{1/2}\ , \qquad (1)$$

where θ is the angle between a portion of the bilayer and the field direction and τ is the relaxation time for the charging of the membrane capacitance, i.e. the classical Maxwell-Wagner dielectric dispersion (see e.g. Kell and Harris 1985a; Pethig and Kell 1987). Generally these values of $\Delta\psi$ are tiny, and whilst the dipole moments μ of membrane proteins may be reasonably large, some several hundred Debyes (Neumann 1986; Pethig and Kell 1987), it is not expected that the Langevin factor $\mu E_0/kT$ (which determines the extent of any thermally based biological response) will attain any significant magnitude. Indeed, it is well known that induced DC values of $\Delta\psi$ of at least 150 mV are required to drive the synthesis of ATP by membranes incorporating an H^+-ATP synthase (see Kell 1986; Tsong and Astumian 1986). Since rather weak AC fields have been shown to elicit a plethora of biological responses in membranous systems (see e.g. Adey 1981; Pilla et al. 1983; Sepersu and Tsong 1984), however, it is evident that the idea of a membrane protein as a passive dipole is inadequate. In any event, both lipids and proteins are known not to undergo "flip-flop" motions (rotation about an axis in the plane of the membrane) on any time-scale of biological significance. By contrast, all types of proteins possess a marked fluctuational mobility, necessarily accompanied by (permanent and induced) dipolar changes, on a time-scale from picoseconds upwards (e.g. Gurd and Rothgeb 1979; Englander and Kallenbach 1984; Welch et al. 1982; Somogyi et al. 1984; Pethig and Kell 1987). Thus any membrane protein will be able to change its conformation(s) in response to an exogenous electrical field of virtually arbitrary frequency and magnitude, with a concomitant change in its enzymatic activity.

One explicit possibility, developed by Astumian, Tsong, Westerhoff and colleagues (see Westerhoff et al. 1986, 1987; Tsong and Astumian 1986) considers an ion-motive membrane protein as a machine which can exist in (at least) four dipolar states, each state being populated (and differentially liganded) to an extent dependent upon the number and nature of its bound ligands. Provided that these states are asymmetric, the protein will succeed in "harvesting" the free energy in the electrical field so as to drive a cyclic ion-pumping process even though the free energy in the field is apparently insufficient.

This simple example indicates one means by which a membranous enzyme might be affected by an electrical field. Similarly, whilst the fact that a protein is embedded in a membrane causes an "amplification" of the macroscopic electric field (see Tsong and Astumian 1986), this type of mechanism can work perfectly well for a "soluble" protein, particularly if its Debye-like rotation is restricted by its being bound to a cytoskeletal structure (or a cognate arrangement in a prokaryote). There has been some discussion of the ability of such structures to participate in energy transfer in vivo (e.g. Welch and Berry 1985), but whilst the existence and importance of such cellular organization is becoming well established (e.g. Clegg 1984; Welch 1985; Welch and Clegg 1987), experimental evidence for an energy-coupling role of these structures remains elusive.

That the activity of an enzyme may be affected by an exogenous parameter, such as the imposition of an electrical field, does not of itself mean that the flux of metabolites through a pathway of which the enzyme is a part will be similarly affected. This is because not all enzymes contribute to an equal extent to the control of flux.

Following the work of Kacser, Burns, Heinrich and Rapoport, the appropriate relationships between enzyme activities and pathway fluxes have been formalized in the so-called Metabolic Control Theory (for reviews see Westerhoff et al. 1984b; Kell and Westerhoff 1986a,b). This analysis provides a quantitative mechanistic description of the effects of an external variable on a steady-state metabolic system, and it is to be hoped that workers studying the effects of electrical fields on cells will adopt the formalism in the description of their experiments.

4 Concluding Remarks

What I hope to have been able to convey in this short overview is (a) that proteins, as complex macromolecules, must exhibit collective motions by virtue of their structures and the ambient thermal energy; (b) that non-thermally excited states may also be achieved and persist for periods that are long relative to those exhibited by simple molecules in viscous media; (c) that the distinction between these types of states is conceptually difficult and to an extent self-defining; (d) that both soluble and membranous proteins may absorb electromagnetic radiation which may affect both their kinetics and energetics, even when the free energy in the exciting field is miniscule; and (e) that the imposition of exogenous electrical fields of virtually arbitrary magnitudes and frequencies will lead to biological effects, but not necessarily in proportion to the change in turnover number of the "target" enzymes. On a more applied note, this opens up the possibility, despite a long history of empiricism (Rowbottom and Susskind 1984), of what really represents an entirely novel approach to affecting selectively the metabolism of living systems (Kell 1987c), both for good and for ill (Becker and Marino 1982; Becker and Selden 1985; Marino and Ray 1986), and one which we may hope might lack the side-effects of current chemotherapies.

References

Adey WR (1981) Tissue interactions with non-ionising electromagnetic radiation. Physiol Rev 61: 435–514

Ansari A, Berendzen J, Bowne S, Frauenfelder H, Iben IET, Sauke TB, Shyamsunder E, Young RD (1985) Protein states and proteinquakes. Proc Natl Acad Sci USA 82:5000–5004

Astumian RD, Chack PB, Tsong TV, Chen Y, Westerhoff HV (1987) Can free energy be transduced from electric noise? Proc Natl Acad Sci USA 84:434–438

Barrow JD, Silk J (1983) The left hand of creation. The origin and evolution of the expanding universe. Unwin, London

Bechtold R, Kuehn C, Lepre C, Isied SS (1986) Directional electron transfer in ruthenium-modified horse heart cytochrome c. Nature 322:286–287

Becker RO, Marino AA (1982) Electromagnetism and life. State University of New York Press, Albany

Becker RO, Selden G (1985) The body electric; electromagnetism and the foundation of life. Morrow, New York

Bilz H, Büttner H, Fröhlich H (1981) Electret model for the collective behaviour of biological systems. Z Naturforsch 36B:208–212

Blumenfeld LA (1983) Physics of bioenergetic processes. Springer, Berlin Heidelberg New York

Careri G, Wyman J (1984) Soliton-assisted unidirectional circulation in a biochemical cycle. Proc Natl Acad Sci USA 81:4386–4388

Careri G, Geraci M, Giansanti A, Rupley JA (1985) Protonic conductivity of hydrated lysozyme powders at megahertz frequencies. Proc Natl Acad Sci USA 82:5342–5346
Clegg JS (1984) Properties and metabolism of the aqueous cytoplasm and its boundaries. Am J Physiol 246:R133–R151
Cooper A (1984) Protein fluctuations and the thermodynamic uncertainty principle. Prog Biophys Mol Biol 44:181–214
Davydov AS (1983) Energy transfer along alpha-helical proteins. In: Clementi E, Sarma RH (eds) Structure and dynamics; nucleic acids and proteins. Adenine, New York, pp 377–387
Del Giudice E, Doglia S, Milani M, Fontana MP (1984) Raman spectroscopy and order in biological systems. Cell Biophys 6:117–129
Englander SW, Kallenbach NR (1984) Hydrogen exchange and structural dynamics of proteins and nucleic acids. Q Rev Biophys 16:521–655
Ferguson SJ (1985) Fully delocalised chemiosmotic or localised proton flow pathways in energy coupling? A scrutiny of experimental evidence. Biochim Biophys Acta 811:47–95
Ferguson SJ, Sorgato MC (1982) Proton electrochemical gradients and energy transduction processes. Annu Rev Biochem 51:185–217
Fersht AR (1985) Enzyme structure and mechanism, 2nd edn. Freeman, San Francisco
Fersht AR, Leatherbarrow RJ, Wells TNC (1986) Quantitative analysis of structure-activity relationships in engineering proteins by quantitative linear free energy relationships. Nature 322: 284–286
Fröhlich H (1968) Long-range coherence and energy storage in biological systems. Int J Quantum Chem 2:641–649
Fröhlich H (1969) Quantum mechanical concepts in biology. In: Marois M (ed) Theoretical physics and biology. North Holland, Amsterdam, pp 13–22
Fröhlich H (1980) The biological effects of microwaves and related questions. Adv Electronics Phys 53:85–152
Fröhlich H (1986) Coherence and the action of enzymes. In: Welch GR (ed) The fluctuating enzyme. Wiley, New York, pp 421–449
Fröhlich H, Kremer F (eds) (1983) Coherent excitations in biological systems. Springer, Berlin Heidelberg New York
Gurd FRN, Rothgeb TM (1979) Motions in proteins. Adv Protein Chem 33:73–165
Harold FM (1986) The vital force; a study of bioenergetics. Freeman, Oxford
Harris CM, Kell DB (1985) On the dielectrically observable consequences of the diffusional motions of lipids and proteins in membranes. 2. Experiments with microbial cells, protoplasts and membrane vesicles. Eur Biophys J 13:11–24
Herweijer MA, Berden JA, Slater EC (1986) Uncoupler-inhibitor titrations of ATP-driven reverse electron transfer in bovine submitochondrial particles provide evidence for direct interaction between ATPase and NADH: Q oxidoreductase. Biochim Biophys Acta 849:276–287
Jaenicke R (1984) Protein folding and protein organisation. Angew Chem Int Ed Engl 23:395–413
Jardetzky O, King R (1983) Soliton theory of protein dynamics. Ciba Found Symp 93:295–309
Kell DB (1979) On the functional proton current pathway of electron transport phosphorylation; an electrodic view. Biochim Biophys Acta 549:55–99
Kell DB (1986) Localized protonic coupling; overview and critical evaluation of techniques. Methods Enzymol 127:538–557
Kell DB (1987a) Forces, fluxes and the control of microbial growth and metabolism. J Gen Microbiol 133:1651–1665
Kell DB (1987b) Non-thermally excited modes and free energy transduction in proteins and biological membranes. In: Barrett TW, Pohl HA (eds) Energy transfer dynamics. Springer, Berlin Heidelberg New York, pp 237–246
Kell DB (1987c) Bioelectrochemical phenomena; their role and exploitation in science and technology. Univ Wales Rev Sci Technol 1:64–71
Kell DB (1988) Protonmotive energy-transducing systems; some physical principles and experimental approaches. In: Anthony CJ (ed) Bacterial energy transduction. Academic Press, London, in press
Kell DB, Harris CM (1985a) Dielectric spectroscopy and membrane organisation. J Bioelectricity 4:317–348

Kell DB, Harris CM (1985b) On the dielectrically observable consequences of the diffusional motions of lipids and proteins in membranes. 1. Theory and overview. Eur Biophys J 12:181–197

Kell DB, Hitchens GD (1983) Coherent properties of the membranous systems of electron transport phosphorylation. In: Fröhlich H, Kremer F (eds) Coherent excitations in biological systems. Springer, Berlin Heidelberg New York, pp 178–198

Kell DB, Westerhoff HV (1985) Catalytic facilitation and membrane bioenergetics. In: Welch GR (ed) Organized multienzyme systems; catalytic properties. Academic Press, New York, pp 63–139

Kell DB, Westerhoff HV (1986a) Towards a rational approach to the optimisation of flux in microbial biotransformations. Trends Biotechnol 4:137–142

Kell DB, Westerhoff HV (1986b) Metabolic control theory; its role in microbiology and biotechnology. FEMS Microbiol Rev 39:305–320

Kemeny G (1974) The second law of thermodynamics in bioenergetics. Proc Natl Acad Sci USA 71:2655–2657

Lomdahl PS (1984) Nonlinear dynamics of globular protein. In: Adey WR, Lawrence AF (eds) Nonlinear electrodynamics in biological systems. Plenum, New York, pp 143–154

Lomdahl PS, Layne SP, Bigio IJ (1984) Solitons in biology. Los Alamos Science 10:2–31

Lumry R (1980) Dynamical aspects of small-molecule protein interactions. In: Braibanti A (ed) Bioenergetics and thermodynamics; model systems. Reidel, Dordrecht, pp 435–452

Marino AA, Ray J (1986) The electric wilderness. San Francisco Press, San Francisco

McClare CWF (1971) Chemical machines, Maxwell's demon and living organisms. J Theor Biol 30:1–34

Neumann E (1986) Elementary analysis of chemical electric field effects in biological macromolecules. In: Gutmann F, Keyzer H (eds) Modern bioelectrochemistry. Plenum, New York, pp 97–175

Nicholls DG (1982) Bioenergetics; an introduction to the chemiosmotic theory. Academic Press, London

Pethig R, Kell DB (1987) The passive electrical properties of biological systems; their role in physiology, biophysics and biotechnology. Phys Med Biol 32:933–970

Petronilli V, Azzone GF, Pietrobon D (1987) Analysis of mechanisms of free energy coupling and uncoupling by inhibitor titrations; theory, computer modeling and experiments. Biochemistry: in press

Pietrobon D, Caplan SR (1986a) Double-inhibitor and inhibitor-uncoupler titrations. I. Analysis with a linear model of chemiosmotic coupling. Biochemistry 25:7682–7690

Pietrobon D, Caplan SR (1986b) Double-inhibitor and inhibitor-uncoupler titrations. II. Analysis with a non-linear model of chemiosmotic coupling. Biochemistry 25:7690–7696

Pilla AA, Sechaud P, McLeod BR (1983) Electrochemical and electrical aspects of low-frequency electromagnetic current induction in biological systems. J Biol Phys 11:51–58

Ringe D, Petsko GA (1985) Mapping protein dynamics by X-ray diffraction. Progress Biophys Mol Biol 45:197–235

Rowbottom M, Susskind C (1984) Electricity and medicine; history of their interaction. San Francisco Press, San Francisco

Scott AC (1983) Solitons on the alpha-helix protein. In: Clementi E, Sarma RH (eds) Structure and dynamics; nucleic acids and proteins. Adenine, New York, pp 389–404

Sepersu EH, Tsong TY (1984) Activation of electrogenic Rb^+ uptake in human erythrocytes by an electric field. J Biol Chem 259:7155–7162

Somogyi B, Welch GR, Damjanovich S (1984) The dynamic basis of energy transduction in enzymes. Biochim Biophys Acta 768:81–112

Tsong TY (1983) Voltage modulation of membrane permeability and energy utilisation in cells. Biosci Rep 3:487–505

Tsong TY, Astumian RD (1986) Absorption and conversion of electric field energy by membrane-bound ATPase. Bioelectrochem Bioenerg 15:457–476

Webb SJ (1980) Laser-Raman spectroscopy of living cells. Phys Rep 60:201–224

Welch GR (ed) (1985) Organized multienzyme systems; catalytic properties. Academic Press, New York

Welch GR (ed) (1986) The fluctuating enzyme. Wiley, New York

Welch GR, Berry MN (1985) Long-range energy continua and the coordination of multienzyme sequences in vivo. In: Welch GR (ed) Organized multienzyme systems; catalytic properties. Academic Press, New York, pp 419–447

Welch GR, Clegg JS (eds) (1987) The organization of cell metabolism. Plenum, New York

Welch GR, Kell DB (1986) Not just catalysts. The bioenergetics of molecular machines. In: Welch GR (ed) The fluctuating enzyme. Wiley, New York, pp 451–492

Welch GR, Somogyi B, Damjanovich S (1982) The role of protein fluctuations in enzyme action; a review. Prog Biophys Mol Biol 39:109–146

Westerhoff HV, Kell DB (1987) A control theoretic analysis of inhibitor titrations of metabolic channeling. Comments Mol Cell Biophys (in press)

Westerhoff HV, Melandri BA, Venturoli G, Azzone GF, Kell DB (1984a) A minimal hypothesis for membrane-linked free-energy transduction. The role of independent, small coupling units. Biochim Biophys Acta 768:257–292

Westerhoff HV, Groen AK, Wanders RJA (1984b) Modern theories of metabolic control and their application. Biosci Rep 4:1–22

Westerhoff HV, Tsong TY, Chock PB, Chen Y, Astumian RD (1986) How enzymes can capture and transmit free energy from an oscillating electric field. Proc Natl Acad Sci USA 83:4734–4738

Westerhoff HV, Kamp F, Tsong TY, Astumian RD (1987) Interactions between enzyme catalysis and non-stationary electric fields. In: Blank M, Findl E (eds) Interactions of electromagnetic fields with living systems. Plenum, New York, pp 203–216

Coherence in the Cytoskeleton: Implications for Biological Information Processing

S. R. HAMEROFF [1]

1 Introduction

Fröhlich's model of coherent excitations in biological systems can provide mechanisms for long-range order and cooperativity, factors useful for biomolecular communication and information processing. Fröhlich's coherence is based on oscillating dipoles in a voltage field, and has been applied primarily to biological membranes (Fröhlich 1970, 1975, 1986). Relatively overlooked, the interiors of living cells contain parallel networks of dynamic protein filamentous polymers which organize and regulate intracellular activities and whose properties strongly suggest communication and intelligence. These networks are collectively termed the cytoskeleton because they were originally thought to provide merely structural bone-like support to living cells. It is now recognized that the cytoskeleton is a dynamic information-processing system capable of organizing cell movement, division, growth and behavior. Thus the cytoskeleton may be described as the nervous system within all cells, ranging from single cell organisms like amoeba and paramecium, to nerve cells (neurons) within the human brain. In neurons, the cytoskeleton accounts for formation and rearrangements of neuronal form and synaptic connections, factors implicated in wide ranges of cognitive functions including neural networking, learning, memory, and consciousness. Components of the cytoskeleton (microtubules, actin and intermediate filaments, microtrabecular lattice) are oriented assemblies of "polar" subunits and have been accordingly described as electrets. The characteristics of polar electrets have been considered sufficient to support Fröhlich-type coherent excitations in the cytoskeleton (Del Giudice et al. 1983c). The presence of coherent excitations within the cytoskeleton would provide an intracellular dimension of molecular cognition. Several models of information representation and processing could take advantage of coherent "clocking" mechanisms in the cytoskeleton. These include parallel switching (neural network) matrices, reaction-diffusion patterns, cellular automata, nonlinear wave propagation, and coherent interference (holography). This chapter will review characteristics of the cytoskeleton, application of Fröhlich coherence mechanisms to cytoskeletal components, and theories of cytoskeletal information processing for which coherent oscillations may be important (Hameroff 1987).

---

1 Department of Anesthesiology, University of Arizona, Health Sciences Center, 1501 N. Campbell, Tucson, Arizona 85724, USA

2 The Cytoskeleton

Living organisms are collective assemblies of cells which contain collective assemblies of organized material called protoplasm. In turn, protoplasm consists of membranes, organelles, nuclei and the bulk interior medium of living cells: cytoplasm. Dynamic rearrangements of cytoplasm within living cells account for their changing shape, repositioning of internal organelles, and in many cases, movement from one place to another. We now know that the cytoskeleton, a dynamic network of filamentous proteins, is responsible for cytoplasmic organization (Fig. 1).

2.1 The Nature of Cytoplasm

Nineteenth century microscopists recognized that cytoplasm had structural, contractile, motile, elastic and rigid properties which permitted elaborate and dynamic forms. They concluded that cytoplasm is not merely a liquid nor an emulsion nor an aqueous suspension of life-bearing granules. Rather, cytoplasmic architecture and contractility could be explained by the proposal of a mesh-like reticular or fibrous substructure whose interstices were filled with fluid (Burnside 1974). Biologists began to consider cytoplasm as a gel in which rod-shaped filaments formed cross-linkages. Gel aptly describes the mechanical properties of cytoplasm: an elastic intermeshing of linear crystalline units giving elasticity and rigidity to a fluid while allowing it to flow.

Development of the electron microscope through the 1960's initially did not illuminate the substructures of cytoplasm. Portions of cells which were optically empty by light microscopy persisted in being empty in electron micrographs and the cell was perceived by many to be a "bag of watery enzymes". Ironically, the fixative then used in electron microscopy, osmium tetroxide, had been dissolving filamentous elements so that their presence was observed only sporadically. Later, with the advent of glutaraldehyde fixation, delicate tubular structures were found to be present in virtually all cell types and they came to be called microtubules. Subsequent characterization of actin, intermediate filaments, microtrabecular lattice and the structure of centrioles led to the recognition that cell interiors were comprised of dynamic networks of connecting filaments and brought the cytoskeleton out of the closet.

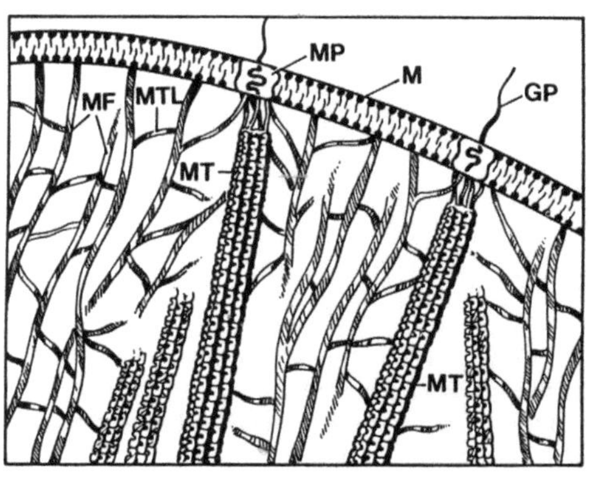

Fig. 1. Schematic of cellular cytoskeleton/membrane. *M* cell membrane; *MP* membrane protein; *GP* glycoprotein extending into extra-cellular space; *MT* microtubules; *MF* microfilaments (actin filaments or intermediate filaments); *MTL* microtrabecular lattice. Cytoskeletal proteins which connect MT and membrane proteins include spectrin, fodrin, ankyrin, and others. (Hameroff 1987)

2.2 Microtubules

Soifer (1986):

"When microtubules are required by a cell for a particular function, microtubules assemble in the appropriate part of the cell, with the necessary orientation. As microtubules are no longer needed, they depolymerize."

The most visible and widely studied cytoskeletal elements are microtubules (MT), slender cylinders intimately involved in the real time execution of important cell functions. Microtubules (MT) are hollow cylinders about 25 nanometers (1 nm = 10 Å = 10^{-9} m) in diameter whose walls are polymerized arrays of protein subunits. Their lengths may range from tens of nanometers during early assembly, to possibly meters in nerve axons within large animals. The protein subunits assemble in longitudinal strings called protofilaments; 13 parallel protofilaments laterally align to form the hollow tubules. The subunits are barbell-shaped proteins (dimers) which in turn consist of two globular proteins (monomers) known as alpha and beta tubulin (Fig. 2). Alpha and beta tubulin monomers are similar, but slightly different molecules with identical orientation within protofilaments and tubule walls. Each monomer consists of about 500 amino acids, weighs about 55 kilodaltons, and has a polarity or charge orientation. MT which grow from cell centers have a plus end (beta tubulin) which extends outward from the cell center (centrosome) into the cell periphery. The minus end (alpha tubulin) remains anchored to a microtubule organizing center (MTOC) within the centrosome. Each dimer, as well as each MT, appears to have an electrical polarity or dipole, with the negative end oriented towards the alpha monomer and cell center, and the positive end towards the beta monomer and cell periphery. The dimers are held together by relatively weak Van der Waals hydrophobic forces such as dipole coupling. Dimer neighbors form hexagonal lattices with a "leftward" tilt and several helical patterns occur in the relations among dimers (Dustin 1984).

MT from different life forms have marked similarities, but subtle differences. Comparison of MT from nerve cells of earthworms and mammals shows that the more

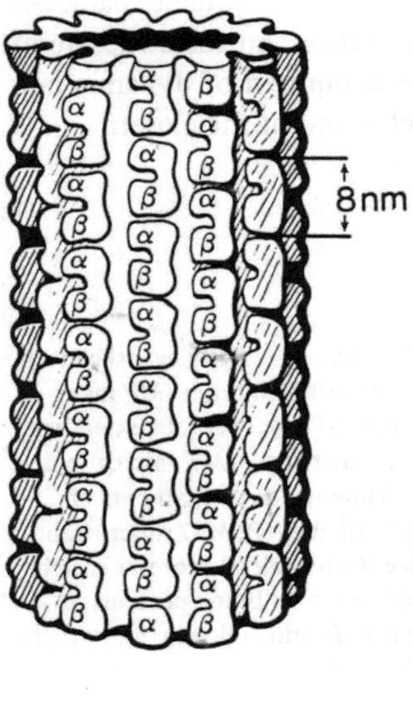

Fig. 2. Microtubules are cylinders whose walls are 13 protofilaments, each a string of 8-nm tubulin dimers. Alpha and beta tubulin monomers form the dimers. Each dimer has six neighbors. (Hameroff 1987)

primitive worm MT are more variable in geometric structure with MT ranging from 9 to 11 protofilaments, whereas mammalian MT are generally more consistent with 13. Tubulins from among different species including mammals and plants bind to common antibodies, and tubulins from different species may coassemble into hybrid MT. Despite these common traits, the diversity of tubulin gene expression has proved far greater than imagined years ago. Analysis of tubulin by amino acid sequencing and advanced electrophoretic techniques has shown that different alpha and beta tubulins exist concurrently, with the greatest diversity shown by beta tubulin. For example, Lee et al. (1986) have shown that as many as 11 different tubulin forms exist in rat thyroid microtubules and 17 different forms exist in rat brain microtubules. Thus alpha and beta tubulin are families of "isozymes," each of which may have specific functions or binding of microtubule-associated proteins (MAPs). Another tubulin variable, detyrosination, occurs in the cytoplasm subsequent to DNA transcription. Detyrosination is the removal of the terminal amino acid, tyrosine, from the polypeptide chain which comprises beta tubulin. Removal of tyrosine exposes an acidic amino acid, glutamate. Local factors in the cytoplasm beyond genetic programming determine whether or not individual tubulin subunits are "tyrosinated" or "glutamated." Thus both genetic and nongenetic coding mechanisms exist for the structure of tubulin subunits assembled within MT.

Since early electron microscopy studies, microtubules have been invariably described as being surrounded by a "clear zone" which gives the impression of a halo around them when they are viewed in cross-section. A 5–10 nanometer distance from the surface of MT is free of cytoplasmic ground substance or any other material normally seen elsewhere throughout the cell. Stebbings and Hunt (1982) have studied the "clear zone" and point out that MT surfaces are strongly "anionic" since tubulin is an acidic protein due to its high content of acidic amino acids such as glutamate and aspartate. These amino acids give up positively charged hydrogen ions to solution, leaving MT with excess electrons. Stebbings and Hunt propose that anionic, or electronegative fields at MT surfaces can explain the clear zones as well as the staining of MT by positively charged dyes, binding to MT of positively charged proteins, cations such as calcium, metals and other compounds. Electronegative fields surrounding MT may be heterogeneous or patterned due to detyrosinated subunits (exposing anionic glutamate), or other factors and act as excitable ionic charge layers (Debye layers) which are also thought to occur immediately adjacent to cell membranes (Green and Triffet 1985). Excitable clear zone charge layers and their ordered water and ions next to MT could facilitate collective communicative mechanisms occurring due to coherent excitations within the cytoskeleton.

The question of the hollow core within MT is even more mysterious; it too appears devoid of ground substance. It is unknown whether the interiors of MT are also electronegative zones, or perhaps positive ones which would create voltage gradients across MT walls. Del Giudice et al. (1986) have suggested electromagnetic focusing and possibly superconductivity within microtubule cores. Insulated from aqueous surroundings and held together by water-excluding hydrophobic forces, MT and the rest of the cytoskeleton comprise a solid state network within living cells.

What are the functions of microtubules? They do provide skeletal support, being the most rigid structures in most cells. To establish the pattern of the cytoskeleton and the form and function of living cells, MT assemble from subunits at the proper time, place, and direction. Biochemical energy in the form of GTP plays an important role in MT assembly (and ATP for actin polymerization). Tubulin subunits bound to GTP add on to polymerizing MT; however, energy is not utilized via hydrolysis until after the GTP tubulin is incorporated. Phosphate bond hydrolysis energy from GTP and ATP may be providing energy to cytoskeletal structures for coherent excitations and/or lattice solitons, polarization waves etc.

MT are often anchored and guided by MT-organizing centers (MTOC) containing centrioles. Centrioles are cylinders comprised of nine MT triplets which exist in perpendicularly oriented pairs. Centrioles provide guidance and directional nagivation for nearly all cytoplasmic movements including cell division, or mitosis (Fig. 3). Once in place, MTOC and MT direct and participate in movement of cytoplasm, organelles and materials, growth, reproduction, synaptic plasticity, and nearly all examples of dynamic cytoplasmic activity.

The full range of MT function is achieved by the actions of various proteins which may bind in precise fashions at specific patterns of tubulin dimers in MT lattices. These *microtubule-associated proteins* (MAPs) include electromechanical enzymes which generate force and movement, communicative crossbridges to other cytoskeletal filaments and organelles, MAPs which enhance MT assembly and a class of MAPs within neurons whose functions are not understood. In many cases, the attachment patterns of MAPs to MT lattice walls have a precise geometrical configuration which appears related to the function of the MAP-MT complex (Fig. 4). Motion producing MAP "arms" which consume ATP hydrolysis energy to generate force are analogous to myosin bridges of skeletal muscle. Moving arms attached to MT are made of pro-

Fig. 3

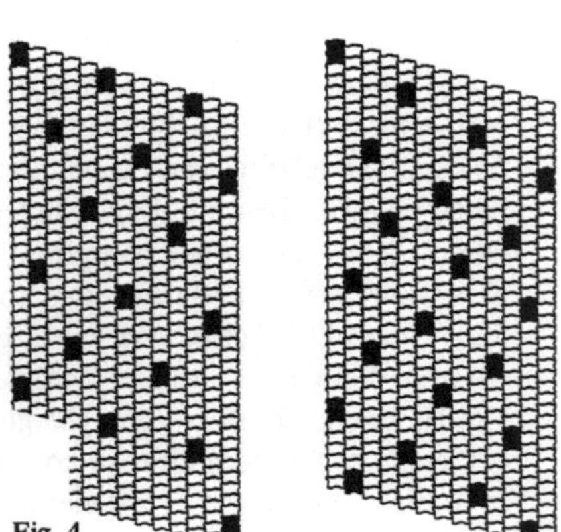

Fig. 4

Fig. 3. Centrioles in cell division. *1* Cross-section of centriole microtubule triplet. *2* Cross section of a centriole with 9 microtubule triplets, 9 satellite bodies, and central "pinwheel" structure. *3* Centriole pairs near cell nucleus, prior to cell division. *4* Centriole pairs have separated and migrated; chromosomes ready for separation. *5* Mitotic spindles, composed of microtubules, have formed from centriole centers (MTOCs) with chromosomes in middle. *6* Centriole anchored microtubules separating a pair of duplicate chromosomes. (Paul Jablonka in: Hameroff 1987)

Fig. 4. Patterns of MAP attachment to microtubules observed by electron microscopy. (Burns 1978; Kim et al. 1986; and others, from Hameroff 1987)

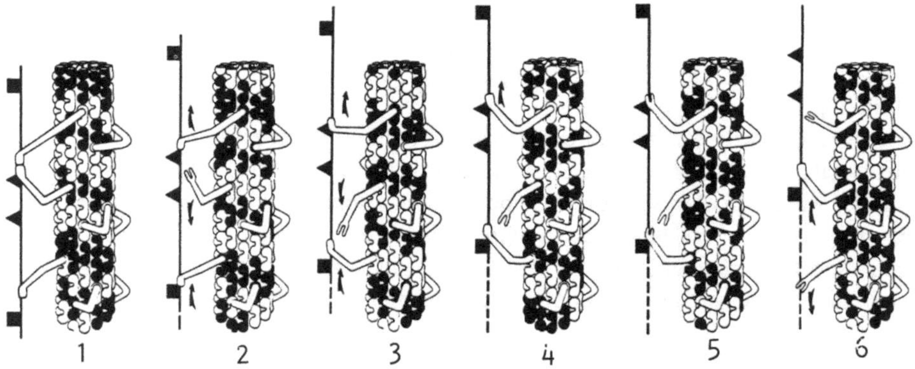

Fig. 5. Axoplasmic transport occurs by coordinated activities of microtubule attached sidearm, contractile proteins ("dynein") which cooperatively pass material in a "bucket brigade." The orchestration mechanism is unknown, but shown here as the consequence of signaling by "soliton" waves of microtubule subunit conformational states. By Fred Anderson

teins called kinesin and dynein. Dynein arms on MT which contract in organized sequences to produce collective movements were first described and characterized in cilia and flagella by (Gibbons 1968). The arms reside at periodic intervals along the outer MT within cilia and flagella and contract conformationally due to hydrolysis of ATP. Many MT-related activities generate force, locomotion and movement of vesicles and other material by a similar mechanism; axoplasmic transport is one well-studied example (Lasek 1981; Ochs 1982). Enzymes, neurotransmitter precursors, and organelles synthesized in neuronal cell bodies are transported along the lengths of axons by contractile dynein arms attached to axonal MT which act collectively to pass material in "bucket brigade" axoplasmic transport. The contractility mechanism is understood; however, the orchestration, signaling and coordination of the dynein MAPs appears to require some form of MT information processing (Fig. 5).

MT also participate in sensory perception of the cell's external environments. Many sensory receptors are modified cilia, assemblies of microtubules similar in structure to centrioles. Sensory transduction, centriole-based guidance and alignment, and axoplasmic transport are "intelligent" MT activities which are vital to biological growth, embryological development, secretion, synapse formation and many other important biological functions. MT are the scaffolding, conveyor belts, nagivators, and computers of living cells.

2.3 Intermediate Filaments, Actin and the Microtrabecular Lattice

The major filamentous components of the cytoskeleton are MT, actin filaments, intermediate filaments (IF) and a class of delicate interconnecting fibrils called the microtrabecular lattice (MTL). Actin is the most versatile component of the cytoskeleton. In conjunction with other proteins, it can polymerize in string-like filaments, form dynamic branching nanoscale meshworks or geodesic polyhedrons. Even more evanescent than labile MT, assembly of actin and associated proteins creates transient configurations of cytoplasm for specific purposes. Dynamic cytoplasmic expressions such as contractile actin-based rings which divide the cytoplasm in cell division, probing amoeboid lamellipodia, and dendritic spines and synapses in neurons may be the roots of intelligence.

Interaction with a variety of other proteins (actin-binding proteins) unleash actin's full capabilities. When crosslinked by proteins such as fimbrin, actin can form rigid bundles which provide structural support. When associated with myosin (a mechano-enzyme), useful muscle-like contraction occurs. Also important are proteins such as talin, spectrin, vinculin, ankyrin and fodrin which connect the cytoskeleton with membrane proteins, and calmodulin which mediates effects of calcium ion fluxes on the cytoskeleton. Proteins such as alpha actinin, troponin, and filamin link actin in networks which cause a gelatinous consistency to cytoplasm – a "gel" (Fig. 6). In the presence of calcium ions and actin-fragmenting proteins such as villin and gelsolin, the gel network liquefies to a solution – "sol." Other actin regulatory proteins include actin capping proteins, which stabilize polymerized actin and promote a gel condition, and profilin, which binds actin subunits, prevents polymerization and maintains sol conditions (Satir 1984). Actin gel polymers, like microtubules, are structurally coherent assemblies which consume sufficient phosphate bond hydrolysis energy (ATP) to fuel coherent oscillations, depolarization waves or solitons-phenomena which would help explain the complex and dynamic functions of actin gels.

The "unknown" members of the cytoskeleton, intermediate filaments represent the most nebulous and chemically variable subgroup among the cytoskeleton. Lazarides (1980) has shown that different types associate with specific types of cells. For example, subunit structure defines five major classes of intermediate filaments: (1) keratin (tonofilaments), which are found in epithelial cells, (2) desmin filaments, predominantly found in smooth, skeletal and cardiac muscle cells, (3) vimentin filaments, found in mesenchymal cells, (4) neurofilaments, found in neurons, and (5) glial filaments, found in glial cells. Often two or more of these classes co-exist in the same cell.

MT and intermediate filaments are not the finest texture of cytoplasmic organization. Smaller, more delicate structures made of actin branch and interconnect in gel state networks which comprise the microtrabecular lattice, or ground substance of living material. Techniques in electron microscopy developed by Porter and colleagues at the University of Colorado (Porter and Tucker 1981) led to observation of an ir-

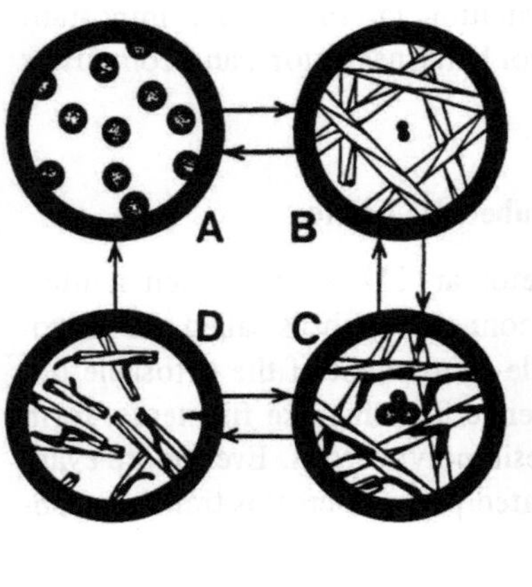

Fig. 6. Components of the cytomatrix. Assembly of actin and related proteins form cytoplasmic "gel" states. *A* monomeric actin; *B* assembled actin filaments which are helical pairs of monomeric strands; center-cross section of actin filament; *C* the presence of filamin (dark strands) causes actin cross-linking and dense "gel state;" *D* calcium and/or actin fragmenting proteins cause liquification to a "sol state." (By Paul Jablonka, from Hameroff 1987)

regular three-dimensional "trabecular" lattice of slender strands throughout the cytoplasm, interconnecting nearly everything in the cell. Slender strands of the microtrabecular lattice range in diameter from 4 to 10 nanometers with lengths of 10 to 100 nanometers and divide the ground substance of cytoplasm into two phases, a protein-rich polymerized phase and a water-rich fluid phase that fills the intratrabecular spaces.

MT, actin, intermediate filaments and the MTL comprise a dynamic cytoskeletal, solid state network within cytoplasm. Gershon et al. (1985) found that the cytoskeleton/MTL comprised from 16 to 21% of cytoplasmic volume with a surface area from 69,000 to 91,000 square microns per cell (69 to 91 billion nm^2). Biologist James Clegg (1981) has accrued evidence regarding the role of water surrounding cytoskeletal structures. Using neutron diffraction and other techniques, Clegg has found that water adjacent to the cytoskeleton is "ordered," that is aligned with polar bonds on the protein surfaces. Thus a layer of ordered water extends at least 3 nm from the billions of square nanometers of solid state surface area within each cell. This ordered water may be dipole coupled to coherent dynamics of the solid state, inhibit the thermal dissipation of protein oscillation energy from within the solid state, and shield calcium and other ions from random solid state interactions. Far from being "bags of watery enzymes," the interiors of living cell are highly ordered networks of dynamic, structurally coherent polymers.

3 Protein Conformational Dynamics

Proteins are the structural and organizational elements of the living state. Their essential functions are intrinsically linked to their structure and dynamic switching among different conformational states. Proteins such as enzymes, ion channels, receptors, cytoskeletal filaments, muscle myosin and hemoglobin undergo important conformational changes in response to a variety of stimuli. Coherent conformational states among cytoskeletal subunits may serve to represent information, exert control over routine biological functions and provide a basis for biomolecular communication.

3.1 Protein Conformation

Most protein molecules are able to shift reversibly between several different but related stable conformations due to several alternative sets of hydrogen bonds, disulfide bridges and hydrophobic Van der Waals forces of about equal energy among their constituent amino acid side chains. Thus step-like, nonlinear jumping occurs between specific conformations so that only a discrete number of alternative conformations exist for a given protein, and random switching from conformation to conformation is limited. Fröhlich (1975) proposed that dipole oscillations within water-excluding hydrophobic protein pockets were coupled to the protein's conformational state.

Different frequencies of conformational changes coexist cooperatively in the same protein (Table 1). The vast majority of processes of biological interest are in the time scale greater than 10^{-9} s, or one ns, which also lies in the collective mode realm for protein dynamics (Karplus and McCammon 1983). Protein conformational changes in the nanosecond time frame are coupled to stimuli by three fundamental features:

Table 1. Protein conformational dynamics – motions of globular proteins at physiological temperature. (After Karplus and McCammon 1983; from Hameroff 1987)

Amplitude of motions	0.001 to 10 nm
Energy	0.1 to 100 kcal
Time range	10^{-15} s (fs) to 10^{3} s (many minutes)
Time for collective functional steps	10^{-9} s (ns)
Types of motion	Local atom fluctuations
	Side chain oscillations
	Displacements of loops, arms, helices, domains and subunits
	Collective elastic body modes, coupled atom fluctuations, solitons and other nonlinear motions, coherent excitations

hydrophobic interactions, charge redistribution and hydrogen bonding. Appropriate stimuli including neurotransmitters, voltage alterations, hormones, calcium and other ions, and enzyme substrates can induce conformational changes within specific proteins (Fig. 7). Conformational transduction is the sensitive link in the mechanism of anesthetic gas molecules, and indicates that cognitive functions relating to consciousness depend in some way on protein conformational regulation. Dynamic conformational changes of proteins are the dynamics of living organisms.

3.2 Hydrophobic Interactions

The polymerization of cytoskeletal polymers and other biomolecules appears to flow against the tide of order proceeding to disorder dictated by the second law of thermodynamics. This apparent paradox is explained by the hydrophobic forces which bind protein assemblies including cytoskeletal polymers and which exclude ordered water from their protein sites. Thus the net effect is order proceeding to disorder because of the hydrophobic exclusion and disordering of previously ordered water.

The attractive forces which bind hydrophobic groups are called Van der Waals forces after the Dutch chemist who described them in 1873 (Baron 1961). At that time it had been experimentally observed that gas molecules failed to follow behavior predicted by the "ideal gas laws" regarding pressure, temperature and volume relationships. Van der Waals attributed this deviation to the volume occupied by the gas molecules and by attractive forces among the gas molecules. These same attractive forces are vital to the assembly of organic molecules, including protein assemblies and can explain cooperativity and coherent interaction among polymer subunits.

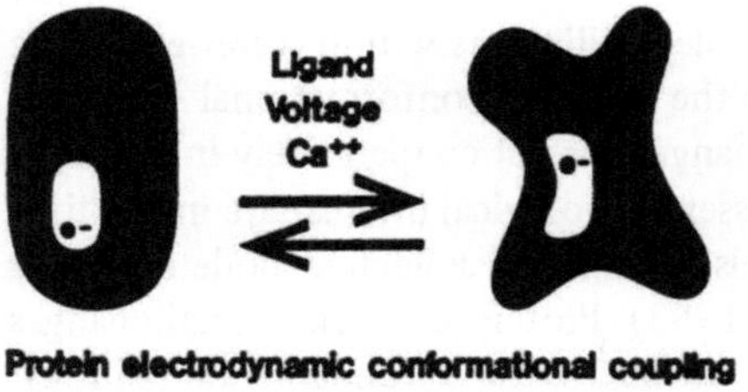

Fig. 7. Protein switching between two different conformational states induced by binding of ligand, calcium ion, or voltage change. One mechanism proposed by Fröhlich is that dipole oscillations (e^-) within hydrophobic regions of proteins may be a trigger, or switch for the entire protein. (Hameroff 1987)

Hydrophobic Van der Waals forces which are essential to the assembly and function of important biomolecules consist of dipole-dipole attraction, induction effect, and London dispersion forces.

Dipole-dipole attractions occur among molecules with permanent dipole moments. Alignments in which attractive, low energy arrangements occur are favored and net attraction between two polar molecules can result if their dipoles are properly configured. The induction effect occurs when a permanent dipole in one molecule can polarize and distort electrons in a nearby molecule so that dipole interactions are attractive. London dispersion forces explain why all molecules, even those without intrinsic dipoles, attract each other. Recognized by F. London in 1930 (Barrow 1961), the effect depends on quantum mechanical motion of electrons. Electrons in atoms without permanent dipole moments (and "shared" electrons in molecules) have, on the average, a zero dipole. However, dynamic instantaneous dipoles can be recognized which can induce dipoles in neighboring polarizable atoms or molecules. The strength of London forces can be significant only when two or more atoms or molecules are very close together (Barrow 1961). By dynamically coupling dipoles and conformational states among neighboring proteins, Van der Waals hydrophobic interactions can result in cooperativity and coherent oscillations of protein assemblies.

3.3 Electret, Piezo, and Pyroelectric Effects

Assemblies whose microscopic subunits and macroscopic whole both possess permanent electric dipoles are known as electrets. They exhibit properties known as piezoelectricity, which may help explain biological activities. In these crystals, dipolar elementary subunits are arranged in such a way that all the positive dipole ends point in one direction and all the negative dipole ends are oriented in the opposite direction. In microtubules, the positive ends of tubulin dimer subunits point away from microtubule organizing centers (MTOC) toward the cell periphery, and the negative ends point towards MTOC. Electrets can store charge and polarization and have now been identified in a variety of nonbiological materials such as ionic crystals, molecular solids, polymers, glasses, ice, liquid crystals and ceramics. Biological tissues demonstrating electret properties include bone, blood vessel wall materials, keratin, cellulose, collagen, gelatin, artificial polypeptides, keratin, DNA, cellulose and microtubules (Athenstaedt 1974). An electret effect accounts for specific properties such as anti-blood clotting in biomaterials and the non-stickiness of Teflon. Sources of polarization or charge storage in macromolecules are dipoles, ionic space charges, or ordered surface water.

Electret materials are piezoelectric (Gubkin and Sovokin 1960). Piezoelectric materials change their shape or conformation in response to electrical stimuli, and change their electrical state in response to mechanical stimuli. Koppenol (1980) has shown that the dipole moment orientation of an enzyme changes in concert with its functional activity. Electrets are also pyroelectric, in that any change in temperature alters the electrical and conformational characteristics of the molecule. The requisites for such behavior are permanent dipole moments in molecules bound in parallel alignment. Microtubules and other cytoskeletal structures appear to be appropriately configured electret, pyroelectric, piezoelectric devices.

The electret state within bone has been well studied and is able to store large amounts of polarization of the order of 10^{-8} coulombs/cm^2 (Mascarenhas 1974, 1975). The limit of charge separation (equivalent to maximal information density) has been calculated (Gutman 1986) to be about 10^{17} electronic charges/cm^3, while there may be about 10^{21} total molecules/cm^3. One cm^3 of densely arrayed parallel microtubules 100 nm apart contains about 10^{17} tubulin subunits and therefore may contain 10^{17} dipoles equivalent to the maximal density of charge! Electret and related properties can impart interesting and potentially useful coherent properties to biomolecules including cytoskeletal polymers.

3.4 Coherent Excitations (Fröhlich)

Proteins are vibrant, dynamic structures in physiological conditions. A variety of recent techniques (nuclear magnetic resonance, X-ray diffraction, fluorescence depolarization, infrared spectroscopy, Raman and Brillouin laser scattering) have shown that proteins and their component parts undergo conformational motions over a range of time scales from femtoseconds (10^{-15} s) to many minutes. The most significant conformational vibrations are suggested by Harvard's Karplus and McCammon (1983) to be in the middle of this range: nanoseconds. Such fluctuations are appropriate for conformational motion of globular proteins (4 to 10 nm diameter), consistent with enzymatic reaction rates and coupled modes like solitons.

Collective cooperativity of dynamic conformational states have been elegantly woven in a theory of coherent protein excitations by Professor Herbert Fröhlich, who has come to several profound conclusions. One is that changes in protein conformation in the nanosecond time scale may be triggered by a charge redistribution such as a dipole oscillation within hydrophobic regions of proteins (Fröhlich 1975). Another Fröhlich (1970) concept is that a set of proteins connected in a common voltage gradient field (membrane or polymer electret) would oscillate coherently at nanosecond periodicity if energy such as biochemical ATP or GTP hydrolysis were supplied. Far-reaching biological consequences may be expected from such coherent excitations, particularly when applied to the cytoskeleton.

Biomolecular excited states with very high dipole moments may, according to Fröhlich, become stabilized (metastable states) through internal and external deformations and through displacement of counter ions like calcium. Metastable states, which can correlate with functional conformations, are thus collective effects involving the molecule and its surroundings. A molecule might be lifted into a metastable state through the action of electric fields, binding of ligands or neurotransmitters, or effects of neighbor proteins. Thus rapid, nanosecond oscillations may become "locked" in specific modes which correspond to useful conformations of a protein. For example, an ion channel, receptor, enzyme, or tubulin subunit may stay in metastable conformational states for relatively long periods, on the order of milliseconds.

Expanding on Fröhlich's work, Wu and Austin (1978) concluded that coherently oscillating dipoles within a narrow band of resonance frequencies with large enough coupling constants may be expected to cause strong long-range (about 1 μ) attractive forces among dipoles. In a dense microtubule array, 1 μ^3 (one billion nm^3) would encompass about 160,000 tubulin subunits – an array sufficiently large for collective effects.

In a further extension of Fröhlich's work, a group of scientists from the University of Milan (Del Giudice et al. 1986) have viewed living matter as a sea of electric dipoles. The Milan group sees order in living systems to be induced by reduction of tridimensional symmetry to a rotational alignment along filamentous electrets such as cytoskeletal structures. In their view, quantum field theory and the Goldstone theorem require that the symmetry breaking (Bose condensation) results in long range interactions among system components (dipoles) conveyed by massless particle/waves (Goldstone bosons). The Milan group argues that the energy required to generate massless bosons is invested in the electret states of biomolecules and correlated fluctuations of their surrounding water and ions.

Celaschi and Mascarenhas (1977) showed that electret activation energy of biomolecules (0.2-0.4 eV) is equivalent to the hydrolysis of one ATP or GTP molecule and what Soviet scientist A.S. Davydov predicted for initiation of biological solitons. Consequently solitons, massless bosons, and Fröhlich's coherent polarization waves may be synonymous.

Pursuing their quantum field approach, the Milan group arrived at an astounding concept of self-focusing of electromagnetic energy within cytoskeletal filaments. Electromagnetic energy exceeding a threshold originating within, or penetrating into, cytoplasm would be confined inside filaments whose diameters depended on the original symmetry breaking (Bose condensation) of ordered dipoles. Any electric disturbance produced by thermally fluctuating dipoles or by any other source would be confined inside filamentous regions whose diameter depends on the polarization density, or ordering of biological water. The Milan group calculated a self focusing diameter of about 15 nm (precisely the inner diameter of microtubules) and argues the cytoskeleton is the material consequence of dynamic self-focusing of polarization waves in the cytoplasm. The Milan group concludes that focusing occurs in cytoplasm of living cells due to the spatial coherence and ordering imparted by cytoskeletal electret behavior.

4 Coherence and Models of Cytoskeletal Information Processing

Cooperative, collective effects of coherent protein conformational oscillations are a likely substrate for biological intelligence ranging from cytoplasmic probing to human consciousness. The activities, functions, and structures of microtubules and other cytoskeletal components have led at least a dozen author groups to publish theoretical models of information processing within MT and the cytoskeleton. The concepts include passive MT signal transduction, descriptive patterns among MT subunit states, dynamic cooperative automaton effects among coherent oscillations of microtubule subunits, and cytoplasmic/cytoskeletal sol-gel field effects utilizing holographic imagery. Coherent oscillations of cytoskeletal protein subunits such as tubulin are relevant to several of these communicative modes.

Experimental findings which support the notion of cytoskeletal communication include electrical signaling along microtubules (Vassilev et al. 1985), fluorescence resonance transfer among MT subunits and membranes (Becker et al. 1975), parallel alignment of MT in applied electric and magnetic fields (Vassilev et al. 1982) and involvement of intraneuronal MT in nerve membrane excitability and synaptic transmission (Matsumoto and Sakai 1979; Alvarez and Ramirez 1979). The following models of cytoskeletal information processing are consistent with these findings and would be strengthened by coherent protein oscillations.

4.1 MT Sensory Transduction (Atema)

Biologist Jelle Atema (1973) proposed that sensory cilia transduced environmental information to the rest of the cell by propagated conformational changes in the microtubule subunits which constituted these cilia. Atema reviewed conformational changes in MT tubulin subunit dimers observed by a number of authors, and concluded that sequences of subunit conformational changes are likely to occur in microtubules coupled to wavelike or whiplike motions of cilia and flagella. Atema's microtubule theory assumed that distortion of tubulin conformation by any number of sources was sufficient to propagate a conformational wave. Thus light energy, chemical bond energy, and mechanical forces could be transduced by sensory cilia.

Atema's view of MT information processing was an all-or-none propagation of conformational changes along tubulin protofilaments. His was the first theory to look beyond the global behavior of cilia to consider conformational effects in tubulin components. Subsequent theories became more elaborate to consider localized analog functions, switching, and collective neighbor interactions among MT subunits.

4.2 MT Mechano-Ionic Transducers (Moran and Varela)

Expanding on Atema's suggestion, Harvard biologists Moran and Varela (1971) suggested that compression or bending of MT would cause release and flux of bound ions from MT subunits (16 calcium ions per tubulin). Moran and Varela saw these ionic currents capable of generating membrane depolarization much like the coordinated release of calcium ion waves by muscle cell sarcoplasmic reticulum triggers actin-myosin contractile shortening and like the calcium waves thought to regulate bending and waving of cilia and flagella as well as calcium wave coupling to sol-gel state in the cytoskeletal ground substance. Moran and Varela's contribution was to observe that MT could release ions such as calcium in a controlled and modifiable manner useful for intracellular communication.

4.3 Cytomolecular Computing (Conrad and Liberman)

Wayne State University computer scientist Michael Conrad and Soviet information scientist E.A. Liberman (Conrad and Liberman 1982) have viewed a molecular analog within the cytoskeleton as a representation of the external world.

Conrad (1985) notes that
"highly parallel signal processing and vibratory behavior on the part of microtubules and other cytoskeletal elements could play a significant role".

4.4 MT Signal Processing (DeBrabander)

Cell biologist Marc DeBrabander and colleagues (Geuens et al. 1987) have developed a method of labeling individual tubulin subunits within MT which are either tyrosinated or glutamated. Tyrosine, the terminal amino acid in tubulin, may be removed from polymerized MT to expose the more acidic glutamate as the new terminal amino acid. Thus all tubulins within polymerized MT are either tyrosinated, or glutamated which may serve as a convenient programming or memory function. DeBrabander and colleagues developed a double labeling technique in which immunogold particles bind to tubulin subunits. Large immunogold particles (10 nm) identify glutamated tubulin and smaller particles (5 nm) bind to tyrosinated tubulin. Patterns of tyrosinated/glutamated tubulin within MT show heterogenous distributions, and suggest the potential for a coding mechanism. MAP attachment, tubulin conformation, calcium binding and other factors could be coupled to MT function via such patterns. DeBrabander and colleagues have provided the first direct evidence of modifiable patterns of tubulin variability in intact microtubules. Consequently, MT appear capable of signal processing in addition to signal transduction.

4.5 Cytoskeletal String Processors (Barnett)

Brooklyn College computer and information scientist Michael P. Barnett has pursued the design of computer components suited to molecular scale devices and artificial intelligence (AI). Unlike most AI researchers, Barnett scoured the subcellular biological realm in search of molecular scale information processing concepts and "discovered" cytoskeletal microtubules and neurofilaments! He has proposed information representation as patterns in the subunits of cytoskeletal polymer subunits which may operate like information strings analogous to word processors.

In Barnett's conceptualization (Fig. 8), information strings move from right to left along processing channels which run parallel to one-dimensional memory channels in which character strings can be stored. String transformers can perform global replacements on sequences of characters like common word-processors. Barnett's model assumes the existence of processing channels (MT) along which strings of information can move, and memory channels (neurofilaments) which consist of a succession of locations, each of which can hold a single character. Parallel array and lateral interconnectedness of MT and neurofilaments could qualify these cytoskeletal elements as string processors, assuming that information may be represented in the polymers. Barnett's model is thus compatible and complementary with other models of conformational patterns within MT and the cytoskeleton. The actual implementation of Barnett's string processing would be greatly facilitated by clocking, or coherent oscillations of subunits in both MT and neurofilaments.

4.6 Microtubule "Gradions" (Roth, Pihlaja, Shigenaka)

Roth et al. (1970), and Roth and Pihlaja (1977) have considered information processing in biomolecular assemblies. Citing cooperative effects among adjacent proteins or protein subunits, they proposed that conformational gradients in microtubules and other protein arrays represent information by patterns of conformational states (gradions) among near neighbors in protein lattices.

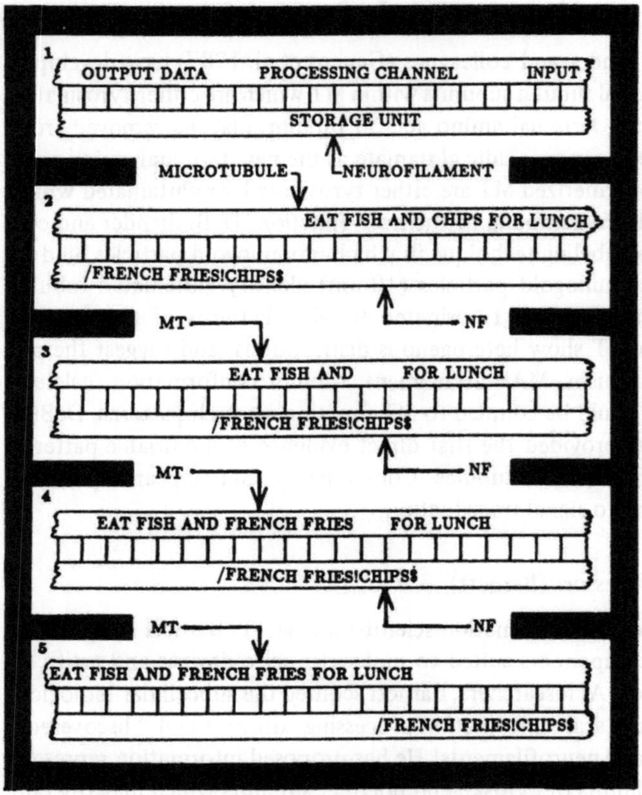

Fig. 8. Barnett's string processor model of microtubule-neurofilament memory storage. *1* a processing channel (microtubule-MT) is laterally connected to a parallel storage unit (neurofilament-NF); *2* information traverses the processing channel from right to left; on the storage channel, synonym pairs move from left to right; *3* as the word "chips" moves by its own self in storage, it is deleted; *4* "chips" is replaced by its synonym "french fries", and *5* the replacement is complete. (By Paul Jablonka, from Hameroff 1987)

Conformational states of individual tubulin subunits are thought to be controlled by factors which include ligand induced conformation, tubulin isozyme, detyrosination, or binding of MAPs including inter-microtubule bridges. Discrete gradion fields within microtubule surfaces were defined as the neighborhood areas around inter-microtubule MAP bridges. Roth, Pihlaja and Shigenaka assumed five possible conformational states for each dimer determined by inter-MT linkage sites and their cooperative allosteric effects, as well as dimer binding by a number of different native substrates or foreign molecules. The pattern of tubulin dimer conformational states in a region of MT lattice governed by attachment of inter-MT linkages (or other MAPs) is defined as a gradion, or conformational field which includes about seventeen dimers (Fig. 9). Consequently, 5^{17} different gradion patterns could exist if the conformation of each subunit were independent of all others. Such deterministic patterns would be generally useful when coupled to coherent dynamics in the cytoskeleton.

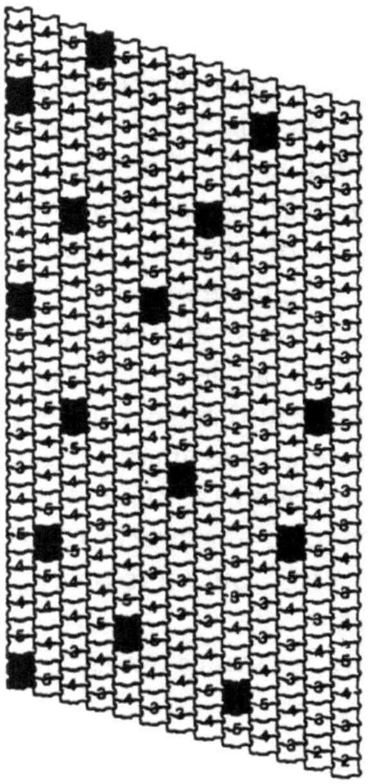

Fig. 9. "Gradions" within microtubule lattice. Dark shaded dimers are MAP attachment patterns; numbered dimers are determined by proximity to MAP attachment sites. MAP binding and other "gradionators" are thought to induce coded patterns representing information. (By Paul Jablonka from Roth 1970)

4.7 Gyroscopic Centrioles (Bornens)

French molecular biologist Michel Bornens (1979) argues for a dynamic stability based on organizational properties of centrioles (Fig. 3) and the cytoskeleton. Specifically, Bornens proposes that centrioles are animated by rapid oscillatory rotation about their longitudinal axis, which results in a dynamic stability and inertia analogous to a spinning top or gyroscope.

The rate of rotation required for gyroscopic inertia has been calculated as 2.3 million revolutions per minute (Albrecht-Buehler 1981). To account for such rapid activity, Bornens suggests a submicroscopic mechanism "allowing more independence of the centriole with respect to surrounding material." Factors which could support this contention include ordered water coupled to centriolar oscillation, some unknown property of the "pericentriolar material," ionic charge layer, or even superconductivity as suggested by Fröhlich and Del Giudice's group. Rapid rotatory oscillations in protein assemblies would be consistent with the findings of Michels et al. (1985), who found collective energy absorption in the range of 10^7 per second by protein assemblies such as virus coats. Bornens also suggests that ATP-generated centriolar rotation triggers propagating impulses along microtubules by transitory contact/stimulus of the nine rotating MT doublet/triplets in the centriole wall with their surrounding satellite bodies. This would lead to coherent rhythmic signals throughout the cytoskeleton with a frequency nine times greater than that of the centriole's rotation. Occurrence of coherent, rhythmic signals throughout the cytoskeleton would be mechanisms close to the nature of life itself.

4.8 Centriole-MT Signaling (Albrecht-Buehler)

Northwestern University biologist Guenter Albrecht-Buehler has considered the general question of intelligence in cytoplasm as well as two specific models of cytoskeletal information processing: centriole signal detection, and propagation of MT impulses.

Albrecht-Buehler (1985) has described a mechanism for impulse propagation along microtubules. He considers that each MT protofilament is a chain of alpha-beta tubulin dimers: AB,AB,AB,...,AB. Within the wall of a microtubule, each monomer is in contact with other monomers of the same protofilament and with those of adjacent protofilaments. Each monomer is consequently subject to attractive Van der Waals forces from surrounding tubulin monomers which hold together the protofilaments and MT cylinders. Albrecht-Buehler proposed that each of these interactions must weaken the A-B dimer bond; consequently the wall of a microtubule exists in a "state of resonance as to the relative strengths of the intermonomeric bonds. For example, (A-B)(A-B)...(A-B)(A-B) could resonate with (A)(B-A)(B-A)...(B-A)(B). Such a resonating chain could propagate information at close to the speed of light".

Albrecht-Buehler is suggesting a coherent communicative resonance among protein conformational states within MT assemblies.

4.9 Dynamic Tensegrity (Heidemann and Jarosch)

Buckminister Fuller proposed an interesting architecture he called tensegrity, constructed from an assembly of tension and compression members (Fuller 1975). The compression members of solid struts are isolated from each other, held together by the tension members. Joshi et al. (1985) have shown that cytoplasm has both compressive and tensile elements. Semi-rigid microtubules are under compression, presumably due to tension generated by actin filaments and actin based microtrabecular lattice. In general, MT do not contact each other so that the self-supporting capability of cytoplasm may indeed stem from tensegrity.

Robert Jarosch (1986a,b) has published a series of papers describing actin-MT interactions which suggest that (1) contractile actin filaments are spirally wound around microtubules, (2) coordinated contraction of the actin filaments imparts a rotational torque to MT, somewhat like a spinning top, (3) actin filaments wound in opposite directions on the same MT can cause rotational oscillations of the MT. These two models fit together to provide a picture of a dynamic cytoplasmic tensegrity network in which the cytoskeletal elements may be coherently twisting back and forth. Perturbation of any part of such a tensegrity network could have dynamic consequences throughout its domain. Transient changes in tension, compression, or oscillatory rhythm caused by a variety of factors would be transmitted and possibly amplified throughout the cytoskeleton.

4.10 Dynamic MT Probing (Kirschner and Mitchison)

Strong evidence supports the concept of dynamic instability in microtubule assembly. Kirschner and Mitchison (1986) ask: "how can a peripheral clue lead to reorganization deep within a cell?" One possibility is that a signal is relayed to the microtubule-organizing center leading to a change in its structure, orientation, and directed nucleation of microtubules. This would be consistent with a primary role for information integration and decision-making within the MTOC. A simpler non-hierarchical idea is that a signal at the periphery affects distribution directly. Since the whole cytoskeletal array is very dynamic, it would only be necessary to stabilize or destabilize a particular subset of microtubules for the entire cytoskeleton to rapidly transform.

4.11 Sphere Packing Screw Symmetry (Koruga)

There are 32 possible symmetry arrangements of packed spheres in a cylindrical crystal. Erickson (1973) used hexagonal packing of protein monomers to explain the form and patterns of viruses, flagella and microtubules. Djuro Koruga (1986) of the University of Belgrade's Molecular Machines Research Unit has analyzed the symmetry laws which describe cylindrical sphere packing and the structure of microtubules.

Koruga (1986):

"The sixfold symmetry and dimer configuration lead to screw symmetry on the cylinder: a domain may repeat by translocating it in a spiral fashion on the cylinder. From coding theory, the symmetry laws of tubulin subunits suggest that 13 protofilaments are optimal for the best known binary error correcting codes with 64 code words. Symmetry theory further suggests that a code must contain about 24 monomer subunits or 12 dimers".

Koruga's symmetry arguments may be compared with the gradion concept of Roth, Pihlaja, and Shigenaka in which a field of about 17 monomers is thought to represent a basic information unit. Koruga concludes that microtubule symmetry and coherent structure are optimal for information processing.

4.12 Cytoskeletal Self-Focusing (Del Giudice)

Starting from Fröhlich's model of coherent excitations, Del Giudice and colleagues (Del Giudice et al. 1983a,b, 1986; Del Giudice 1986) have used quantum field theory to describe the electret state of biological systems (ordered water surrounding linear biomolecules) and determined that there exists a strong likelihood for the propagation of particle-like waves in biomolecules. Further, the ordering of water should lead to self-focusing of electromagnetic energy into filamentous beams excluded by the ordered symmetry. For ordered cytoplasm, they calculate the diameter for the confinement and propagation of particle-like waves (massless bosons, polarization waves, or solitons) in biomolecules to be about 15 nanometers, exactly the inner diameter of microtubules.

Cytoskeletal polymers may thus be capable of capturing and utilizing ambient or biologically generated electromagnetic energy. One possible example is infrared energy which is routinely generated by dipoles in biological molecules. This energy is generally believed to be dissipated into heat within the aqueous cytoplasm, however self-focusing within a solid-state cytoskeleton could utilize this energy productively in a communicative medium.

4.13 MT Automata, Holography (Hameroff, Watt, Smith)

The self-focusing of electromagnetic energy described by the Milan group is thought to occur by an electret induced increase in the refractive index of cytoplasm. A similar concept was proposed (Hameroff 1974) in which microtubules were thought to act like dielectric waveguides for electromagnetic photons. It was also proposed that the periodic array of MT subunits leaked or diffracted energy with 8 nanometer periodicity, resulting in a source of coherent energy (or calcium ions) from MT. Cytoplasmic interference of the coherent sources from among multiple MT could lead to holographic imaging in cytoplasm. Coupling of calcium concentrations to cytoplasmic sol-gel states could "hardwire" holographic interference patterns into the microtrabecular lattice. In parallel arrays of MT within nerve fibers, graded potentials or traveling action potentials were suggested to activate planes of cytoplasm perpendicular to the long axis of the MT and nerve fibers (Fig. 10). These traveling planes may be likened to image screens as in TV sets. In a TV picture tube, the screen is motionless and electron beams move to create a pattern by their intersection with the screen. Perhaps imaging within neurons occurs on travelling screens generated by action potentials moving through parallel MT arrays. The content of such images would depend on programming mechanisms in the conformation of tubulin subunits which comprise the MT walls, and which would update with each successive action potential. Hameroff and Watt (1982) described a method of MT tubulin programming in which charge carriers (calcium ions, electrosolitons) or conformational waves such as phonons or solitons were steered through MT lattices by genetically or cytoplasmically programmed tubulins and specific MAP-binding sites.

Hameroff et al. (1984, 1986) have utilized principles of cellular automata to explain information processing in MT. Cellular automata are dynamical systems which can generate and process patterns and information, and are capable of computing. Cellular automata require a lattice structure of like neighbors with discrete states and neighbor rules, and a universal clock to which all neighbors are timed. Adopting Fröhlich's model of coherent nanosecond dipole oscillations coupled to conformational

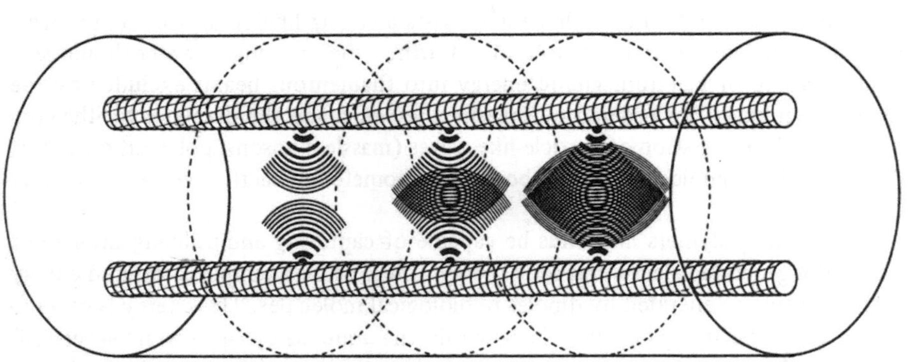

Fig. 10. Interference patterns in cytoplasm caused by coherent waves (e.g. Ca^{2+}, sol-gel state, MTL) generated by dynamic activities in microtubules may be a basis for holographic information imagery. (Hameroff 1987)

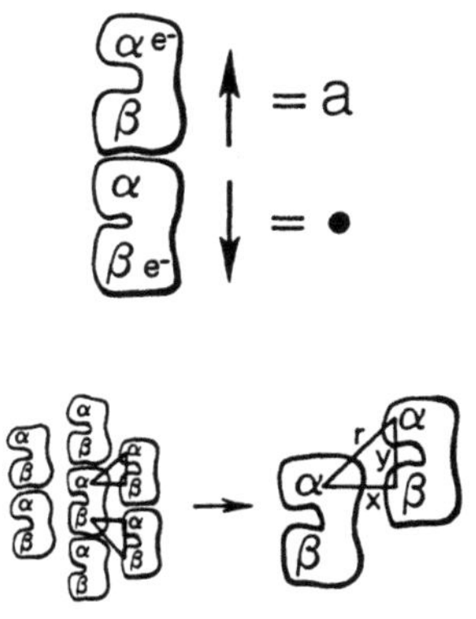

Fig. 11. *Top* MT-tubulin dimer subunits comprised of α and β monomers. Relative electron occupancy of either monomer may correlate with coherent dipole oscillations in the nanosecond time domain. *Bottom* screw symmetry hexagonal packing leads to unequal neighbor rules based on lattice distances and Van der Waals dipole interactions. For dimers shown at right, x = 5 nm, y = 4 nm, r = 6.4 nm. The relative strength of each neighbor dimer dipole state is y/r^3

states as a clocking mechanism, the authors calculated MT lattice neighbor Van der Waals dipole interactions as neighbor rules for an MT-automaton computer simulation. Each tubulin dimer was considered in one of two possible states at each nanosecond generation. The two states were related to Fröhlich's concept of dipole oscillation so that the dipole can be oriented either toward the alpha tubulin end (a, Fig. 11) or toward the beta tubulin end (represented by a dot in Fig. 11). The polarity and electret behavior of MT indicate that in the resting state, tubulin dimer dipoles should be oriented toward the beta monomer.

Dimer states at each "clock tick," or generation were determined by neighbor states at the previous generation.

$$\text{state} = \alpha,\ \text{if}\ \sum_{i=1}^{n} f(y) > 0,\quad \text{state} = \beta,\ \text{if}\ \sum_{i=1}^{n} f(y) < 0\ ,$$

where n = 7 as the number of neighbors, and f(y) the force from the *i*th neighbor in the y direction.

$$f(y) \propto \frac{\sin\theta}{r^2},\ \sin\theta = \frac{y}{r};\ f(y) \propto \frac{y}{r^3}\ .$$

The dipole state of any particular dimer at each clock tick thus depends on the summation of the dimer's neighbor dipole states (including its own) at the previous clock tick. The neighbor influences are unequal because of the screw symmetry of the MT lattice. Distant dimers (more than one neighbor away) would be expected to have little influence because of the dropoff in force intensity (y/r^3) with distance. However, collective influences from many like oriented dimers could augment long-range cooperativity. Using only near neighbor influences, computer simulation of an MT automaton yields interesting patterns and behavior of dipole/conformational states. These include both stable and traveling interactive patterns capable of computing and regulation of cytoskeletal activities. For example, Fig. 12 shows such patterns traveling through MT lattices, leaving altered wakes, or memory. Assuming nanosecond generations, these traveling patterns would travel at 8 nm per ns (8 m/s), a velocity consistent with propagating action potentials, or solitons. Variability in individual tubulin dimer isozymes, ligand binding, or MAP attachments could program and read out information in routine cellular functions. As one example, propagation

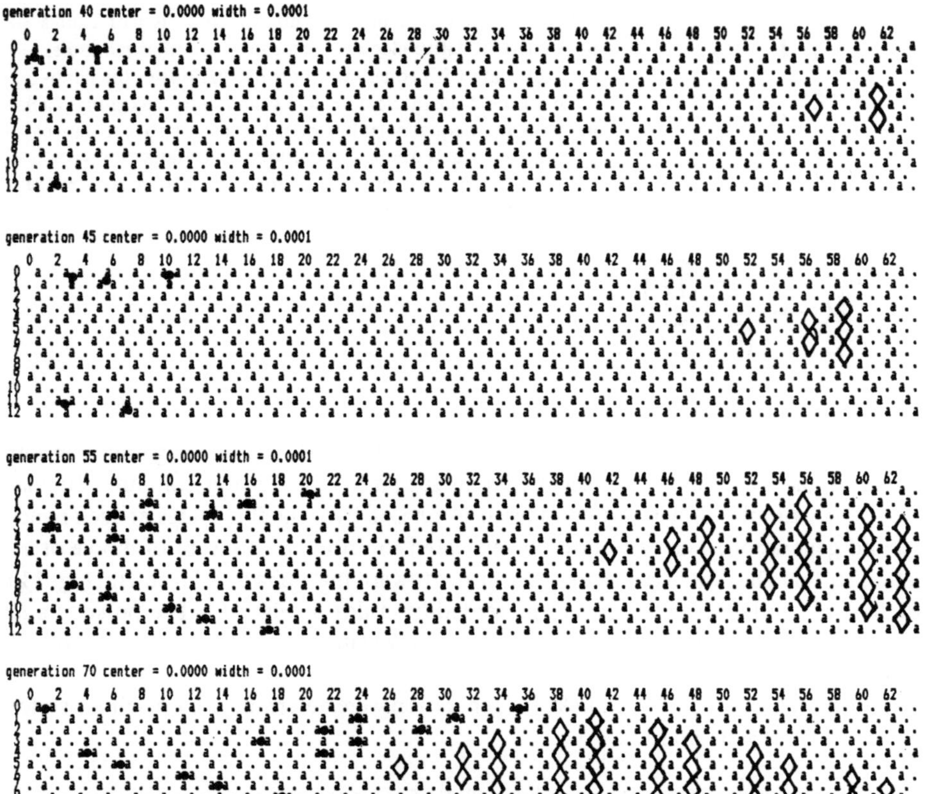

Fig. 12. 30 "nanoseconds" of computer simulation of MT cellular automaton based on Fröhlich coherent oscillations and tubulin subunit neighbor rules based on dipole-dipole interactions. Thirteen protofilaments are arranged horizontally, 64 rows are shown vertically yielding a flat MT lattice. "Alpha" states of tubulin (Fig. 11) are shown by "a," beta states by a *dot*. Four contiguous beta states are highlighted by *diamond patterns*. Four contiguous alpha states by *darkened dots*. The stimulation sequence (generations 40, 45, 55 and 70) shows movement of patterns at velocity of 8 nm/ns (8 m/s). (Hameroff et al. 1984)

of MT conformational patterns could coordinate the activities of contractile MAP sidearms in axoplasmic transport (Fig. 5). In a general sense, MT automata may be the information substrate for biological activities ranging from ciliary bending to human consciousness. Specific automata patterns distributed throughout wide volumes of cytoskeletal arrays within the brain could lead to cooperative resonance and collective effects resulting in images, thoughts, or ideas similar to coherence-induced phase transitions in certain metals yielding emergent collective properties such as superconductivity.

5 The Cytoskeletal Connection

The current prevalent model of brain function and artificial intelligence is the neural network, a collective dynamical system whose output is determined by the states of component neuronal synapses. In turn, the state of each synapse is determined by other neurons, modulated by the interneuronal cytoskeletons. In turn, the cytoskeleton is a collective dynamical system whose net activity is determined by the states of its component subunits. Their states, in turn, are determined by another level of organization including cytoplasmic factors, ordered water and ions, genetic isozymes of individual subunits, other cytoskeletal elements such as the microtrabecular lattice, and cooperative effects – coherence – from among collective assemblies of subunits. Cognitive processes which have been ascribed to neural net concepts may be more accurately interpreted as fractal hierarchies of dynamical systems which are highly parallel, highly interconnected, and of increasing capacity as they become more microscopic. Coherent oscillations of these subunits would permit collective effects to occur at each level and at successively more macroscopic levels with a net effect of consciousness.

References

Albrecht-Buehler G (1981) Does the geometric design of centrioles imply their function? Cell Motility 1:237–245

Albrecht-Buehler G (1985) Is the cytoplasm intelligent too? Cell and Muscle Motility 6:1–21

Alvarez J, Ramirez BU (1979) Axonal microtubules: their regulation by the electrical activity of the nerve. Neurosci Lett 15:19–22

Atema J (1973) Microtubule theory of sensory transduction. J Theor Biol 38:181–190

Athenstaedt H (1974) Pyroelectric and piezoelectric properties of vertebrates. Ann NY Acad Sci 238:68–93

Barrow GM (1961) Physical chemistry. McGraw-Hill, New York

Becker JS, Oliver JM, Berlin RD (1975) Flourescence techniques for following interactions of microtubule subunits and membranes. Nature 254:152–154

Bornens M (1979) The centriole as a gyroscopic oscillator: implications for cell organization and some other consequences. Bio Cell 35 (11):115–132

Burnside B (1974) The form and arrangement of microtubules: an historical, primarily morphological review. Ann NY Acad Sci 253:14–26

Celaschi S, Mascarenhas S (1977) Thermal stimulated pressure and current studies of bound water in lysozyme. Biophys J 29, 2:273–277

Clegg JS (1981) Intracellular water, metabolism, and cellular architecture. Collect Phenom 3: 289–312

Conrad M (1985) On design principles for a molecular computer. Communications of the ACM 28(5):464–480

Conrad M, Liberman EA (1982) Molecular computing as a link between biological and physical theory. J Theor Biol 98:239–252

Del Giudice E (1986) Collective properties of biological systems. In: Gutmann F, Keyzer H (eds) Modern bioelectrochemistry. Plenum, New York

Del Giudice E, Doglia S, Milani M (1983a) Actin polymerization in cell cytoplasm. In: Earnshaw JC, Stear MW (eds) The application of laser light scattering to the study of biological motion. Plenum, New York, pp 493–497

Del Giudice ES, Doglia S, Milani M, Vitiello G (1983b) Spontaneous symmetry breakdown and boson condensation in biology. Phys Rev Lett vol 95 (a):508–510

Del Giudice E, Doglia S, Milani M (1983c) Self focusing and ponderomotive forces of coherent electric waves: a mechanism for cytoskeleton formation and dynamics. In: Fröhlich H, Kremer F (eds) Coherent excitations in biological systems. Springer, Berlin Heidelberg New York, pp 123–127
Del Giudice E, Doglia S, Milani M, Vitiello G (1986) Collective properties of biological systems – solitons and coherent electric waves in a quantum field theoretical approach. In: Gutmann F, Keyzer H (eds) Modern bioelectrochemistry. Plenum, New York, pp 263–287
Dustin P (1984) Microtubules (2nd Revised Edition). Springer-Verlag, Berlin, p 442
Erickson RO (1973) Tubular packing of spheres in biological fine structure. Science 181:705–716
Fröhlich H (1970) Long-range coherence and the actions of enzymes. Nature 228:1093
Fröhlich H (1975) The extraordinary dielectric properties of biological materials and the action of enzymes. Proc Natl Acad Sci, USA 72(11):4211–4215
Fröhlich H (1986) Coherent excitations in active biological systems. In: Gutmann F, Keyzer H (eds) Modern bioelectrochemistry. Plenum Press, New York, pp 241–261
Fuller RB (1975) Synergetics. Macmillan, New York
Gershon ND, Porter KR, Trus BL (1985) The cytoplasmic matrix: its volume and surface area and the diffusion of molecules through it. Proc Natl Acad Sci, USA 82:5030–5034
Geuens G, Gundersen GG, Nuydens R, Cornelissen F, Bulinski VC, DeBrabander M (1986) Ultrastructural colocalization of tyrosinated and nontyrosinated alpha tubulin in interphase and mitotic cells. J Cell Biol 103(5), 1883–1893
Gibbons IR (1968) The biochemistry of motility. Annu Rev Biochem 37:521–546
Green HS, Triffet T (1985) Extracellular fields within the cortex. J Theor Biol 115(1):43–64
Gubkin A, Sovokin W (1960) Trans Bull Acad Sci, USSR 24:246
Gutmann F (1986) Some aspects of charge transfer in biological systems. In: Gutmann F, Keyzer H (eds) Modern bioelectrochemistry. Plenum, New York, pp 177–197
Hameroff SR (1974) Chi: a neural hologram? Am J Chin Med 2(2):163–170
Hameroff SR (1987) Ultimate computing: biomolecular consciousness and nanotechnology. Elsevier-North Holland, Amsterdam
Hameroff SR, Watt RC (1982) Information processing in microtubules. J Theor Biol 98:549–561
Hameroff SR, Smith SA, Watt RC (1984) Nonlinear electrodynamics in cytoskeletal protein lattices. In: Adey WR, Lawrence AF (eds) Nonlinear electrodynamics in biological systems. Plenum, New York, pp 567–583
Hameroff SR, Smith SA, Watt RC (1986) Automaton model of dynamic organization in microtubules. Ann NY Acad Sci 466:949–952
Jarosch R (1986a) A model for the molecular basis of filament contractility and sliding as demonstrated by helix models. Cell Motil 6(2):229–236
Jarosch R (1986b) The mechanical behaviour of doublet microtubules simulated by helical models. Cell Motil 6(2):209–216
Joshi HC, Chu D, Buxbaum RE, Heidemann SR (1985) Tension and compression in the cytoskeleton of PC 12 neurites. J Cell Biol 101:697–705
Karplus M, McCammon JA (1983) Protein ion channels, gates, receptors. In: King J (ed) Dynamics of proteins: elements and function. Annu Rev Biochem. Benjamin/Cummings Menlo Park, 53: 263–300
Kirschner M, Mitchison T (1986) Beyond self-assembly: from microtubules to morphogenesis. Cell 45:329–342
Koppenol WH (1980) Effect of a molecular dipole on the ionic strength dependence of a bimolecular rate constant. Identification of the site of reaction. Biophys J 29:493–507
Koruga D (1986) Microtubule screw symmetry: packing of spheres as a latent bioinformation code. Ann NY Acad Sci 466:953–955
Lasek RJ (1981) The dynamic ordering of neuronal cytoskeletons. Neurosci Res Prog Bull 19(1): 7–13
Lazarides E (1980) Intermediate filaments as mechanical integrators of cellular space. Nature 283: 249–256
Lee JC, Field DJ, George HJ, Head J (1986) Biochemical and chemical properties of tubulin subspecies. Ann NY Acad Sci 466:111–128

Mascarenhas S (1974) The electret effect in bone and biopolymers and the bound water problem. Ann NY Acad Sci 238:36–52

Mascarenhas S (1975) Electrets in biophysics. J Electrostat 1:141–146

Matsumoto G, Sakai H (1979) Microtubules inside the plasma membrane of squid giant axons and their possible physiological function. J Membr Biol 50:1–14

Michels B, Dormoy Y, Cerf R, Schulz JA (1985) Ultrasonic absorption in tobacco mosaic virus and its protein aggregates. J Mol Biol 181:103–110

Moran DT, Varela FG (1971) Microtubules and sensory transduction. Proc Natl Acad Sci USA 68:757–760

Ochs S (1982) Axoplasmic transport and its relation to other nerve functions. Wiley Interscience, New York

Porter KR, Tucker R (1981) The ground substance of the living cell. Sci Amer 56–61

Roth LE, Pihlaja DJ (1977) Gradionation: hypothesis for positioning and patterning. J Protozool 24:1, 2–9

Roth LE, Pihlaja DJ, Shigenaka Y (1970) Microtubules in the heliazoon axopodium. I. The gradion hypothesis of allosterism in structural proteins. J Ultrastruct Res 30:7–37

Satir P (1984) Cytoplasmic matrix: old and new questions. Cell Biol 99(1):235–238

Soifer D (1986) Factors regulating the presence of microtubules in cells. In: Soifer D (ed) Dynamic aspects of microtubule biology. Ann NY Acad Sci 466:1–7

Stebbings H, Hunt C (1982) The nature of the clear zone around microtubules. Cell Tissue Res 227:609–617

Vassilev PM, Dronzine RT, Vassileva MP, Georgiev GA (1982) Parallel arrays of microtubules formed in electric and magnetic fields. Biosci Rep 2:1025–1029

Vassilev P, Kanazirska M, Tien HT (1985) Intermembrane linkage mediated by tubulin. Biochem Biophys Res Commun 126(1):559–565

Wu TM, Austin S (1978) Bose-Einstein condensation in biological systems. J Theor Biol 71:209–214

Subject Index

MIX
Papier aus verantwortungsvollen Quellen
Paper from responsible sources
FSC® C105338

If you have any concerns about our products,
you can contact us on
ProductSafety@springernature.com

In case Publisher is established outside the EU,
the EU authorized representative is:
Springer Nature Customer Service Center GmbH
Europaplatz 3, 69115 Heidelberg, Germany

Printed by Libri Plureos GmbH
in Hamburg, Germany